Second Edition of Municipal Wastewater Treatment

Second Edition of Municipal Wastewater Treatment

Guest Editors

Yung-Tse Hung
Hamidi Abdul Aziz
Issam A. Al-Khatib

MDPI

Basel • Beijing • Wuhan • Barcelona • Belgrade • Novi Sad • Cluj • Manchester

Guest Editors
Yung-Tse Hung
Civil and Environmental Engineering
Cleveland State University
Cleveland
United States

Hamidi Abdul Aziz
Civil Engineering
Universiti Sains Malaysia
Nibong Tebal
Malaysia

Issam A. Al-Khatib
Institute of Environmental and Water Studies
Birzeit University
Birzeit
West Bank

Editorial Office
MDPI AG
Grosspeteranlage 5
4052 Basel, Switzerland

This is a reprint of the Special Issue, published open access by the journal *International Journal of Environmental Research and Public Health* (ISSN 1660-4601), freely accessible at: www.mdpi.com/journal/ijerph/special_issues/second_municipal_wastewater.

For citation purposes, cite each article independently as indicated on the article page online and using the guide below:

Lastname, A.A.; Lastname, B.B. Article Title. *Journal Name* **Year**, *Volume Number*, Page Range.

ISBN 978-3-7258-3506-5 (Hbk)
ISBN 978-3-7258-3505-8 (PDF)
https://doi.org/10.3390/books978-3-7258-3505-8

Contents

About the Editors

Yung-Tse Hung

Prof. Dr. Yung Tse Hung, Ph.D., P.E., DEE, fellow-ASCE, served as a professor of Civil Engineering at Cleveland State University from 1981 to 2024. He earned his B.S. and M.S. in Civil Engineering from Cheng Kung University, Taiwan, and his Ph.D. from the University of Texas at Austin. Prof. Hung has taught at 16 universities across 8 countries and started the public health engineering program at the University of Canterbury, New Zealand, in 1972. He has served on the faculties of numerous universities globally, including in New Zealand, the USA, Hong Kong, UAE, Singapore, Australia, Russia, and Kyrgyzstan. Prof. Hung's research focuses on biological wastewater treatment, industrial water pollution control, and municipal wastewater treatment. He has published approximately 40 books, 242 book chapters, 198 refereed publications, and 352 other scholarly works, totaling around 811 publications and presentations. He is a fellow of ASCE, a diplomate of AAEE, a fellow of the Ohio Academy of Science, a member of AEESP, and a life member of WEF. He serves as editor-in-chief for several international journals and books and is a registered professional engineer in Ohio and North Dakota.

Hamidi Abdul Aziz

Prof. Dr. Hamidi Abdul Aziz is a distinguished professor in Environmental Engineering at the School of Civil Engineering, Universiti Sains Malaysia. He earned his Ph.D. in Civil Engineering (Environmental) from the University of Strathclyde in 1992. Currently, he heads the Solid Waste Management Cluster (SWAM) at Universiti Sains Malaysia. With 29 years of teaching and research experience in environmental engineering, Professor Aziz has made significant contributions in areas such as solid waste management, landfill technology, water and wastewater treatment, leachate treatment, bioremediation, pollution control, and environmental impact assessment. To date, he has mentored over 100 Ph.D. and MSc students and has published over 200 ISI papers and several books. Additionally, he serves as an editor and editorial board member for several international journals. In recognition of his outstanding research contributions, the Malaysia Academy of Sciences named him a Top Research Scientist of Malaysia in 2012. In 2020, he was listed among the top 2% of scientists in his field globally, as compiled by the prestigious Stanford University.

Issam A. Al-Khatib

Prof. Dr. Issam A. Al-Khatib is a faculty member at the Institute of Environmental and Water Studies, Birzeit University, Palestine. His expertise spans across water resources management, environmental assessment, wastewater management, and climate change, with a focus on environmental health and sustainable development. Dr. Al-Khatib has led numerous research and evaluation projects on topics such as medical waste management, solid waste, water and sanitation, and urban environments. He is dedicated to promoting public environmental awareness through training, workshops, and community campaigns. He has contributed to national committees on environmental standards, including swimming pool water quality regulations. Dr. Al-Khatib has supervised over 40 M.Sc. theses, focusing on these critical environmental issues, and is an expert in academic quality evaluation. Currently, he is working on a project to evaluate the attitudes, perceptions, and behaviors of Birzeit University students toward the management of single-use plastic waste. Prof. Al-Khatib is ranked among the top 0.5% of researchers globally according to Scholar GPS in 2025, recognized for his exceptional contributions to environmental science, monitoring, and waste management.

Preface

Water scarcity has become a pressing global issue. The primary goal of municipal wastewater treatment is to eliminate contaminants from wastewater prior to releasing it back into the environment. The treated effluent can then be repurposed for various water reuse applications and resource recovery initiatives. This reprint, including 17 papers, addresses various aspects of wastewater treatment, highlighting advancements and challenges in the field. One study investigates the influence of salts on the photocatalytic degradation of formic acid in wastewater, while another assesses the impact of COD in toxic industrial wastewater on nitrification efficiency. Another explores the efficiency of side stream nitrification in modern two-stage activated sludge plants. One study conducts a risk analysis of heavy metals migration from sewage sludge, and another discusses municipal wastewater connections related to water crises and jaundice outbreaks in Shimla City. Another evaluates magnetic nanomaterials for wastewater treatment using biochemical methane potential tests. One review focuses on advanced oxidation technology in sludge conditioning and dewatering, while another analyzes micro- and macro-elements in plants used for landfill leachate treatment. Another study examines the effects of temperature shifts in anaerobic digestion for municipal wastewater sludge. One investigation looks into the influence of particle size and zeta potential in treating highly colored landfill leachate. Another examines heavy metals in harvested rainwater for domestic use in Palestine. Another assesses the impact of treated wastewater irrigation on barley metabolism in arid regions. One review discusses the status of chlorine and UV irradiation in wastewater treatment, while another explores the synthesis and application of 11Å-TA from coal fly ash. One study investigates the potential of Nephelium lappaceum seed as a coagulant in landfill leachate treatment. Another analyzes seasonal and technological shifts of multi-resistant pathogens in municipal wastewater treatment. This reprint contributes to the understanding and improvement of wastewater treatment technologies, addressing environmental and public health concerns.

Yung-Tse Hung, Hamidi Abdul Aziz, and Issam A. Al-Khatib
Guest Editors

International Journal of *Environmental Research and Public Health*

MDPI

Communication

Influence of Salts on the Photocatalytic Degradation of Formic Acid in Wastewater

Azzah Nazihah Che Abdul Rahim [1,2], Shotaro Yamada [1], Haruki Bonkohara [1], Sergio Mestre [3], Tsuyoshi Imai [1], Yung-Tse Hung [4] and Izumi Kumakiri [1,*]

1 Graduate School of Sciences and Technology for Innovation, Yamaguchi University, 2-16-1 Tokiwadai, Ube 755-8611, Japan
2 Department of Oil and Gas Engineering, School of Chemical Engineering, College of Engineering, Universiti Teknologi MARA, Shah Alam 40450, Selangor, Malaysia
3 Chemical Engineering Department, University Institute of Ceramic Technology, Universitat Jaume I. Avda, Vicent Sos Baynat, 12071 Castellon, Spain
4 Department of Civil and Environmental Engineering, Cleveland State University, Cleveland, OH 44115, USA
* Correspondence: izumi.k@yamaguchi-u.ac.jp

Abstract: Conventional wastewater treatment technologies have difficulties in feasibly removing persistent organics. The photocatalytic oxidation of these contaminants offers an economical and environmentally friendly solution. In this study, TiO_2 membranes and Ag/TiO_2 membranes were prepared and used for the decomposition of dissolved formic acid in wastewater. The photochemical deposition of silver on a TiO_2 membrane improved the decomposition rate. The rate doubled by depositing ca. 2.5 mg of Ag per 1 g of TiO_2. The influence of salinity on formic acid decomposition was studied. The presence of inorganic salts reduced the treatment performance of the TiO_2 membranes to half. Ag/TiO_2 membranes had a larger reduction of ca. 40%. The performance was recovered by washing the membranes with water. The anion adsorption on the membrane surface likely caused the performance reduction.

Keywords: photocatalysis; Ag/TiO_2; inorganic salts; immobilization; isoelectric point

Citation: Che Abdul Rahim, A.N.; Yamada, S.; Bonkohara, H.; Mestre, S.; Imai, T.; Hung, Y.-T.; Kumakiri, I. Influence of Salts on the Photocatalytic Degradation of Formic Acid in Wastewater. *Int. J. Environ. Res. Public Health* **2022**, *19*, 15736. https://doi.org/10.3390/ijerph192315736

Academic Editor: Paul B. Tchounwou

Received: 28 October 2022
Accepted: 24 November 2022
Published: 26 November 2022

1. Introduction

A large variety of persistent organic molecules are found in the wastewater from municipal and industrial activities of a city. Some of these organics are bio-accumulative and can be toxic, even in low concentrations. However, current technologies have difficulties in treating these dilute components in an economic and environmentally friendly manner. In addition, municipal wastewater contains salinity in the coastal area, which can change the properties of wastewater treatment technology [1–3]. For example, salts can hinder the growth of microorganisms and change the performance of activated sludge treatment [1]. Concentrations of 10 to 100 mmol/L NaCl in water changed the adhesion energy of humic substances and affected their removal by adsorption [3]. Coexisting NaCl in water is also reported to slightly reduce catalytic wet-air oxidation activity [2].

Catalytic wet-air oxidation decomposes various types of organics in wastewater that are difficult to degrade by biological treatment [4,5]. Immobilizing catalyst particles on membranes eliminate the process required to separate spent catalysts from the treated water. In addition, the configuration of catalytic membrane contactors facilitates oxygen supply to the reaction field, which enhances the decomposition rate of dissolved organics by catalytic oxidation [6,7]. For example, platinum-based catalytic membranes decomposed dissolved phenols in seawater under mild conditions, showing the potential for membranes to treat wastewater [6]. However, the catalytic performance needs to be improved to reduce the footprint of the membrane unit. As photocatalysts can decompose various types of organics in water [8], immobilizing these photocatalysts on membranes is a possible solution to improve the catalytic membrane performance.

Titanium dioxide (TiO_2) is a commonly used photocatalyst as it is robust, nontoxic, and cost-effective. However, TiO_2 requires UV light irradiation to activate due to the wide band gap energy of 3.2 and 3.0 eV for the anatase and rutile phases, respectively [9,10]. Large efforts have been made to narrow the band gap by modifying TiO_2 with dye, noble metals, such as Au, Ag, and Pt, transition metals, such as Fe and Cu, and other materials [10]. Successful modifications made the TiO_2-based catalysts visible-light active and the high performance of decomposing dissolved organics in water is reported [11,12].

Many studies evaluating photocatalytic activity use solutions prepared by dissolving a controlled amount of target organic compounds into distilled water. While salinity can exist in wastewater, its influence on photocatalytic activity is not fully understood. For example, Chen et al. reported a negative influence of anions, such as chloride, on the photocatalytic oxidation of dichloroethane, as anions inhibit the adsorption of dichloroethane to a TiO_2 surface [13]. On the contrary, Makita et al. reported a positive influence of cations on the decomposition rate of dye [14].

In this study, the influence of water salinity was examined using TiO_2 and Ag/TiO_2 membranes. Formic acid was used as a model organic compound in water as its degradation process is simple and easy to follow [15,16]. Salinity in municipal wastewater can increase in the coastal area due to the infiltration of seawater. To simulate the seawater contamination, sodium chloride (NaCl) and magnesium sulfate ($MgSO_4$) were used as salts. Magnesium chloride ($MgCl_2$) and potassium sulfate (K_2SO_4) were also used for comparison.

2. Materials and Methods

Porous flat discs were prepared using kaolin (ER, Caobar S.A., Guadalajara, Spain), alumina (AR12B5, Aluminium Pechiney, Salindres, France), potato starch (Sigma Aldrich Inc., St. Louis, MO, USA) and polyvinyl alcohol (PVA, Mowiol 4-88, Sigma Aldrich Inc., USA) as a ligand. After sintering the supports at 1673 K for four hours, XRD analysis identified mullite, corundum, and cristobalite (Figure S1). TiO_2 particles (P25, Evonik Industries) were applied to the surface of the disc supports, having a 47 mm diameter. A detailed procedure will be found elsewhere [15]. Then, silver was photochemically applied to some of the prepared TiO_2 membranes using a silver acetate solution of 0.01–1 mmol/L. The concentration of silver in the immersed solution was measured by inductively coupled plasma (ICP, SII Nano Technology Co. Ltd., Tokyo, Japan) before and after applying UV light to the membrane. The amount of silver deposited on the TiO_2 membranes was calculated from the concentration decrease in silver ions in the solution and the weight of the solution used. Three pieces of black lamps (Toshiba, maximum light emission at 352 nm) were used as the UV source and the light intensity was adjusted to 3.3 mW/cm^2 by changing the distance between the membrane and the lamps. The light strength was measured by a photometer (C10427H102428, Hamamatsu Photonics, Tsukuba, Japan). Prepared TiO_2 and Ag/TiO_2 membranes were analyzed by scanning electron microscope (FE-SEM, JSM-633F, JSM-7600FG, JEOL Ltd., Tokyo, Japan) and X-ray diffraction (XRD, XRD-6100, SHIMADZU Co., Kyoto, Japan) with Cu-Kα radiation.

The photocatalytic activity of the obtained TiO_2 and Ag/TiO_2 membranes was evaluated by the decomposition of formic acid (Fujifilm Wako Pure Chemical Corporation, Osaka, Japan) in water. Formic acid was used in this study. The flat disc-shaped membrane was soaked in formic acid solution and kept under dark conditions for 20 min before applying UV light with a light intensity of 3.3 mW/cm^2 at room temperature. This light intensity was used to simulate the UVA part of the sunlight [17]. The same black lamps being used in the silver deposition were used as the UV light source in the photocatalytic activity tests. Figure 1 illustrates the experimental setup. Humidity was not controlled nor measured. The starting concentration of the formic acid solution was adjusted to 200 mg/L, a concentration similar to other studies [15]. The concentration of the formic acid solutions was measured using a UV–Vis spectrophotometer UV-1800 (Shimadzu, Columbia, MD, USA) at a 205.6 nm wavelength. Calibration curves to determine the formic acid concentration were prepared for each salt concentration.

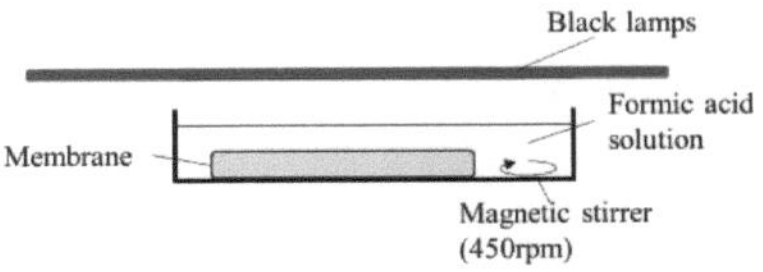

Figure 1. The experimental setup for photocatalytic formic acid decomposition.

Inorganic salts, NaCl, $MgSO_4$, $MgCl_2$, and K_2SO_4, were purchased from Fujifilm Wako Pure Chemical Corporation, Japan, and dissolved into the formic acid solution at the concentration of 0.6, 6, 60 mmol/L for NaCl, $MgSO_4$ and K_2SO_4, and 0.3, 3, 30 mmol/L for $MgCl_2$. The maximum concentration of NaCl studied was about 1/10 of that in seawater. After each test, the membranes were washed with water.

3. Results and Discussion

3.1. Influence of Silver Deposition

Figure 2 shows the membrane morphology. The surface of the porous ceramic disc (Figure 2a) was completely covered with TiO_2 particles (Figure 2b). The thickness of the TiO_2 layer was about 10–30 μm (Figure 2c). The variation is due to the large pores in the ceramic disc, which were plugged by the TiO_2 particles. Element mapping of the cross-sectional view of the TiO_2 membrane is shown in Figure S2. Al and Si mapping images show a void in the flat support. The Ti mapping showed that such voids can be filled with TiO_2 particles applied to the surface. As a result, the surface of the TiO_2 membrane became smoother compared to the support surface (Figure 2a,b). The surface morphology of the $AgTiO_2$ membrane was similar to that of the TiO_2 membrane. No peaks relating to Ag or AgO_x were found in the XRD pattern, probably due to its small amount and size [18] (Figure S1). X-ray photoelectron spectroscopy (XPS) analyses showed that silver was deposited as Ag^0, Ag^+, and Ag^{2+} (Figure S3).

The concentration of formic acid did not change in the dark, showing little influence on the adsorption of formic acid on the membranes. On the contrary, the concentration decreased when UV light was irradiated, as shown in Figure 3. The concentration at time t (C_t) was normalized with the initial concentration (C_0) in the figure. The results obtained with a TiO_2 membrane and four pieces of Ag/TiO_2 membrane having different deposition amounts of silver are shown in the same figure. The formic acid concentration became about 75% and 50% compared to the initial concentration after 80 min with TiO_2 membrane and Ag/TiO_2 (AgT2) membrane, respectively. Silver deposition improved the photocatalytic activity of the membrane. The membranes showed reproducible performance, as indicated with smaller keys.

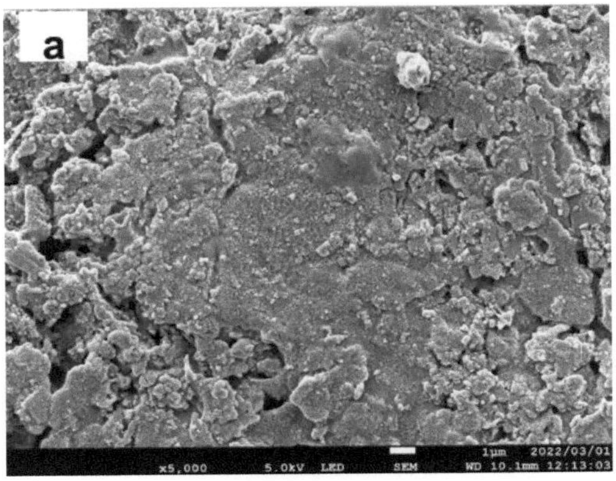

Figure 2. *Cont.*

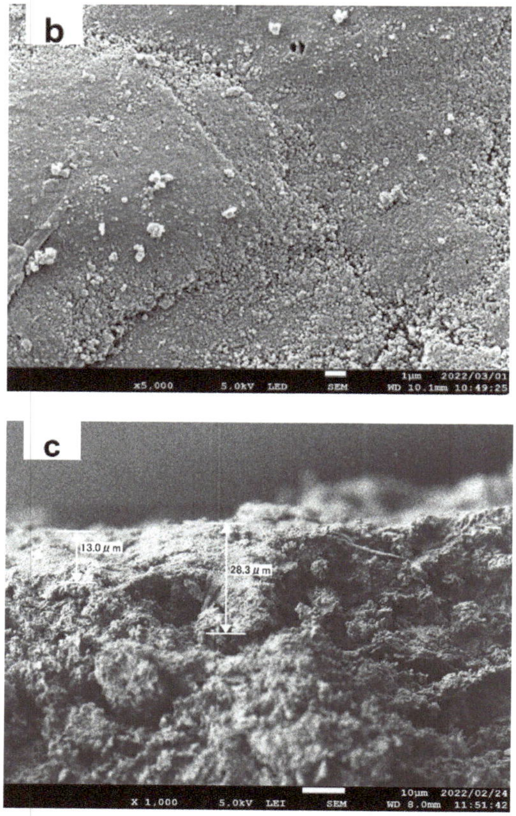

Figure 2. Membrane morphologies (**a**) surface view of the porous ceramic disc, (**b**) surface view of the TiO_2 membrane, (**c**) cross-sectional view of the TiO_2 membrane.

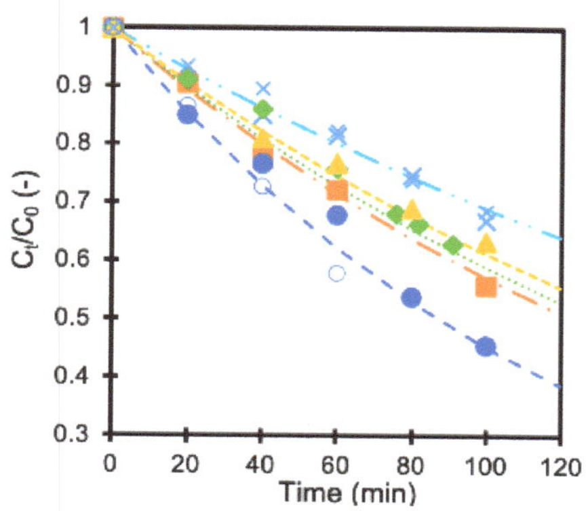

Figure 3. Normalized formic acid concentration as a function of time (x: TiO_2 membranes, x: TiO_2 membranes 2nd run, ■ AgT1 membrane, ● AgT2 membrane, ○ AgT2 membrane 2nd run, ◆ AgT3 membrane, ▲ AgT4 membrane).

The concentration change was fitted (lines in Figure 3) with the first-order equation (Equation (1)) as reported earlier [16].

$$C_t/C_0 = e^{-kt} \tag{1}$$

where k is a rate constant, and $C_{0,t}$ are the formic acid concentrations at the start and at a time, t, respectively. The calculated rate constant values are summarized in Table 1. Applying small amounts of silver to the TiO_2 membranes improved the formic acid decomposition rate as reported earlier, for it reduces the recombination of an electron-hole pair that facilitates the formation of oxidants [16,19]. However, an optimum amount of silver maximizes the rate constant. An excessive amount of silver may act as the center for electron-hole recombination, which is considered to be one of the reasons for its negative influence [20].

Table 1. Influence of silver on the rate constant values obtained in formic acid decomposition.

Membrane No.	Mass of Silver (mg)	$k \times 10^2$ (min^{-1})	Coefficient of Determination (R^2)
TiO_2	0	0.39	0.942
AgT1	0.03	0.57	0.962
AgT2	0.15	0.78	0.976
AgT3	0.37	0.50	0.963
AgT4	3.1	0.48	0.975

3.2. Effect of Salinity on the Photocatalytic Performance

The TiO_2 membrane and the Ag/TiO_2 (AgT2) membrane showing the fastest decomposition rate were used to examine the influence of coexisting salts. Figure 4a,b show the influence of adding different types of salts to the formic acid solution in which TiO_2 and Ag/TiO_2 membranes were immersed, respectively. The decomposition of formic acid became slower when the salts were added to the solutions. For example, the decomposition rate became about half when 60 mmol/L NaCl coexisted in the formic acid solution (Figure 4a). Adding 60 mmol/L NaCl and 30 mmol/L $MgCl_2$ to the formic acid solutions caused almost the same reduction in the decomposition rate for both TiO_2 and Ag/TiO_2 membranes. Adding $MgSO_4$ and K_2SO_4 further hindered the formic acid decomposition. Interestingly, the performances were recovered by washing the membranes with water (open keys in the figure as control). These results suggest that anions, $SO_4{}^{2-}$ and Cl^-, inhibit formic acid decomposition. A similar reduction by the inorganic salts was observed with TiO_2 and Ag/TiO_2 membranes prepared on alumina supports. Consequently, the support materials seem to have a negligible influence on the effect of salts.

Figure 5 shows the influence of NaCl and $MgSO_4$ concentrations on the rate constant of formic acid decomposition. Adding $MgSO_4$ to the formic acid solution reduced the photocatalytic activity of both TiO_2 and Ag/TiO_2 membranes. The $MgSO_4$ concentration did not affect the decomposition rate. Both TiO_2 and Ag/TiO_2 membranes showed almost the same rate constant when $MgSO_4$ was added, showing no relevant enhancement of depositing silver to the TiO_2 membrane. A NaCl concentration greater than 6 mmol/L in the solution hindered the decomposition of formic acid by the TiO_2 membrane to the same degree as $MgSO_4$. On the contrary, NaCl addition showed less influence on the Ag/TiO_2 membrane performance. The decomposition rate with the latter was higher compared to that with the TiO_2 membrane when membranes were applied in solutions of the same NaCl concentration.

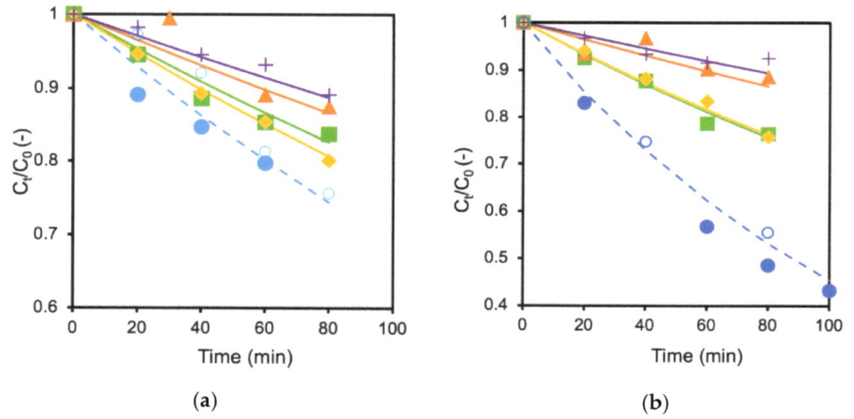

Figure 4. Influence of different salts toward formic acid decomposition; (**a**) TiO_2 membrane, (**b**) Ag/TiO_2 membrane; • No salt, ■ with 60 mmol/L NaCl, ◆ with 30 mmol/L $MgCl_2$, ▲ with 60 mmol/L $MgSO_4$ and + with 60 mmol/L K_2SO_4, ○ membrane performance after washing (no salt).

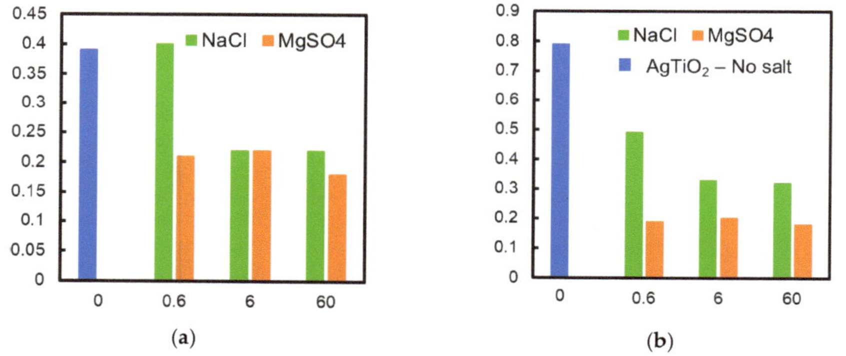

Figure 5. Influence of salt concentration on the rate constant (**a**) TiO_2 and (**b**) Ag/TiO_2 membrane; results obtained without any salt addition and addition of NaCl and $MgSO_4$ at different concentrations are shown in the figure.

Since chloride enhances silver dissolution [21], this process under experimental conditions was checked. Membranes were prepared under the same conditions as for the AgT2 membrane in Table 1. The membranes were soaked in a 40 g solution for one hour and the concentration of silver was measured by ICP. Silver dissolution was enhanced about 8 times in the NaCl-containing solution under dark conditions. The dissolution was almost negligible under UV light (Table 2).

Table 2. Silver dissolution from Ag/TiO_2 membrane in different solutions.

Solutions	Silver Concentration (mg/L)	
	Dark	**UV Irradiation**
Formic acid	0.011	0.0038
Formic acid + NaCl	0.088	0.0037
Formic acid + $MgSO_4$	0.015	0.0034

Figure 6 shows the zeta potential of TiO_2 (P25) particles and Ag/TiO_2 particles measured using a Zetasizer Ultra (Malvern Instruments Ltd., Malvern, UK). HNO_3 and NaOH were used to adjust the pH of the solution. The Ag/TiO_2 particles were prepared in the same manner as for membrane preparation. The amount of silver on TiO_2 particles, 0.89 and 8.9 mg Ag/g TiO_2, were comparable to the membranes. For example, the AgT2 membrane in Table 2 had a silver amount of ca. 2.5 mg Ag/g TiO_2.

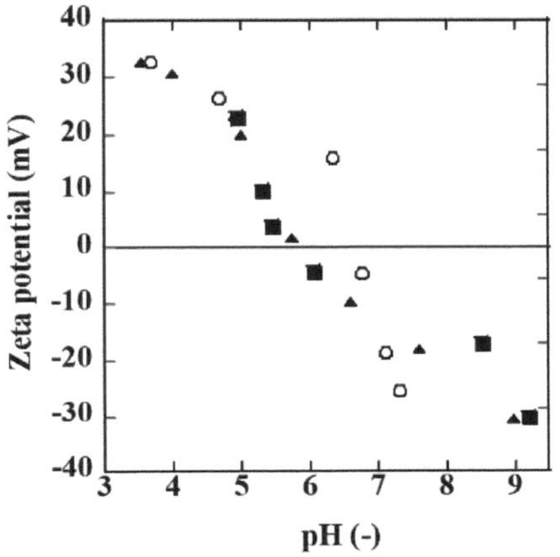

Figure 6. Zeta potential as a function of pH (○: TiO_2 (P25), ■: Ag/TiO_2 (0.89 mg Ag/g TiO_2), ▲: Ag/TiO_2 (8.9 mg Ag/g TiO_2).

The isoelectric point (IEP) of TiO_2 was about 6.5, which is a value reported similarly in [22]. The IEP value became smaller with silver deposition on TiO_2. The difference in the silver amount did not affect the IEP. The pH of the formic acid solution used to evaluate the photocatalytic property of the membranes changed from ca. 3.1 to 3.3 during the decomposition of the acid. Accordingly, the surface of both the TiO_2 and Ag/TiO_2 membranes was positively charged when immersed in the formic acid solution. The SO_4^{2-} and Cl^- anions coexisting in the solution can adsorb on the membrane surface and may hinder the photocatalytic reactions by hindering the adsorption of hydroxide to the surface which reduces the formation of oxidants, for example.

4. Conclusions

Ag/TiO_2 membranes were prepared by the photochemical deposition of silver on TiO_2 membranes. A small amount of silver deposition improved the photocatalytic decomposition rate of formic acid dissolved in wastewater. The excessive addition of silver reduced the decomposition rate. The largest enhancement was with 2.5 mg Ag/g TiO_2. Salinity in water reduced the decomposition property of both the TiO_2 and Ag/TiO_2 membranes, but the membrane performance recovered by washing the membrane with water. The addition of NaCl influenced the Ag/TiO_2 membranes less than the TiO_2 membranes. The reduction in oxidation performance can be attributed to the anion adsorption on the membrane surface, which is positively charged in a formic acid solution. Ag deposition on TiO_2 shifts the IEP value to lower pH, which may reduce the anion influence at around a pH of 6. The zeta potential cannot explain the different influences of SO_4^{2-} and Cl^- on the Ag/TiO_2 and TiO_2 membranes and further study is required to understand this mechanism. Even though the decomposition rate became about half when using the coexisting salts, photocatalytic membranes decomposed dissolved formic acid. The results suggest a wider application potential of photocatalytic membranes.

Supplementary Materials: The following supporting information can be downloaded at: https://www.mdpi.com/article/10.3390/ijerph192315736/s1. Figure S1: XRD patterns (a) porous ceramic disk, (b) TiO_2 membrane and (c) $AgTiO_2$ membrane; Figure S2: SEM-EDS images of $AgTiO_2$ membrane (cross sectional view) (a) cross sectional image, (b) Al mapping, (c) Si mapping, (d) Ti mapping; Figure S3: XPS of TiO_2 membrane and $AgTiO_2$ membrane (a) wide scan spectra, (b) O_{1s} spectra, (c) Ti_{2p} spectra, and (d) Ag_{3d} spectrum of $AgTiO_2$ membrane with fitting curves.

Author Contributions: Conceptualization, I.K.; validation, A.N.C.A.R., S.Y., H.B. and I.K.; investigation, A.N.C.A.R., S.Y., H.B. and I.K.; resources, S.M. and I.K.; data curation, A.N.C.A.R. and I.K.; writing—original draft preparation, A.N.C.A.R. and I.K.; writing—review and editing, I.K., S.M., T.I. and Y.-T.H.; visualization, A.N.C.A.R. and I.K.; project administration, I.K.; funding acquisition, I.K. All authors have read and agreed to the published version of the manuscript.

Funding: This research was funded by the Japan Science and Technology Agency, JST SICORP, grant number JPMJSC18C5, Japan.

Institutional Review Board Statement: Not applicable.

Informed Consent Statement: Not applicable.

Data Availability Statement: Not applicable.

Conflicts of Interest: The authors declare no conflict of interest.

References

1. He, H.; Chen, Y.; Li, X.; Cheng, Y.; Yang, C.; Zeng, G. Influence of salinity on microorganisms in activated sludge processes: A review. *Int. Biodeterior. Biodegrad.* **2016**, *119*, 520–527. [CrossRef]
2. Béziat, J.-C.; Besson, M.; Gallezot, P.; Durécu, S. Catalytic Wet Air Oxidation of Carboxylic Acids on TiO_2-Supported Ruthenium Catalysts. *J. Catal.* **1999**, *182*, 129–135. [CrossRef]
3. Xie, L.; Lu, Q.; Mao, X.; Wang, J.; Han, L.; Hu, J.; Lu, Q.; Wang, Y.; Zeng, H. Probing the intermolecular interaction mechanisms between humic acid and different substrates with implications for its adsorption and removal in water treatment. *Water Res.* **2020**, *176*, 115766. [CrossRef] [PubMed]
4. Rodríguez, A.; Ovejero, G.; Romero, M.; Díaz, C.; Barreiro, M.; García, J. Catalytic wet air oxidation of textile industrial wastewater using metal supported on carbon nanofibers. *J. Supercrit. Fluids* **2008**, *46*, 163–172. [CrossRef]
5. Zou, L.Y.; Li, Y.; Hung, Y.-T. Wet Air Oxidation for Waste Treatment. In *Advanced Physiochemical Treatment Technologies*; Wang, L.K., Hung, Y.-T., Shammas, N.K., Eds.; Handbook of Environmental Engineering; Humana Press: Totowa, NJ, USA, 2007.
6. Kumakiri, I.; Hokstad, J.; Peters, T.A.; Melbye, A.G.; Ræder, H. Oxidation of aromatic components in water and seawater by a catalytic membrane process. *J. Pet. Sci. Eng.* **2011**, *79*, 37–44. [CrossRef]
7. Iojoiu, E.E.; Walmsley, J.C.; Raeder, H.; Miachon, S.; Dalmon, J.-A. Catalytic membrane structure influence on the pressure effects in an interfacial contactor catalytic membrane reactor applied to wet air oxidation. *Catal. Today* **2005**, *104*, 329–335. [CrossRef]
8. Furtado, R.X.D.S.; Sabatini, C.A.; Zaiat, M.; Azevedo, E.B. Perfluorooctane sulfonic acid (PFOS) degradation by optimized heterogeneous photocatalysis (TiO_2/UV) using the response surface methodology (RSM). *J. Water Process. Eng.* **2021**, *41*, 101986. [CrossRef]
9. Lee, S.-Y.; Park, S.-J. TiO_2 photocatalyst for water treatment applications. *J. Ind. Eng. Chem.* **2013**, *19*, 1761–1769. [CrossRef]
10. Etacheri, V.; Di Valentin, C.; Schneider, J.; Bahnemann, D.; Pillai, S.C. Visible-light activation of TiO_2 photocatalysts: Advances in theory and experiments. *J. Photochem. Photobiol. C Photochem. Rev.* **2015**, *25*, 1–29.
11. Khdary, N.H.; Alkhuraiji, W.S.; Sakthivel, T.S.; Khdary, D.N.; Salam, M.A.; Alshihri, S.; Al-Mayman, S.I.; Seal, S. Synthesis of Superior Visible-Light-Driven Nanophotocatalyst Using High Surface Area TiO_2 Nanoparticles Decorated with Cu_xO Particles. *Catalysts* **2020**, *10*, 872. [CrossRef]
12. Sanzone, G.; Zimbone, M.; Cacciato, G.; Ruffino, F.; Carles, R.; Privitera, V.; Grimaldi, M. Ag/TiO2 nanocomposite for visible light-driven photocatalysis. *Superlattices Microstruct.* **2018**, *123*, 394–402. [CrossRef]
13. Chen, H.; Zahraa, O.; Bouchy, M. Inhibition of the adsorption and photocatalytic degradation of an organic contaminant in an aqueous suspension of TiO_2 by inorganic ions. *J. Photochem. Photobiol. A Chem.* **1997**, *108*, 37–44. [CrossRef]
14. Makita, M.; Harata, A. Photocatalytic decolorization of rhodamine B dye as a model of dissolved organic compounds: Influence of dissolved inorganic chloride salts in seawater of the Sea of Japan. *Chem. Eng. Process. Process Intensif.* **2008**, *47*, 859–863. [CrossRef]
15. Kumakiri, I.; Diplas, S.; Simon, C.; Nowak, P. Photocatalytic Membrane Contactors for Water Treatment. *Ind. Eng. Chem. Res.* **2011**, *50*, 6000–6008. [CrossRef]
16. Pipelzadeh, E.; Derakhshan, M.V.; Babaluo, A.A.; Haghighi, M.; Tavakoli, A. Formic Acid Decomposition Using Synthesized Ag/TiO_2 Nanocomposite in Ethanol-Water Media Under Illumination of Near UV Light. *arXiv* **2018**, arXiv:physics/1811.12508.

17. Tran, T.T.H.; Bui, T.T.H.; Nguyen, T.L.; Man, H.N.; Tran, T.K.C. Phase-pure brookite TiO_2 as a highly active photocatalyst for the degradation of pharmaceutical pollutants. *J. Electron. Mater.* **2019**, *48*, 7846–7861.
18. Kumakiri, I.; Murasaki, K.; Yamada, S.; Azzah Nazihah, C.A.R.; Ishii, H. A Greener Procedure to Prepare TiO_2 Membranes for Photocatalytic Water Treatment Applications. *J. Membr. Sci. Res.* **2022**, *8*. *in progress*. [CrossRef]
19. Chakhtouna, H.; Benzeid, H.; Zari, N.; Bouhfid, R. Recent progress on Ag/TiO_2 photocatalysts: Photocatalytic and bactericidal behaviors. *Environ. Sci. Pollut. Res.* **2021**, *28*, 44638–44666. [CrossRef] [PubMed]
20. Seery, M.K.; George, R.; Floris, P.; Pillai, S.C. Silver doped titanium dioxide nanomaterials for enhanced visible light photocatalysis. *J. Photochem. Photobiol. A Chem.* **2007**, *189*, 258–263. [CrossRef]
21. Levard, C.; Mitra, S.; Yang, T.; Jew, A.D.; Badireddy, A.R.; Lowry, G.V.; Brown, G.E. Effect of Chloride on the Dissolution Rate of Silver Nanoparticles and Toxicity to *E. coli. Environ. Sci. Technol.* **2013**, *47*, 5738–5745. [CrossRef]
22. Lee, H.-S.; Hur, T.; Kim, S.; Kim, J.-H.; Lee, H.-I. Effects of pH and surface modification of TiO_2 with SiOx on the photocatalytic degradation of a pyrimidine derivative. *Catal. Today* **2003**, *84*, 173–180. [CrossRef]

International Journal of
Environmental Research and Public Health

MDPI

Article

Influence of COD in Toxic Industrial Wastewater from a Chemical Concern on Nitrification Efficiency

Iwona B. Paśmionka [1,*], Piotr Herbut [2,3], Grzegorz Kaczor [4], Krzysztof Chmielowski [4], Janina Gospodarek [1], Elżbieta Boligłowa [1], Marta Bik-Małodzińska [5] and Frederico Márcio C. Vieira [3]

[1] Department of Microbiology and Biomonitoring, Faculty of Agriculture and Economics, University of Agriculture in Krakow, 31-120 Krakow, Poland
[2] Department of Rural Building, Faculty of Environmental Engineering and Land Surveying, University of Agriculture in Krakow, 31-120 Krakow, Poland
[3] Biometeorology Study Group (GEBIOMET), Universida de Tecnológica Federal do Paraná (UTFPR), Estrada para Boa Esperança, km 04, Comunidade São Cristóvão, Dois Vizinhos 85660-000, Brazil
[4] Department of Sanitary Engineering and Water Management, Faculty of Environmental Engineering and Land Surveying, University of Agriculture in Krakow, 31-120 Krakow, Poland
[5] Institute of Soil Science, Engineering and Environmental Management, University of Life Sciences in Lublin, 20-069 Lublin, Poland
* Correspondence: iwona.pasmionka@urk.edu.pl; Tel.: +48-126624400

check for updates

Citation: Paśmionka, I.B.; Herbut, P.; Kaczor, G.; Chmielowski, K.; Gospodarek, J.; Boligłowa, E.; Bik-Małodzińska, M.; Vieira, F.M.C. Influence of COD in Toxic Industrial Wastewater from a Chemical Concern on Nitrification Efficiency. *Int. J. Environ. Res. Public Health* **2022**, *19*, 14124. https://doi.org/10.3390/ijerph192114124

Academic Editors: Yung-Tse Hung, Hamidi Abdul Aziz, Issam A. AlKhatib and Paul B. Tchounwou

Received: 3 October 2022
Accepted: 27 October 2022
Published: 29 October 2022

Abstract: COD is an arbitrary indicator of the content of organic and inorganic compounds in wastewater. The aim of this research was to determine the effect of COD of industrial wastewater on the nitrification process. This research covered wastewater from acrylonitrile and styrene–butadiene rubbers, emulsifiers, polyvinyl acetate, styrene, solvents (butyl acetate, ethyl acetate) and owipian® (self-extinguishing polystyrene intended for expansion) production. The volume of the analyzed wastewater reflected the active sludge load in the real biological treatment system. This research was carried out by the method of short-term tests. The nitrification process was inhibited to the greatest extent by wastewater from the production of acrylonitrile (approx. 51%) and styrene–butadiene (approx. 60%) rubbers. In these wastewaters, nitrification inhibition occurred due to the high COD load and the presence of inhibitors. Four-fold dilution of the samples resulted in a two-fold reduction in the inhibition of nitrification. On the other hand, in the wastewater from the production of emulsifiers and polyvinyl acetate, a two-fold reduction in COD (to the values of 226.4 $mgO_2{\cdot}dm^{-3}$ and 329.8 $mgO_2{\cdot}dm^{-3}$, respectively) resulted in a significant decrease in nitrification inhibition. Wastewater from the production of styrene, solvents (butyl acetate, ethyl acetate) and owipian® inhibited nitrification under the influence of strong inhibitors. Lowering the COD value of these wastewaters did not significantly reduce the inhibition of nitrification.

Keywords: chemical wastewater; activated sludge nitrifiers; inhibition of nitrification

1. Introduction

Municipal and industrial wastewater is a carrier of many dangerous organisms and substances, which include parasites and pathogenic organisms [1–3], toxic substances and compounds disturbing the natural biological balance in treated sewage receivers by increasing the saprobicity and eutrophication of these waters [4–6]. Such wastewater can be neutralized by mechanical–biological or mechanical–biological–chemical methods of treatment with the use of activated sludge. Effective detection of any disruptions and instabilities in the processes taking place during wastewater treatment enables quick and effective actions to be taken to eliminate them, thus ensuring safety for the environment and human health [7–10].

The effectiveness of wastewater treatment by the activated sludge method depends on the appropriate balance between the number of nitrifying and heterotrophic bacteria, among other factors [11]. The dominant group of microorganisms in the activated sludge is determined by the quality and quantity of substrates contained in raw sewage and the availability of nutrients [12]. Nitrification is the most sensitive part of biological nitrogen removal from wastewater, with autotrophic nitrifying biomass being about 10 times more sensitive to inhibitory factors than its oxygen heterotrophic counterpart [13]. In the nitrification process, chemolithoautotrophic nitrosobacteria of the genera *Nitrosomonas*, *Nitrosovibrio*, *Nitrosococcus*, *Nitrosospira* and *Nitrosolobus* transform ammonium nitrogen to nitrite [14,15]. Then, nitrites are oxidized to nitrates with the participation of nitrobacteria of the genera *Nitrobacter*, *Nitrospira*, *Nitrococcus* and *Nitrospina* [16]. The critical nitrogen removal step is initiated by the conversion of ammonia to nitrate by the nitrifying microorganisms. Nitrification is considered to be a key step in biological wastewater treatment as it converts the toxic ammonia into nitrates, which can be further converted by denitrification into environmentally harmless gaseous forms of nitrogen. The nitrification process is sensitive to the presence of toxic substances in the wastewater. This is of particular importance in the case of industrial wastewater. These wastewaters, depending on the type of industry, are characterized by a different quantitative, and especially qualitative, composition. Many chemical compounds in this wastewater can inhibit the nitrification process. These compounds include, among others: butadiene, styrene, alkylbenzenesulfonic acid, acetic acid, butylcatechol, tricresol, acetonitrile, alkyldimethylbenzylammonium chloride, alkyldimethylbenzylammonium bromide and many other substances, which have been widely documented in the literature. Many of these compounds are quaternary ammonium compounds that found widespread use as disinfectants during COVID-19. They induce a reduction in the intensity of nitrification or even put a complete stop to this process [17,18].

Inhibition of nitrification in industrial wastewater can be attributed to the presence of compounds with hydrophobic properties, a high molecular weight and aromatic structures [19]. According to Pagg et al. [20], inhibition of nitrification depends on the level of biodegradability of the potential inhibitory compound and is less severe in the case of biodegradable substances. The biodegradability of wastewater may be indicated by the ratio of COD to BOD_5 [21]. In order to consider raw sewage as susceptible to biodegradation, the value of the COD/BOD_5 relationship, according to Siwiec et al. [22], Samudro et al. [23] and Bader et al. [24], should be lower than 2.2 or even 2.0. On the other hand, the ratio of the concentration of easily digestible organic compounds to the concentration of total nitrogen and total phosphorus is decisive for the possible effectiveness of nutrient removal from municipal wastewater [25].

The growth rate of nitrifying bacteria is influenced by temperature, substrate concentration, oxygen content, pH and the presence of toxic inhibitory compounds, i.e., inhibitors. In a study conducted by Bawiec [26], free ammonia was a factor that inhibited the metabolism of bacteria of the genus *Nitrobacter*. The amount of free ammonia in sewage depends on the concentration of ammonium ions, temperature and pH of the sewage [5]. The production of excessive ammonia nitrogen emissions is a common problem in many types of industrial wastewater [19].

Heavy metals are considered to be very strong inhibitors of nitrification. The influence of cadmium, copper and mercury on *Nitrosomonas europaea* in quasi-stationary batch reactors is well-known [27]. Ouyang et al. [28] demonstrated in their research that the increased concentration of Cu in wastewater inhibited the oxidation of ammonium nitrogen rather than nitrite.

Inhibition of nitrification in mechanical–biological municipal and industrial wastewater treatment plants may cause problems resulting from the non-conversion of the ammonium form of nitrogen and thus a high concentration of this ion in the wastewater. Inhibition problems may sometimes occur periodically, which makes it very difficult to find the source of substances inhibiting the nitrification process [29]. The factor influencing the inhibition of nitrification may also be excessive salinity of the wastewater. It causes plasmolysis or a reduction in the activity of organisms. In the presence of salt, there is a lower efficiency of biological nitrogen removal processes, including nitrification, and especially denitrification. According to Dincer and Karga [30], a salt concentration above 2% causes a significant reduction in the efficiency of both nitrification and denitrification, and the authors confirmed in their research that denitrification was more sensitive to salinity than nitrification.

In the case of industrial wastewater from coking plants, although the activated sludge process has been adapted to their treatment, nitrification is often disturbed by high concentrations of chemical compounds, expressed in the COD value. Kim et al. [31] investigated the inhibitory effect of ammonia, thiocyanate, iron cyanide, phenol and *p*-cresol on nitrification in the activated sludge treatment technology. The authors showed that ammonia in a concentration below 350 mg·dm^{-3} did not limit the substrate for nitrifying bacteria. On the other hand, a concentration of thiocyanate above 200 mg·dm^{-3} significantly inhibited nitrification, but iron cyanide with a concentration below 100 mg·dm^{-3} did not. Nitrification was also inhibited by phenol with a concentration above 200 mg·dm^{-3} and *p*-cresol with a concentration above 100 mg·dm^{-3}. In the studies on the treatment of wastewater from chemical production, carried out by Paśmionka and Gospodarek [17], it was found that the nitrification process was inhibited to the greatest extent (72%) by wastewater from styrene–butadiene rubber production. On the other hand, wastewater from the production of methacrylate (polymethyl methacrylate) had the lowest degree of inhibition (16%). The investigated sewage also had a toxic effect on the entire biocenosis and adversely affected the structure of activated sludge flocs. According to Ge et al. [32], measuring the effect of potential inhibitors on the nitrification rate is important for maintaining the appropriate efficiency and effectiveness of mechanical–biological wastewater treatment plants. Juliastuti et al. [33], on the basis of research carried out on industrial wastewater containing selected organic compounds, found that the degree of nitrification inhibition was reduced by the compounds in the following order: chlorobenzene > trichlorethylene > phenol > ethylbenzene. Chlorobenzene, even at the level of 0.25 mg·dm^{-3}, reduces the autotrophic biomass. The nitrification process is completely inhibited by chlorobenzene at the concentration of 0.75 mg·dm^{-3}. Trichlorethylene (TCE) has a less inhibitory effect on the nitrification process. The increase to 50% of inhibition is observed only at the concentration of 0.75 mg·dm^{-3} TCE. The nitrification process is completely inhibited at the concentration of 1 mg·dm^{-3} TCE. Phenol inhibits nitrification by 50% at 3 mg·dm^{-3}. The inhibitory effect of phenol is almost constant in the range of 4–10 mg·dm^{-3}, and the complete inhibition of this process is achieved at 50 mg·dm^{-3}. The inhibitory effect of ethylbenzene is 50% at 8 mg dm^{-3}, and the autotrophic biomass is completely inactive at 50 mg·dm^{-3}. In a study by Park et al. [34], the effect of phenol on nitrification was assessed. After 400 mg·dm^{-3} of phenol was added to the sewage, the NO_3-N concentration in the sewage outflow decreased from 69.24 and 51.24 mg·dm^{-3} to 1.89 and 1.51 mg·dm^{-3}, respectively, within 14 days. After lowering the phenol concentration to 60 mg·dm^{-3}, the efficiency of nitrification gradually increased.

The aim of this research was to determine the effect of the concentration of organic and inorganic pollutants (expressed in COD value) in industrial wastewater on the degree of inhibition of nitrification. The study covered wastewater from acrylonitrile and styrene–butadiene rubbers, emulsifiers, polyvinyl acetate, styrene, solvents (butyl acetate, ethyl acetate) and owipian® (self-extinguishing polystyrene intended for expansion) production.

2. Materials and Methods

2.1. Description of the Research Object

The research was conducted in the Municipal and Industrial Sewage Treatment Plant, located in the city of Oświęcim in the Monowice district (50°02′17.1″ N 19°19′13.8″ E) in Lesser Poland. The sewage treatment plant is supplied with municipal sewage from the city and commune of Oświęcim as well as industrial sewage from the Synthos chemical plant.

The planned capacity of the sewage treatment plant is 53,400 m^3·d^{-1}, with industrial sewage amounting to 26,400 m^3·d^{-1}, while municipal sewage amounts to 27,000 m^3·d^{-1}. The facility's technology is based on the processes of mechanical, chemical and biological treatment of industrial and municipal wastewater, using the activated sludge method.

Municipal and industrial wastewater is pretreated in separate technological lines consisting of grates, sand traps, degreasers and primary sedimentation tanks. Its purpose is to separate floating and dragged pollutants, sand, fats, organic suspension from sewage and to correct the pH to a value that allows for their further biological treatment. Industrial wastewater is additionally neutralized and coagulated in a mixer system. Pretreated wastewater streams are mixed and directed to biological treatment. The biological treatment system uses an anaerobic chamber, four aeration chambers, three secondary radial settling tanks, a blower station and an activated sludge-pumping station. Wastewater treatment takes place in two stages:

I. The first stage (anaerobic) is carried out in an anaerobic chamber to which sewage, free of dissolved oxygen, is supplied. This stage guarantees the occurrence of favorable conditions conducive to the development of anaerobic organisms responsible for the reduction processes.
II. The second stage (aerobic) takes place in aeration chambers equipped with agitators and an installation for fine-bubble aeration with compressed air. The concentration of dissolved oxygen in the aerobic chambers ranges from 1.5 to 2.5 mg·dm^{-3}, which favors the development of organisms responsible for the aerobic decomposition of pollutants.

Treated sewage is separated from activated sludge in secondary radial settling tanks. After the sedimentation process, the activated sludge is collected in the central hopper of the secondary radial settling tank, from where it then flows into the chamber, from which it is returned to the process as recirculated sludge or is pumped out as excess sludge. The treated wastewater is collected through the overflow troughs on the outskirts of the settling tanks, and from there it is discharged to the receiver, which is the Macocha stream, a tributary of the Vistula.

Active sludge cultivated in the presence of pretreated industrial sewage from the Synthos chemical plant, mixed with municipal sewage, was used in the conducted research. Table 1 presents the average values of technological parameters of the activated sludge used during the research.

Table 1. Technological parameters of the activated sludge used for the research at the Municipal and Industrial Sewage Treatment Plant in Oświęcim.

Parameter	Unit	Value
temperature	°C	14.6
oxygenation	mgO_2·dm^{-3}	1.5–3
aeration time	h	4–5
total suspension of the activated sludge	mg·dm^{-3}	3200–5300
total suspension in activated excess sludge	mg·dm^{-3}	6000–8000
increase in excess activated sludge	m^3·d^{-1}	300–420
recirculation	% in relation to the incoming sewage	95–120
dry mass of activated sludge	%	1.69
age of the activated sludge	d	13–14
load BOD_5	$kgBOD_5 \cdot kg_{d.m.} \cdot d^{-1}$	0.13–0.28

2.2. Research Description

The tests were carried out in static conditions, during a four-hour aeration process of properly prepared activated sludge and tested sewage. Wastewater from plants producing individual organic compounds was analyzed. Before starting the tests, the activated sludge (in the amount of 1 dm^3) was centrifuged and then washed with an aqueous solution of sodium bicarbonate and ammonium sulfate. The thus-prepared pellet was diluted with distilled water in the ratio 1:10 and centrifuged again. The activated sludge was suspended in the appropriate volume of tap water in order to obtain the concentration required for the tests. Then, the medium was prepared by dissolving 5.04 g of $NaHCO_3$ and 2.64 g of $(NH_4)_2SO_4$ in 1 dm^3 of distilled water. Further details are provided below:

(a) Control flask contents:
- Activated sludge: 125 cm^3;
- Medium: 25 cm^3;
- Distilled water: 100 cm^3.

(b) Comparative flask contents:
- Activated sludge: 125 cm^3;
- Medium: 25 cm^3;
- Reference inhibitor, allylthiourea (ATU): 2.5 cm^3;
- Distilled water: 97.5 cm^3.

(c) Test flask contents:
- Activated sludge: 125 cm^3;
- Medium: 25 cm^3;
- Tested sewage in the amount of 25 cm^3, 50 cm^3, 75 cm^3 and 100 cm^3;
- Distilled water in the amount of 75 cm^3, 50 cm^3, 25 cm^3 and 0 cm^3 (according to the amount of sewage tested).

The volume of the mixture in each flask was 250 cm^3. The systems prepared in this way were aerated for 4 h with moist, compressed air using VEB ELMET TYP Fp 09 pumps. The concentration of dissolved oxygen in the systems was about 2 $mg \cdot dm^{-3}$. Immediately after 4 h of aeration, a sample was taken from each flask, and the concentration of NH_4-N, NO_2-N and NO_3-N was determined after filtering. The test results are expressed as the percentage of nitrification inhibition (IN), calculated in relation to the control flask and to the flask with the comparative inhibitor, according to Formula (1):

$$\%IN = \frac{C_C - C_T}{C_C - C_B} \cdot 100 \quad (1)$$

in which:

C_C—concentration of oxidized forms of nitrogen (NO_2-N + NO_3-N) in the control flask after 4 h of aeration [$mg \cdot dm^{-3}$];
C_T—concentration of oxidized forms of nitrogen (NO_2-N + NO_3-N) in the flask with the tested sewage, after 4 h of aeration [$mg \cdot dm^{-3}$];
C_B—concentration of oxidized forms of nitrogen (NO_2-N + NO_3-N) in a flask with a comparative inhibitor (ATU) after 4 h of aeration [$mg \cdot dm^{-3}$].

Ahlstrom Munktell 389 filters were used to filter the samples.

The study used wastewater inhibiting the nitrification process to a degree higher than 30.0%, coming from the production of styrene–butadiene and acrylonitrile rubbers, styrene, solvents (butyl acetate, ethyl acetate), emulsifiers, polyvinyl acetate and owipian®. Wastewater in the amount of 25 cm^3, 50 cm^3, 75 cm^3 and 100 cm^3 was used for the tests, with a variable COD value.

2.3. Determination of NH_4-N, NO_2-N and NO_3-N Concentrations

The concentrations of NH_4-N, NO_2-N and NO_3-N were determined by the colorimetric method using the CADAS 30S biochemical spectrophotometer (Dr Lange, The Nederland) and appropriate cuvette tests. Measurements were made in triplicate.

2.4. Calculation of the Nitrification Activity of the Activated Sludge and the Specific Rate of Nitrification

The nitrifying activity of the activated sludge was calculated based on the differences in the concentration of oxidized forms of nitrogen (NO_3-N and NO_2-N) in the control flask and the flask with the comparative inhibitor, after 4 h of aeration, using Formula (2):

$$R = \frac{C_C - C_B}{MLVSS \cdot 4} \quad (2)$$

in which:

R—nitrification activity of activated sludge [mg·g·h^{-1}];
C_C—concentration of oxidized forms of nitrogen (NO_2-N + NO_3-N) in the control flask after 4 h of aeration [mg·dm^{-3}];
C_B—concentration of oxidized forms of nitrogen (NO_2-N + NO_3-N) in a flask with a comparative inhibitor (ATU), after 4 h of aeration [mg·dm^{-3}];
$MLVSS$—mixed liquor volatile suspended solids [mg·dm^{-3}].

The specific rate of nitrification was determined on the basis of changes in NH_4-N concentration in the control flask and the flask with the comparative inhibitor, after 4 h of aeration, according to Formula (3):

$$S = \frac{C(NH_4 - N)_B - C(NH_4 - N)_C}{MLVSS \cdot 4} \quad (3)$$

in which:

S—specific rate of nitrification [mg·g·h^{-1}];
$C(NH_4 - N)_B$—ammonium nitrogen concentration in a flask with a comparative inhibitor (ATU) after 4 h of aeration [mg·dm^{-3}];
$C(NH_4 - N)_C$—ammonium nitrogen concentration in the control flask, after 4 h of aeration [mg·dm^{-3}];
$MLVSS$—mixed liquor volatile suspended solids [mg·dm^{-3}].

2.5. Verification of Measurement Data

Before the statistical analysis was performed, the measurement data were verified with regard to outliers. In order to apply appropriate parametric or nonparametric tests of outlier detection, at the beginning, the statistical distributions of individual observational data strings were assessed. The COD values, expressed in mg·dm^{-3}, and the values of the nitrification inhibition parameter (IN), expressed in%, were analyzed. The normality of individual data distributions was tested with the Shapiro–Wilk test. In the null hypothesis H_0 for this test, it was assumed that the given research sample comes from a normally distributed population. The significance level for this test was assumed to be $\alpha = 0.05$. The results of the Shapiro–Wilk test for measurement data, including COD and IN values, are presented in Table 2.

Since for each data string, the assumed significance level $\alpha = 0.05$ is lower than the calculated value of the test probability p-value, there is no reason to reject the null hypothesis that the distribution of individual data is normal. The normality of the distribution of measurement data allowed for the verification of outliers by applying the parametric Grubbs' T-test. The calculations were performed with the statistical package Statistica v. 13.1 (StatSoft, Tulsa, OK, USA). The results of the Grubbs' T-test are summarized in Table 3.

Table 2. The results of the assessment of the normality of the distribution of data including COD and IN values (Shapiro–Wilk test).

Production Line	Shapiro–Wilk Normality Test Results for COD			Shapiro–Wilk Normality Test Results for IN		
	W	α	*p*-Value	W	α	*p*-Value
acrylonitrile rubbers	1.000	0.05	1.000	0.808	0.05	0.133
styrene–butadiene rubbers	1.000		1.000	0.987		0.785
emulsifiers	1.000		0.995	0.993		0.844
polyvinyl acetate	0.990		0.806	0.850		0.240
styrene	1.000		1.000	1.000		1.000
solvents (butyl acetate, ethyl acetate)	1.000		0.999	0.986		0.774
owipian®	1.000		0.630	0.861		0.269

W—the value of the Shapiro–Wilk test statistic. α—significance level, 0.05. *p*-value—the value of the calculated test probability.

Table 3. The results of the Grubbs' T-test for the verification of outliers in the measurement series of COD and IN values.

Production Line	Grubbs' T-Test Results for COD			Grubbs' T-Test Results for IN		
	G	α	*p*-Value	G	α	*p*-Value
acrylonitrile rubbers	1.000	0.05	1.000	1.152	0.05	0.133
styrene–butadiene rubbers	1.000		1.000	1.059		0.785
emulsifiers	1.001		0.995	1.044		0.844
polyvinyl acetate	1.053		0.806	1.146		0.240
styrene	1.000		1.000	1.000		1.000
solvents (butyl acetate, ethyl acetate)	1.000		0.999	1.061		0.774
owipian®	1.006		0.980	1.143		0.269

G—the value of Grubbs' T-test statistic. α—significance level, 0.05. *p*-value—the value of the calculated test probability.

The data in Table 3 show that for each analyzed measurement data sequence, there are no grounds to reject the null hypothesis H_0 stating that there are no outliers in a given data sequence.

2.6. Linear Regression Analysis

Statistical analysis of linear regression was used to determine the strength and direction of the impact of the COD value in industrial wastewater on the process of nitrification inhibition. The independent variable in this analysis was the value of COD in the wastewater, defined in $mgO_2\ dm^{-3}$, while the dependent variable was the parameter IN, given as a percentage, calculated according to Formula (1). The relationship between the tested parameters in individual types of industrial wastewater is described by Equation (4):

$$IN = a \cdot COD + b \quad (4)$$

The parameters of the linear regression model were determined by the least-squares method. The significance of the relationship between the analyzed variables, the significance of the slope coefficient of the regression equation and the significance of the coefficient of determination R^2 were verified by the Fisher–Snedecor F-test. Statistical calculations, along with the verification and estimation of the parameters of the regression equation, were performed with the Statistica v. 13.1 computer program (StatSoft, Tulsa, OK, USA).

3. Results

In studies conducted on wastewater from acrylonitrile and styrene–butadiene rubber production, the inhibition of nitrification occurred due to the concentration of pollutants (expressed in the COD value) contained in both wastewater samples and the presence of inhibitors. In the wastewater from the production of acrylonitrile rubbers, a two-fold reduction in the COD value resulted in a reduction in nitrification inhibition from 50.7% in the initial sample to 19.8%. As the wastewater was diluted further, the inhibitory effect was further reduced (Table 4).

Table 4. Influence of COD of industrial wastewater from acrylonitrile rubber production on the nitrification inhibition in activated sludge under laboratory conditions.

Samples	pH	COD ($mgO_2 \cdot dm^{-3}$)	Concentration NH_4-N ($mg \cdot dm^{-3}$)		Concentration NO_3-N ($mg \cdot dm^{-3}$)		% IN
			Before Aeration	After 4 h of Aeration	Before Aeration	After 4 h of Aeration	
control	7.6	-	56.00	23.00	0.00	15.60	-
ATU	7.6	-	56.00	49.00	0.00	0.00	-
sewage 25 cm^3	7.2	168.4	56.00	23.00	0.00	13.54	13.2
sewage 50 cm^3	7.2	336.7	56.00	29.00	0.00	12.51	19.8
sewage 75 cm^3	7.2	505.1	56.00	31.40	0.00	8.07	48.3
sewage 100 cm^3	7.2	673.4	56.00	29.20	0.00	7.69	50.7

MLVSS—4.700 $g \cdot dm^{-3}$, R—0.45, S—0.75.

In the wastewater from styrene–butadiene rubber production, a four-fold dilution of the initial sample with a COD of 627.2 $mgO_2 \cdot dm^{-3}$ caused only a two-fold reduction in nitrification inhibition (from 59.8% to 28.6%). Therefore, it can be concluded that reducing the concentration of pollutants in the tested sample to the COD value of 156.8 $mgO_2 \cdot dm^{-3}$ does not satisfactorily reduce the inhibition observed in the sample with the maximum concentration of pollutants (Table 5).

Table 5. Influence of COD of industrial wastewater from styrene–butadiene rubber production on the nitrification inhibition in activated sludge under laboratory conditions.

Samples	pH	COD ($mgO_2 \cdot dm^{-3}$)	Concentration NH_4-N ($mg \cdot dm^{-3}$)		Concentration NO_3-N ($mg \cdot dm^{-3}$)		% IN
			Before Aeration	After 4 h of Aeration	Before Aeration	After 4 h of Aeration	
control	7.6	-	56.00	22.60	0.00	14.08	-
ATU	7.6	-	56.00	37.40	0.00	0.00	-
sewage 25 cm^3	7.9	156.8	56.00	26.10	0.00	10.05	28.6
sewage 50 cm^3	7.9	313.6	56.00	27.20	0.00	9.69	31.2
sewage 75 cm^3	7.9	470.4	56.00	29.90	0.00	8.07	42.7
sewage 100 cm^3	7.9	627.2	56.00	33.00	0.00	5.66	59.8

MLVSS—2.910 $g \cdot dm^{-3}$, R—1.21, S—1.27.

Tests conducted on wastewater from emulsifiers and polyvinyl acetate production have shown that in this case, the nitrification process was limited by a high COD load. However, even a two-fold dilution of this wastewater significantly reduced the degree of inhibition, and its further dilution caused its gradual decrease. In the case of wastewater from the production of emulsifiers, a two-fold reduction in COD (to the value of

226.4 $mgO_2 \cdot dm^{-3}$) compared to the initial sample resulted in a decrease in nitrification inhibition from 44.5% to 17.0% (Table 6).

Table 6. Influence of COD of industrial wastewater from emulsifiers production on the nitrification inhibition in activated sludge under laboratory conditions.

Samples	pH	COD ($mgO_2 \cdot dm^{-3}$)	Concentration NH_4-N ($mg \cdot dm^{-3}$)		Concentration NO_3-N ($mg \cdot dm^{-3}$)		% IN
			Before Aeration	After 4 h of Aeration	Before Aeration	After 4 h of Aeration	
control	7.6	-	56.00	22.60	0.00	14.08	-
ATU	7.6	-	56.00	37.40	0.00	0.00	-
sewage 25 cm^3	7.5	113.2	56.00	29.10	0.00	12.06	14.3
sewage 50 cm^3	7.5	226.4	56.00	30.70	0.00	11.69	17.0
sewage 75 cm^3	7.5	339.0	56.00	32.80	0.00	10.02	28.8
sewage 100 cm^3	7.5	452.6	56.00	35.00	0.00	7.82	44.5

MLVSS—2.910 $g \cdot dm^{-3}$, R—1.21, S—1.27.

In the wastewater from polyvinyl acetate production, a two-fold reduction in the COD concentration (to the value of 329.8 $mgO_2 \cdot dm^{-3}$) resulted in a decrease in nitrification inhibition from 38.9% to 16.1% (Table 7).

Table 7. Influence of COD of industrial wastewater from polyvinyl acetate production on the nitrification inhibition in activated sludge under laboratory conditions.

Samples	pH	COD ($mgO_2 \cdot dm^{-3}$)	Concentration NH_4-N ($mg \cdot dm^{-3}$)		Concentration NO_3-N ($mg \cdot dm^{-3}$)		% IN
			Before Aeration	After 4 h of Aeration	Before Aeration	After 4 h of Aeration	
control	7.6	-	56.00	22.60	0.00	14.08	-
ATU	7.6	-	56.00	37.40	0.00	0.00	-
sewage 25 cm^3	7.7	164.9	56.00	23.30	0.00	12.25	13.0
sewage 50 cm^3	7.7	329.8	56.00	25.30	0.00	11.81	16.1
sewage 75 cm^3	7.7	494.6	56.00	25.40	0.00	11.38	19.2
sewage 100 cm^3	7.7	730.0	56.00	16.00	0.00	8.60	38.9

MLVSS—2.910 $g \cdot dm^{-3}$, R—1.21, S—1.27.

Tests conducted on wastewater from styrene, solvents (butyl acetate, ethyl acetate) and owipian® production indicate that the inhibition of the nitrification process was due to the presence of a strong inhibitor or many inhibitors of this process in the tested wastewater. The percentage of nitrification inhibition did not decrease with dilution of this wastewater, but remained more or less constant (despite the continuous reduction in the COD value). In the case of sewage from styrene production, the inhibition of nitrification in the initial sample was slightly lower (45.1%, with a COD of 242.1 $mgO_2 \cdot dm^{-3}$), but after four-fold dilution, it remained at a relatively high level, 37.5%, with a COD of only 60.5 $mgO_2 \cdot dm^{-3}$ (Table 8).

The four-fold dilution of wastewater from solvent (butyl acetate, ethyl acetate) production, despite the decrease in COD from 491.6 $mgO_2 \cdot dm^{-3}$ to 122.9 $mgO_2 \cdot dm^{-3}$, did not remove the inhibition effect. The degree of inhibition of the nitrification process remained at the level of approx. 50.0%; thus, it was comparable to the initial sample (Table 9).

Table 8. Influence of COD of industrial wastewater from styrene production on the nitrification inhibition in activated sludge under laboratory conditions.

Samples	pH	COD ($mgO_2 \cdot dm^{-3}$)	Concentration NH_4-N ($mg \cdot dm^{-3}$)		Concentration NO_3-N ($mg \cdot dm^{-3}$)		% IN
			Before Aeration	After 4 h of Aeration	Before Aeration	After 4 h of Aeration	
control	7.6	-	56.00	21.30	0.00	12.96	-
ATU	7.6	-	56.00	48.90	0.00	0.73	-
sewage 25 cm^3	7.4	60.5	56.00	29.70	0.00	8.51	37.5
sewage 50 cm^3	7.4	121.1	56.00	30.16	0.00	8.32	39.1
sewage 75 cm^3	7.4	181.6	56.00	30.48	0.00	7.96	42.1
sewage 100 cm^3	7.4	242.1	56.00	32.19	0.00	7.61	45.1

MLVSS—5.880 $g \cdot dm^{-3}$, R—0.52, S—1.17.

Table 9. Influence of COD of industrial wastewater from solvent (butyl acetate, ethyl acetate) production on the nitrification inhibition in activated sludge under laboratory conditions.

Samples	pH	COD ($mgO_2 \cdot dm^{-3}$)	Concentration NH_4-N ($mg \cdot dm^{-3}$)		Concentration NO_3-N ($mg \cdot dm^{-3}$)		% IN
			Before Aeration	After 4 h of Aeration	Before Aeration	After 4 h of Aeration	
control	7.6	-	56.00	22.60	0.00	14.08	-
ATU	7.6	-	56.00	37.40	0.00	0.00	-
sewage 25 cm^3	7.5	122.9	56.00	25.70	0.00	7.05	49.9
sewage 50 cm^3	7.5	245.6	56.00	27.20	0.00	6.99	50.4
sewage 75 cm^3	7.5	368.7	56.00	27.70	0.00	6.61	53.1
sewage 100 cm^3	7.5	491.6	56.00	28.20	0.00	6.03	57.2

MLVSS—2.910 $g \cdot dm^{-3}$, R—1.21, S—1.27.

A similar result was obtained when testing industrial wastewater from owipian® production. The reduction in the concentration of organic pollutants in the form of COD from 621.3 $mgO_2 \cdot dm^{-3}$ to 155.3 $mgO_2 \cdot dm^{-3}$ did not reduce the inhibition of nitrification, which still amounted to approx. 50.0% (Table 10).

Table 10. Influence of COD of industrial wastewater from owipian® production on the nitrification inhibition in activated sludge under laboratory conditions.

Samples	pH	COD ($mgO_2 \cdot dm^{-3}$)	Concentration NH_4-N ($mg \cdot dm^{-3}$)		Concentration NO_3-N ($mg \cdot dm^{-3}$)		% IN
			Before Aeration	After 4 h of Aeration	Before Aeration	After 4 h of Aeration	
control	7.6	-	56.00	21.30	0.00	12.96	-
ATU	7.6	-	56.00	48.90	0.00	0.73	-
sewage 25 cm^3	7.2	155.3	56.00	25.46	0.00	6.92	49.4
sewage 50 cm^3	7.2	316.1	56.00	26.97	0.00	6.37	53.9
sewage 75 cm^3	7.2	466.0	56.00	28.24	0.00	6.03	56.7
sewage 100 cm^3	7.2	621.3	56.00	31.72	0.00	5.96	57.2

MLVSS—5.880 $g \cdot dm^{-3}$, R—0.52, S—1.17.

Regression analysis was used to test and statistically confirm the relationship between the COD values in individual industrial wastewaters and the IN parameter. Figure 1 shows the scatterplots of measurement data determined from laboratory tests against the regression lines calculated using the least-squares method.

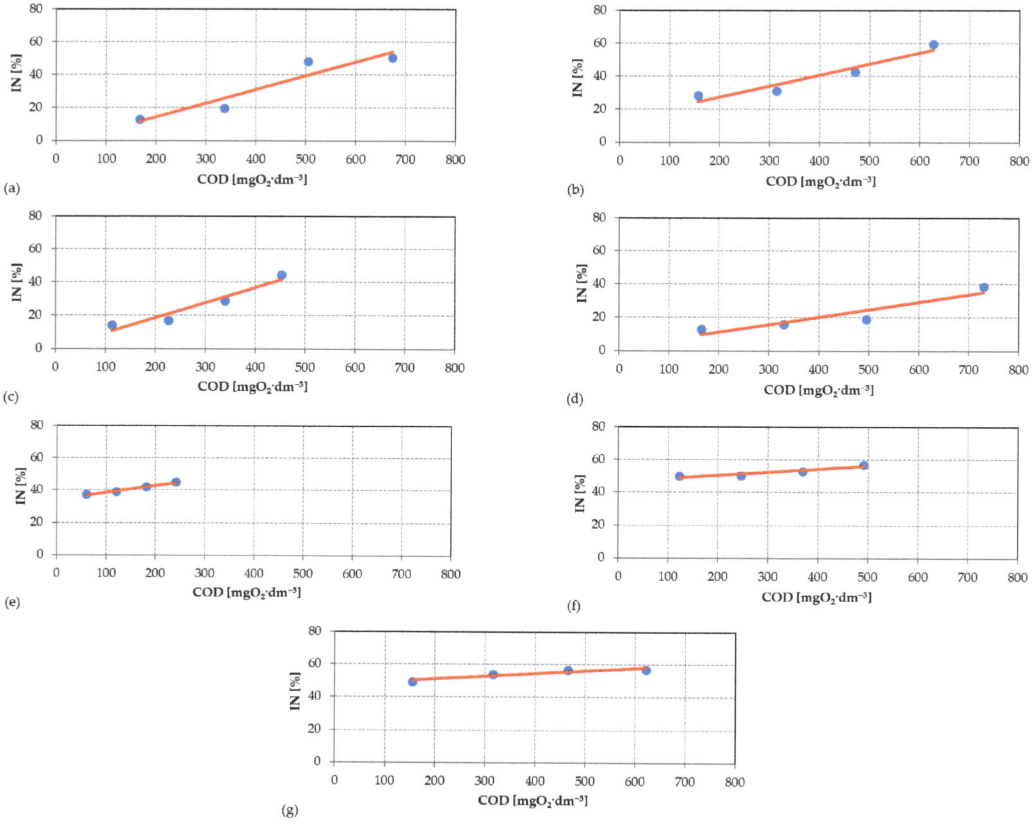

Figure 1. Scatterplots of measurement data from laboratory tests against established regression lines: (**a**) wastewater from acrylonitrile rubber production; (**b**) wastewater from styrene–butadiene rubber production; (**c**) wastewater from emulsifier production; (**d**) wastewater from polyvinyl acetate production; (**e**) wastewater from styrene production; (**f**) wastewater from solvent (butyl acetate, ethyl acetate) production; (**g**) wastewater from owipian® production.

The results of the calculations of the individual parameters of the regression line are summarized in Table 11. The *p*-value is marked in red, which indicates the statistical irrelevance of most often one of the coefficients of the equation for the significance level of 0.05. Although regression models describing the dependence of COD and IN in wastewater from acrylonitrile rubbers, polyvinyl acetate and owipian® production are not statistically certain, high values of R and R^2 confirm a strong correlation between the parameters studied. Additionally, it is worth noting that the greatest statistical uncertainty of the regression model was found to be 6.8% for industrial wastewater from the production of polyvinyl acetate, with the assumed uncertainty level of 5.0%. It follows that the regression line describes 93.2% of the measurement data. The deviation from the assumed materiality level is only 1.8%.

Table 11. Summary of statistical parameters of the linear regression analysis assessing the effect of COD in industrial wastewater on the nitrification inhibition.

Production Line	Equation Parameter Value	R	R^2	F	p-Value
acrylonitrile rubbers	a = 0.0838 b = −2.256	0.945	0.893	16.634	0.055
styrene–butadiene rubbers	a = 0.0670 b = 14.300	0.955	0.912	20.800	0.044
emulsifiers	a = 0.0906 b = 0.537	0.961	0.924	24.144	0.039
polyvinyl acetate	a = 0.0450 b = 2.425	0.931	0.867	13.013	0.068
styrene	a = 0.0430 b = 34.500	0.991	0.983	112.510	0.009
solvents (butyl acetate, ethyl acetate)	a = 0.0200 b = 46.501	0.950	0.903	18.546	0.049
owipian®	a = 0.0170 b = 47.689	0.948	0.900	17.895	0.052

a—the value of the directional coefficient of the regression equation, b—value of the intercept of the regression equation, R—value of the correlation coefficient between the analyzed variables, R^2—value of the determination coefficient, F—the value of the Fisher–Snedecor F statistics, jointly verifying the significance of the relationship between the analyzed variables, the significance of the slope coefficient of the regression equation and the significance of the determination coefficient R^2, p-value—the value of the calculated test probability for the F-test.

Values of the slope of the regression line "a" in the second column of Table 11 indicate the strength of the influence of COD on the value of the IN parameter. This relationship is presented graphically in Figure 2. The individual bars of the graph show by what percent the inhibition of nitrification will decrease, expressed by the value of the IN parameter as a percentage, when the COD value in industrial wastewater drops by 100 $mgO_2 \cdot dm^{-3}$. The presented dependence shows that the pollutants expressed by the COD value most strongly inhibit the nitrification process during the production of emulsifiers and acrylonitrile rubbers, and inhibit it the least during the production of owipian® and solvents.

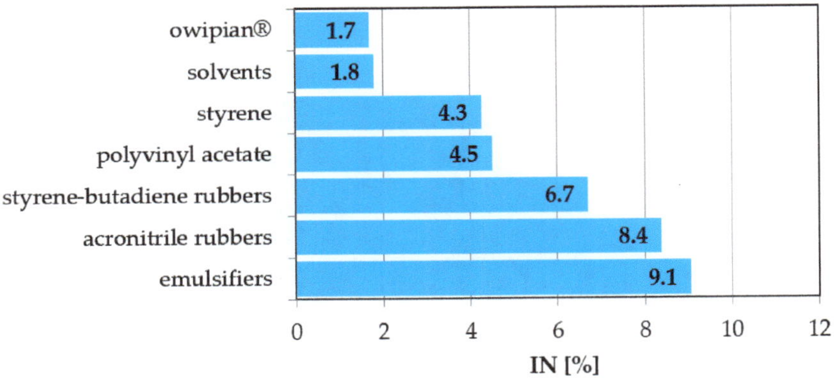

Figure 2. Percentage reduction in nitrification inhibition while reducing the COD value in industrial wastewater by 100 $mgO_2 \cdot dm^{-3}$.

Figure 3 presents the course of the regression lines showing the effect of COD values in industrial wastewater on the inhibition of nitrification expressed by the value of the IN parameter. The presented dependencies indicate that during the production of acrylonitrile rubbers, polyvinyl acetate and emulsifiers, the inhibition of nitrification decreases proportionally with the decrease in the COD value in the wastewater. At very low COD values, the nitrification inhibition process stops. On the other hand, in the case of the production of solvents (butyl acetate, ethyl acetate), with a complete reduction in COD, the inhibition remains constant at 49.9%, and similarly, it remains at 49.4% in the case of the production of owipian® and 37.5% in the production of styrene. It can be concluded from these values that the substances inhibiting the nitrification process are not directly related to the pollutants included in the COD group. These are probably the inhibitory substances found in the tested wastewater.

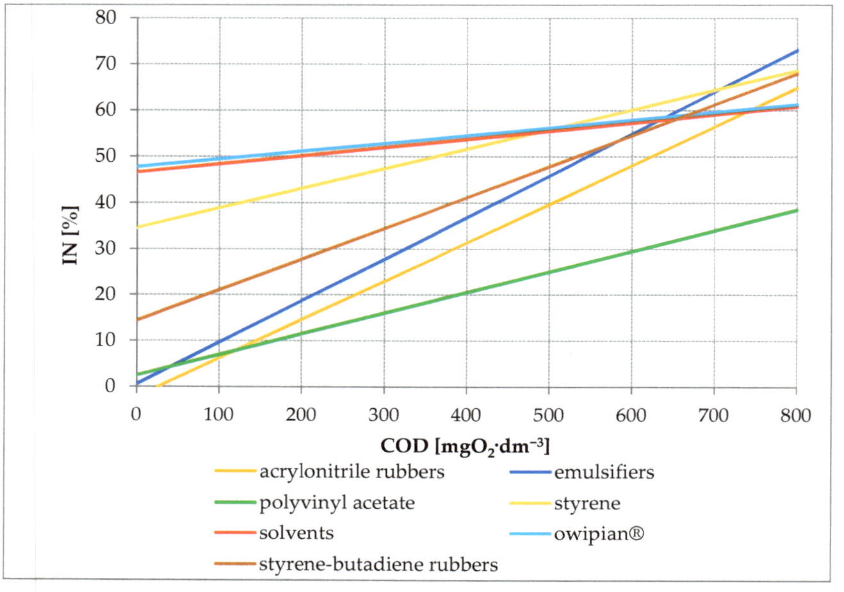

Figure 3. The course of the regression line showing the effect of the COD value in industrial wastewater on the nitrification inhibition expressed by the value of the IN parameter.

4. Discussion

The constantly developing chemical industry affects the degradation of the natural environment, the protection of which requires large financial outlays related to the necessity of neutralizing the resulting wastewater. Pretreated wastewater from production plants is usually directed to a mechanical and biological wastewater treatment plant [35]. Despite pretreatment, this sewage contains numerous organic and inorganic compounds, the concentration of which expresses the COD value. High COD values have a negative impact on purification processes, including nitrification, which was confirmed by our research. COD values exceeding 400 $mgO_2 \cdot dm^{-3}$ caused nitrification inhibition above 50%. COD values below 200 $mgO_2 \cdot dm^{-3}$ only slightly reduced the efficiency of this process, with the exception of sewage from the production of styrene, solvents (butyl acetate, ethyl acetate) and owipian®. In this case, it is likely the factors limiting nitrification were the inhibitors present in the tested chemical sewage. It is also possible that this effect was increased by numerous disinfecting compounds present in municipal wastewater, used during COVID-19 [36].

According to the literature data, the amount of nitrogen removed from the system in the form of biomass ranges from 20 to 30% [12]. These data, however, are not reflected in the biological system of the Municipal and Industrial Sewage Treatment Plant in Oświęcim, in which the amount of nitrogen removed from the system as a result of assimilation is only 12.4% [5]. It should be emphasized that the proportion of carbon to nitrogen incorporated in the biomass in this system is 31:1, which corresponds to the average values given in the available literature. This proves that carbon is not a limiting factor, as the minimum C:N demand ranges from 10:1 to 17:1 [12,21,25]. Limiting the sludge growth and the related decrease in the intensity of nitrification, as observed in our research, is most often the response of activated sludge to dynamic changes in the external environment, which are the cause of a long process of biochemical and physical adaptation in microorganisms, which was also observed by Rahimi et al. [37].

Significant factors limiting or even preventing the course of nitrification in the biological treatment system of the Municipal and Industrial Sewage Treatment Plant in Oświęcim are inhibitors flowing in with industrial sewage and high COD values. The list of organic and inorganic compounds having a negative effect on nitrifying bacteria is very long and includes a number of substances for which the degree of inhibition has been determined [6,13,17,19,20]. Most of these compounds were present in the sewage we analyzed. These substances are much more toxic to nitrifying bacteria than activated sludge heterotrophs. Most industrial wastewater contains various compounds, which, alone or in a mixture, have an inhibitory effect on nitrification [29].

As part of this study, the degree of nitrification inhibition under the influence of various concentrations of chemical pollutants (determined by the COD value) present in selected industrial wastewater was assessed. The research was carried out to determine the COD value in postproduction wastewater, which inhibited the nitrification process. The obtained results clearly show that the high concentration of organic pollutants in most of the tested wastewater was one of the main reasons for a significant reduction in nitrification. This is evidenced by the results obtained during the testing of wastewater from acrylonitrile and styrene–butadiene rubbers, styrene, solvents (butyl acetate, ethyl acetate) and owipian® production. In the tested sewage, sometimes even a four-fold reduction in the concentration of organic pollutants, expressed in the form of COD, did not satisfactorily eliminate the inhibition process. Similar observations were made by Samudro et al. [23], who examined industrial wastewater from the production of urea, urea–formaldehyde adhesives, butanol, 2-ethylhexanol, maleic and phthalic anhydride and adhesive resins.

The conducted research showed that in the biological wastewater treatment system at the Municipal and Industrial Wastewater Treatment Plant in Oświęcim, the nitrification process is clearly disturbed by high COD values and the presence of inhibitory substances in the incoming wastewater.

The diversified production profile in the chemical industry hinders the proper operation of mechanical and biological wastewater treatment plants. In such plants, the amount and composition of sewage as well as the concentration of organic pollutants expressed in the COD value are subject to continuous fluctuations [38,39]. The biocenosis of activated sludge is sensitive to chemical substances [40]; therefore, treatment plants have difficulties in removing toxins and minimizing the concentration of organic pollutants in the incoming sewage [41,42]. This inhibits many processes, including nitrification, which was observed at the WWTP in Oświęcim.

5. Conclusions

Industrial wastewater contains numerous chemicals, including toxic ones, which significantly contribute to the inhibition of the nitrification process. In the Municipal and Industrial Wastewater Treatment Plant in Oświęcim, the most dangerous industrial wastewater comes from the production of acrylonitrile and styrene–butadiene rubbers, emulsifiers, polyvinyl acetate, styrene, solvents (butyl acetate, ethyl acetate) and owipian®. The main reason for the low efficiency of the nitrification process at the WWTP in Oświęcim is the concentration of organic pollutants present in the sewage and the presence of inhibitors. COD values exceeding 400 $mgO_2 \cdot dm^{-3}$ caused nitrification inhibition above 50%. COD values below 200 $mgO_2 \cdot dm^{-3}$ only slightly reduced the efficiency of this process, with the exception of sewage from the production of styrene, solvents (butyl acetate, ethyl acetate) and owipian®. Diluting industrial wastewater with municipal wastewater at a ratio of 1:3, or even 1:1, has a positive effect on the treatment processes, but still does not guarantee the elimination of nitrification inhibition. At present, industrial wastewater destined for biological treatment is mixed with municipal wastewater in a proportion of 2:1 or even 3:1, and it is not possible to dilute it further, due to the insufficient volume of available municipal wastewater. Only a clear advantage of typical domestic sewage in the system would allow for a partial reduction in nitrification inhibition. Nitrifying bacteria present in the activated sludge of the WWTP in Oświęcim are subject to the phenomenon of chronic stress. This is evidenced by the lower nitrification activity of activated sludge than in the literature, which in the period studied was on average about 1.0 $mg \cdot g \cdot h^{-1}$ of oxidized nitrogen. It should be emphasized, however, that under favorable conditions, this activity is sufficient to maintain the parameters of the water-legal permit for the removal of ammoniacal nitrogen from wastewater.

Author Contributions: Conceptualization, I.B.P., P.H. and G.K.; methodology, I.B.P., K.C. and G.K.; software, I.B.P., K.C. and G.K.; validation, P.H., J.G. and F.M.C.V.; formal analysis, I.B.P.; investigation, K.C.; resources, M.B.-M.; data curation, J.G.; writing—original draft preparation, I.B.P.; writing—review and editing, E.B.; visualization, I.B.P., K.C. and G.K.; supervision, E.B.; project administration, P.H., F.M.C.V. and E.B.; funding acquisition, P.H. and E.B. All authors have read and agreed to the published version of the manuscript.

Funding: This research was funded by Ministry of Science and Higher Education of the Republic of Poland and Faculty of Environmental Engineering, University of Agriculture in Krakow, through the project "Scientific subvention D014—Environmental Engineering, Mining and Energy".

Institutional Review Board Statement: Not applicable.

Informed Consent Statement: Not applicable.

Data Availability Statement: Not applicable.

Conflicts of Interest: The authors declare no conflict of interest.

References

1. Paśmionka, I.B.; Bulski, K.; Herbut, P.; Boligłowa, E.; Vieira, F.M.C.; Bonassa, G.; Prá, M.C.D.; Bortoli, M. Evaluation of the Effectiveness of the Activated Sludge Process in the Elimination Both ATB-Resistant and ATB-Susceptible *E. coli* Strains. *Energies* **2021**, *14*, 5868. [CrossRef]
2. Paśmionka, I. Evaluation of the efficiency of removing sanitation indicators in the process of biological wastewater treatment. *Acta Sci. Pol. Form. Circumiectus* **2020**, *19*, 15–22. [CrossRef]
3. Kołecka, K.; Gajewska, M.; Caban, M. From the pills to environment—Prediction and tracking of non-steroidal anti-inflammatory drug concentrations in wastewater. *Sci. Total Environ.* **2022**, *825*, 153611. [CrossRef] [PubMed]
4. Jucherski, A.; Walczowski, A.; Bugajski, P.; Jóźwiakowski, K.; Rodziewicz, J.; Janczukowicz, W.; Wu, S.; Kasprzyk, M.; Gajewska, M.; Mielcarek, A. Long-term operating conditions for different sorption materials to capture phosphate from domestic wastewater. *Sustain. Mater. Technol.* **2022**, *31*, e00385. [CrossRef]
5. Paśmionka, I.B.; Bulski, K.; Herbut, P.; Boligłowa, E.; Vieira, F.M.C.; Bonassa, G.; Bortoli, M.; Prá, M.C.d. Toxic Effect of Ammonium Nitrogen on the Nitrification Process and Acclimatisation of Nitrifying Bacteria to High Concentrations of NH_4-N in Wastewater. *Energies* **2021**, *14*, 5329. [CrossRef]
6. Jakubowicz, P.; Fitobór, K.; Gajewska, M.; Drewnowska, M. Detection and Removal of Priority Substances and Emerging Pollutants from Stormwater: Case Study of the Kołobrzeska Collector, Gdańsk, Poland. *Sustainability* **2022**, *14*, 1105. [CrossRef]
7. Śliz, P.; Bugajski, P. Assessment of the stability and reliability of the water treatment plant in Nowy Sącz using control cards. *J. Water Land Dev.* **2022**, *52*, 251–256. [CrossRef]
8. Bugajski, P.; Nowobilska-Majewska, E.; Majewski, M. The Impact of Atmospheric Precipitation on Wastewater Volume Flowing into the Wastewater Treatment Plant in Nowy Targ (Poland) in Terms of Treatment Costs. *Energies* **2021**, *14*, 3806. [CrossRef]
9. Nowobilska-Majewska, E.; Bugajski, P.M. Concept of a New Technological System of a Biological Reactor in a Wastewater Treatment Plant in Nowy Targ in Terms of the Current Quantity and Quality of Wastewater—Case Study. *J. Ecol. Eng.* **2021**, *22*, 39–46. [CrossRef]
10. Jóźwiakowski, K.; Marzec, M.; Listosz, A.; Gizińska-Górna, M.; Micek, A.; Pytka-Woszczyło, A.; Pochwatka, P.; Rybczyńska-Tkaczyk, K. The Influence of Household Wastewater Treatment Plants with Drainage System on the Quality of Groundwater in the Lublin Province, Poland. *J. Ecol. Eng.* **2021**, *22*, 18–39. [CrossRef]
11. Rajta, A.; Bhatia, R.; Setia, H.; Pathania, P. Role of heterotrophic aerobic denitrifying bacteria in nitrate removal from wastewater. *J. Appl. Microbiol.* **2020**, *128*, 1261–1278. [CrossRef] [PubMed]

12. Gajewska, M.; Jóźwiakowski, K.; Ghrabi, A.; Masi, F. Impact of influent wastewater quality on nitrogen removal rates in multistage treatment wetlands. *Environ. Sci. Pollut. Res.* **2015**, *22*, 12840–12848. [CrossRef] [PubMed]
13. Juliastuti, S.R.J.; Baeyens, J.; Creemers, C. Inhibition of Nitrification by Heavy Metals and Organic Compounds: The ISO 9509 Test. *Environ. Eng. Sci.* **2004**, *20*, 79–90. [CrossRef]
14. Head, I.M.; Hiorns, W.D.; Embley, T.M.; McCarthy, A.J.; Saunders, J.R. The phylogeny of autotrophic ammonia-oxidizing bacteria as determined by analysis of 16S ribosomal RNA gene sequences. *Microbiology* **1993**, *139*, 1147–1153. [CrossRef]
15. Schramm, A.; de Beer, D.; Wagner, M.; Amann, R. Identification and Activities In Situ of *Nitrosospira* and *Nitrospira* spp. as Dominant Populations in a Nitrifying Fluidized Bed Reactor. *Appl. Environ. Microbiol.* **1998**, *64*, 3480–3485. [CrossRef]
16. Khin, T.; Annachhatre, A.P. Novel microbial nitrogen removal processes. *Biotechnol. Adv.* **2004**, *22*, 519–532. [CrossRef]
17. Paśmionka, I.B.; Gospodarek, J. Assessment of the Impact of Selected Industrial Wastewater on the Nitrification Process in Short-Term Tests. *Int. J. Environ. Res. Public Health* **2022**, *19*, 3014. [CrossRef]
18. Lipińska-Ojrzanowska, A.; Walusiak-Skorupa, J. Quaternary ammonium compounds—New occupational hazards. *Med. Pr.* **2014**, *65*, 675–682. [CrossRef]
19. Yang, L.; Liang, X.; Han, Y.; Cai, Y.; Zhao, H.; Sheng, M.; Cao, G. The coupling use of advanced oxidation processes and sequencing batch reactor to reduce nitrification inhibition of industry wastewater: Characterization and optimization. *Chem. Eng. J.* **2019**, *360*, 1577–1586. [CrossRef]
20. Pagga, U.; Bachner, J.; Strotmann, U. Inhibition of nitrification in laboratory tests and model wastewater treatment plants. *Chemosphere* **2006**, *65*, 1–8. [CrossRef]
21. Bugajski, P.M.; Pawełek, J.; Jóźwiakowska, K. The Interdependence of Organic and Biogenic Pollutants Concentrations in the Aspect of their Susceptibility to Biodegradation—A Case Study. *J. Ecol. Eng.* **2021**, *22*, 138–147. [CrossRef]
22. Siwiec, T.; Reczek, L.; Michel, M.M.; Gut, B.; Hawer-Strojek, P.; Czajkowska, J.; Jóźwiakowski, K.; Gajewska, M.; Bugajski, P. Correlations between organic pollution indicators in municipal wastewater. *Arch. Environ. Prot.* **2018**, *44*, 50–57. [CrossRef]
23. Samudro, G.; Mangkoedihardjo, S. Review on BOD, COD and BOD/COD ratio: A triangle zone for toxic, biodegradable and stable levels. *Int. J. Acad. Res.* **2010**, *2*, 235–239.
24. Bader, A.C.; Hussein, H.J.; Jabar, M.T. BOD:COD Ratio as Indicator for Wastewater and Industrial Water Pollution. *Int. J. Spec. Educ.* **2022**, *37*, 2164–2171.
25. Di Capua, F.; de Sario, S.; Ferraro, A.; Petrella, A.; Race, M.; Pirozzi, F.; Fratino, U.; Spasiano, D. Phosphorous removal and recovery from urban wastewater: Current practices and new directions. *Sci. Total Environ.* **2022**, *823*, 153750. [CrossRef]
26. Bawiec, A. Efficiency of nitrogen and phosphorus compounds removal in hydroponic wastewater treatment plant. *Environ. Technol.* **2019**, *40*, 2062–2072. [CrossRef] [PubMed]
27. Park, S.; Ely, R.L. Candidate stress genes of *Nitrosomonas europaea* for monitoring inhibition of nitrification by heavy metals. *Appl. Environ. Microbiol.* **2008**, *74*, 5475–5482. [CrossRef]
28. Ouyang, F.; Zhai, H.; Ji, M.; Zhang, H.; Dong, Z. Physiological and transcriptional responses of nitrifying bacteria exposed to copper in activated sludge. *J. Hazard. Mater.* **2016**, *301*, 172–178. [CrossRef]
29. Kroiss, H.; Schweighofer, P.; Frey, W.; Matsche, N. Nitrification inhibition-a source identification method for combined municipal and/or industrial wastewater treatment plants. *Water Sci. Technol.* **1992**, *26*, 1135–1146. [CrossRef]
30. Dincer, A.R.; Kargi, F. Salt inhibition of nitrification and denitrification in saline wastewater. *Environ. Technol.* **1999**, *20*, 1147–1153. [CrossRef]
31. Kim, Y.M.; Park, D.; Lee, D.S.; Park, J.M. Inhibitory effects of toxic compounds on nitrification process for cokes wastewater treatment. *J. Hazard. Mater.* **2008**, *152*, 915–921. [CrossRef] [PubMed]
32. Ge, S.; Wang, S.; Yang, X.; Qiu, S.; Li, B.; Peng, Y. Detection of nitrifiers and evaluation of partial nitrification for wastewater treatment: A review. *Chemosphere* **2015**, *140*, 85–98. [CrossRef] [PubMed]
33. Juliastuti, S.R.; Baeyens, J.; Creemers, C.; Bixio, D.; Lodewyckx, E. The inhibitory effects of heavy metals and organic compounds on the net maximum specific growth rate of the autotrophic biomass in activated sludge. *J. Hazard. Mater.* **2003**, *100*, 271–283. [CrossRef]
34. Park, S.J.; Oh, J.W.; Yoon, T.I. The role of powdered zeolite and activated carbon carriers on nitrification in activated sludge with inhibitory materials. *Process Biochem.* **2003**, *39*, 211–219. [CrossRef]
35. Jóźwiakowska, K.; Marzec, M. Efficiency and reliability of sewage purification in long-term exploitation of the municipal wastewater treatment plant with activated sludge and hydroponic system. *Arch. Environ. Prot.* **2020**, *46*, 30–41. [CrossRef]
36. Kataki, S.; Chatterjee, S.; Vairale, M.G.; Sharma, S.; Dwivedi, S.K. Concerns and strategies for wastewater treatment during COVID-19 pandemic to stop plausible transmission. *Resour. Conserv. Recycl.* **2021**, *164*, 105156. [CrossRef]
37. Rahimi, S.; Modin, O.; Mijakovic, I. Technologies for biological removal and recovery of nitrogen from wastewater. *Biotechnol. Adv.* **2020**, *43*, 107570. [CrossRef]
38. Mikosz, J. Determination of permissible industrial pollution load at a municipal wastewater treatment plant. *Int. J. Environ. Sci. Technol.* **2015**, *12*, 827–836. [CrossRef]
39. Yang, M.; Qiu, S.; Wang, L.; Chen, Z.; Hu, Y.; Guo, J.; Ge, S. Effect of short-term light irradiation with varying energy densities on the activities of nitrifiers in wastewater. *Water Res.* **2022**, *216*, 118291. [CrossRef]
40. Van den Broeck, R.M.R.; Van Impe, J.F.M.; Smets, I.Y.M. Assessment of activated sludge stability in lab-scale experiments. *J. Biotechnol.* **2009**, *141*, 147–154. [CrossRef]

41. Chhetri, R.K.; Karvelas, S.; Sanchez, D.F.; Droumpali, A.; Kokkoli, A.; Andersen, H.R. A modified nitrification inhibition test for high-salinity wastewater. *Chem. Eng. J.* **2022**, *429*, 132460. [CrossRef]
42. Zhang, M.; Wang, X.; Yang, J.; Wang, D.; Liang, J.; Zhou, L. Nitrogen removal performance of high ammonium and high salt wastewater by adding carbon source from food waste fermentation with different acidogenic metabolic pathways. *Chemosphere* **2022**, *292*, 133512. [CrossRef] [PubMed]

International Journal of
Environmental Research and Public Health

MDPI

Article

Efficiency of Sidestream Nitritation for Modern Two-Stage Activated Sludge Plants

Thomas Baumgartner, Lydia Jahn, Vanessa Parravicini, Karl Svardal and Jörg Krampe *

Institute for Water Quality and Resource Management, TU Wien, 1040 Vienna, Austria
* Correspondence: jkrampe@iwag.tuwien.ac.at

Abstract: The operational costs of wastewater treatment plants (WWTPs) are mainly driven by electric power consumption, making the energy-efficient operation an all-time present target for engineers and operators. A well known approach to reduce the demand for purchased electricity is the operation of an anaerobic sludge stabilisation process. Although anaerobic digesters make it possible to recover large quantities of energy-rich methane gas, additional strategies are required to handle the increased internal return flow of nitrogen, which arises with the sludge dewatering effluent (SDE). SDE treatment increases the oxygen demand and in turn the energy required for aeration. In this study, different SDE treatment processes were compared with regard to the treatment in mainstream, sidestream nitritation, as well as nitritation combined with anammox for two-stage and single-stage WWTPs. Although SDE treatment in sidestream nitritation was found to have no effect on the energy demand of single-stage WWTPs, this concept allows the treatment capacity in the activated sludge tank to be raised, while contributing to a high nitrogen removal under carbon limitation. In contrast, SDE sidestream treatment showed great potential for saving energy at two-stage WWTPs, whereby sidestream nitritation and the further treatment in the first stage was found to be the most efficient concept, with a savings of approx. 11% of the aeration energy.

Keywords: energy efficiency; SDE treatment; nitritation; two-stage WWTP

Citation: Baumgartner, T.; Jahn, L.; Parravicini, V.; Svardal, K.; Krampe, J. Efficiency of Sidestream Nitritation for Modern Two-Stage Activated Sludge Plants. *Int. J. Environ. Res. Public Health* **2022**, *19*, 12871. https://doi.org/10.3390/ijerph191912871

Academic Editors: Yung-Tse Hung, Hamidi Abdul Aziz and Issam A. Al-Khatib

Received: 31 July 2022
Accepted: 4 October 2022
Published: 8 October 2022

1. Introduction

1.1. Energy Assessment

Wastewater treatment includes different cleaning steps that in total require a considerable amount of energy. Wastewater treatment plant (WWTP) concepts usually encompass a mechanical treatment by screens, grit removal, and primary sedimentation to remove particular and organic matter. The subsequent biological treatment serves to remove soluble substances by microbial conversion. Especially the biological treatment is highly energy-intensive, since aeration is required to cover the oxygen demand of the bacteria involved. The energy consumption of WWTPs has been intensively analysed in recent years. Studies from different European countries reveal that the electricity consumption required for wastewater treatment can make up 1.0% of the regional energy consumption [1,2] with a specific energy demand of usually around 0.5 kWh/m^3 wastewater [2,3]. The energy demand depends mainly on the treatment concept and the organic load, normally analysed as chemical oxygen demand (COD). In European countries, the daily amount of COD that is emitted to the sewer system ranges between 100 and 120 g COD/(PE·d). By considering a calorific value of 14 kJ/g COD [4], the energy potential of the organic material in wastewater can be calculated with 140 to 170 kWh/(PE·a). In contrast, the average energy consumption for WWTPs larger than 50,000 PE was found to be 30 kWh/(PE·a) [5], indicating that municipal wastewater contains significantly more energy in the form of organic matter than required for its treatment.

The energy demand for wastewater treatment with activated sludge is mainly driven by the sludge retention time (SRT) and the type of aeration system, whereby the energy required for the aeration can make up 60% of the total energy demand of a plant [6,7]. Moreover, the energy demand correlates with the inflow load and sludge production [8]. Figure 1 shows the proportion of the oxygen utilisation required for carbon removal (OU_C) and the amount of carbon that is incorporated into the excess sludge (COD_{ES}) in correlation to the SRT. The curves by Svardal and Kroiss [4] are based on full-scale experiences that agree with simulation results by Henze et al. [9] and can be derived from the design guideline DWA-A 131 [10]. Figure 1 illustrates that a high SRT is linked to an increased oxygen demand and a lower transfer of carbon to the excess sludge. Since the SRT for WWTPs with aerobic sludge stabilisation is prescribed with 25 d [10], the energy demand for aeration of these plants is clearly increased compared with WWTPs with anaerobic digesters. Anaerobic sludge stabilisation provides a more energy-efficient operation, since this process allows a lower SRT (meaning a lower oxygen demand) for the activated sludge tank. SRTs for WWTPs with anaerobic digesters are mostly between 12 and 14 d, whereby in that range approx. 60% of the COD is oxidised and 40% of the COD is incorporated into the excess sludge and supplied to the digester. WWTP concepts with digesters usually include primary settlers to gain a primary sludge rich in organic matter. Organic compounds of the primary and the excess sludge (hereinafter referred to as raw sludge) are degraded in the digester to energy-rich biogas. The conversion of the methane gained into energy by using combined heat and power plants (CHP) minimises the electric power purchased, making the anaerobic process a cost-effective stabilisation strategy for large-scale WWTPs.

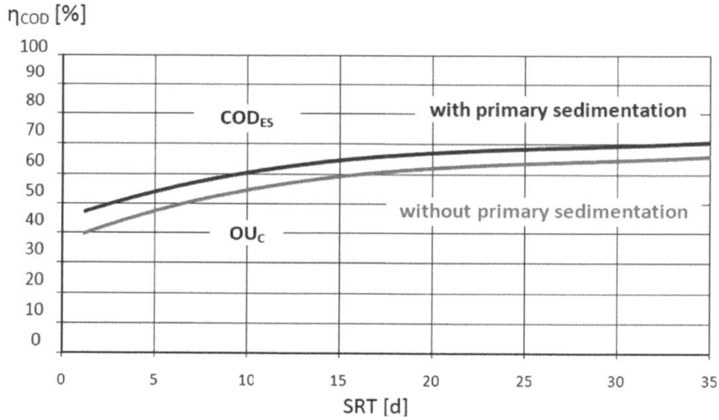

Figure 1. COD in excess sludge (COD_{ES}) and the oxygen utilisation for carbon removal (OU_c) in correlation to the SRT [4].

1.2. SDE Treatment Concepts

The biodegradation of raw sludge leads to the release of the incorporated nitrogen and in turn to high ammonium levels in the sludge dewatering effluent (SDE). NH_4-N levels in SDE are usually above 1000 mg/L. Since the nitrogen load in SDE can make up 20% of the incoming load, it is of high relevance to evaluate different SDE treatment concepts that can be realised on an individual WWTP.

SDE can be treated in mainstream or sidestream, i.e., directly in the activated sludge tank or in a separate tank, before it is recycled to the mainstream. Some plant concepts include a buffer tank that allows the storage and transfer of SDE into the activated sludge tank avoiding nitrogen peak loading. If there is no storage tank available, the SDE is directly returned to the activated sludge tank, where it increases the oxygen and energy demand for the aeration system significantly. The construction of dedicated treatment tanks for SDE represents an essential measure to relieve the aeration system in the biological stage and to create additional treatment capacities. A main benefit of a sidestream SDE treatment is an improved denitrification process due to a better nitrogen to carbon ratio. Since the amount of oxidised carbon is linked to the aerated tank volume, more readily biodegradable carbon is available for denitrification under lower oxygen supply. A side-effect of a reduced COD respiration under anoxic conditions is an increased adsorption of carbon to the excess sludge, which in turn raises the biogas yield during digestion. Depending on the type of SDE treatment applied in sidestream, the N to COD ratio in the influent of the mainstream tank can be also reduced in favour of denitrification.

Biological mainstream or sidestream treatment options for nitrogen rich SDE are, e.g., nitrification/denitrification [11]; nitritation/denitritation [12]; or more specific processes such as deammonification (anammox) [13,14]. The following section briefly summarises the microbial processes involved in different SDE treatment concepts.

1.2.1. Nitrification and Denitrification

Nitrification is a two-step conversion process of ammonium to nitrate. The first step of nitrification is the conversion of ammonium to nitrite (nitritation) by ammonium-oxidising bacteria (AOB). Nitrite-oxidising bacteria (NOB) are responsible for the further oxidation of nitrite to nitrate (nitratation). The oxygen demand for full nitrification (nitritation and nitratation) amounts to 4.57 g O_2/g NO_3-N. Taking into account the biomass growth, the actual oxygen demand for nitrification amounts to 4.33 g O_2/g NO_3-N. Nitrification and denitrification are the conventional nitrogen removal concepts realised in the activated sludge tank (mainstream).

1.2.2. Nitritation and Denitritation

SDE sidestream nitritation relies on the conversion of ammonium only to nitrite, whereby denitritation is usually performed in mainstream. Due to a lower oxygen demand for nitritation (3.25 g O_2/g NO_2-N), the required energy demand for aeration is approx. 25% lower than for nitrification. However, the pre-treatment by nitritation provides less oxygen in the form of nitrite for carbon removal under anoxic conditions in mainstream. Nitritation without a further conversion to nitrate can be achieved by suppressing NOBs, which are known to be more sensitive to high ammonium levels (free ammonia) and temperatures than the AOBs [15,16]. However, to avoid AOB inhibition due to free ammonia, the NH_4-N levels should not exceed 200 mg/L [17]. The conversion of ammonium to nitrite is limited by the alkalinity; thus, only 55% of the incoming ammonium can be converted to nitrite. The pre-treated SDE is returned to the mainstream for the further nitrification of the remaining ammonium and the denitritation of nitrite by heterotrophic bacteria. In this connection, the COD demand for denitritation is approx. 40% lower compared with that for denitrification. Full-scale applications show that nitritation is characterised by a stable process control and low maintenance [18,19].

1.2.3. Nitritation and Deammonification

In addition to conventional nitrogen removal, some bacteria are capable of an anaerobic ammonia oxidation. These so-called anammox bacteria require an optimal proportion of ammonium and nitrite. For that reason, the anammox process is linked to a preceding nitritation under aerobic conditions. In a second step, ammonium and nitrite are converted directly into nitrogen without the need of an organic carbon source. The following equation shows the anammox process according to Strous et al. [20].

$$NH_4^+ + 1.32NO_2^- + 0.06HCO_3^- + 0.13H^+ \rightarrow 1.02N_2 + 0.26NO_3^- + 0.066CH_2O_{0.5}N_{0.15} + 2.03H_2O$$

The oxygen demand for the pre-treatment via nitritation and deammonification is only 1.93 g O_2/g N and thus much lower compared with that for nitrification or nitritation. Moreover, the carbon demand is negligible, making the process attractive for WWTPs with low carbon availability or for technologies aiming for a target COD recovery such as microsieves. However, since anammox bacteria are more sensitive to environmental changes and own longer growth rates, the operation of anammox reactors requires special attention to the retention of the biomass. Operational strategies for full-scale deammonification of dewatering liquors are described by Refs. [21,22].

Table 1 summarises the oxygen demand and oxygen utilisation for carbon removal of the SDE treatment processes considered.

Table 1. Oxygen demand and oxygen utilisation for carbon removal for denitrification (OU_{DN}).

SDE Treatment Concepts	Oxygen Demand ($g\ O_2/g\ N$)	OU_{DN} for TN Removal ($g\ O_2/g\ N$)	Total ($g\ O_2/g\ N$)
nitrification/denitrification	4.33	2.86	~1.5
nitritation/denitritation	3.74	2.23	~1.5
nitritation/anammox	1.93	0.40	~1.5

1.3. Two-Stage WWTP Design

Although most municipal WWTPs are designed as single-stage, several WWTPs are operated as two-stage activated sludge plants. The two-stage plant design relates primarily to the A/B process, which aims to separate carbon and nitrogen removal [23]. Organic carbon is partially removed in a high loaded first stage, whereby the nitrogen removal occurs mainly in the second stage. The first activated sludge tank is generally designed for carbon removal with SRTs in the range of 1 to 5 d. Due to this low SRT, most of the COD is transferred to excess sludge (55% to 60%), resulting in low oxygen consumption. The excess sludge with a high amount of biodegradable carbon ensures an increased methane yield as well as energy production, which contributes towards energy self-sufficiency. The second stage is generally designed for nitrogen removal (nitrification/denitrification) with a higher SRT. Although there is a stable operation reported for the A/B process, an unfavorable N/COD ratio could result in a lack of carbon for denitrification in the second stage. With the aim of overcoming carbon limitation in the second stage, a further development of the two-stage plant design resulted in the hybrid process. This concept relates to a sidestream that can be bypassed from the influent to the second stage in order to avoid carbon limitation and to ensure excellent nitrogen removal [24]. Generally, the operation of two-stage WWTPs is more stable, since the different organic loading rates in the activated sludge tanks can suppress the growth of filaments. A stable two-stage WWTP operation is reported for many large-scale plants (e.g., WWTP Vienna 4.0 Mio. PE; WWTP Munich I+II 3.0 Mio. PE; WWTP Frankfurt Niederrad 1.35 Mio. PE; WWTP Salzburg Siggerwiesen 680,000 PE).

The literature study indicates that there are different technologies available to treat SDE, whereby the concepts were found to affect the energy balance of a plant. Although the treatment concepts are described in earlier publications, there is so far no comparison of SDE treatment concepts available that considers municipal single- and two-stage plant design. However, from an energetic point of view, the optimal SDE treatment concept has a considerable effect on the energy demand and the operational costs. Especially large WWTPs that are often designed as two-stage plants have great potential to save energy. This paper offers valuable advice for planners and operators to decide on the most suitable SDE treatment concept. The study provides a helpful approach that can be applied for an individual plant to calculate an expected energy demand and sludge production.

2. Materials and Methods

2.1. SDE Treatment Scenarios

Different SDE treatment options were evaluated and compared by generating COD and N balances. Table 2 summarises the scenarios considered. Further SDE treatment concepts such as physical processes (e.g., ammonia stripping) or chemical processes (e.g., precipitation of magnesium-ammonium-phosphate) were not considered within this work.

Table 2. Scenarios for the SDE treatment in single- and two-stage WWTPs.

Scenario	Plant Concept	Sidestream	Mainstream
1	single-stage	-	nitrification/denitrification
2	single-stage	nitritation	denitritation
3	single-stage	nitritation/anammox	-
4	two-stage	-	nitrification/denitrification
5	two-stage	nitritation	denitritation in 1st stage
6	two-stage	nitritation/anammox	-

Further assumptions concern the microbial conversion within the different treatment processes. The conversion of ammonium in SDE sidestream is limited by the alkalinity, whereby the effluent after nitritation usually consists of 55% nitrite and 45% ammonium. In the case of nitritation and anammox, 89% of the treated ammonium is converted to nitrogen (N_2), 8% to nitrate and approx. 3% remain as ammonium in the pre-treated SDE. The total COD and nitrogen (TN) removal were assumed to be 95 and 80%. NH_4-N removal was calculated with 100%.

2.2. COD and N Balances

Mass balances are well-known approaches to illustrate differences in the mass flow of a system. Figure 2 shows the COD and N mass balances for the single-stage WWTP concept. The following calculations are based on specific data for municipal WWTPs. The first assumption concerns the inflow loads, which are defined as 120 g COD/(PE·d) and 8 g N/(PE·d). This specific nitrogen load is common for WWTPs > 10,000 PE [25]. All mass balances consider a total nitrogen removal rate of 80%, which corresponds to an effluent load of 1.6 g N/(PE d). COD removal was calculated as 95%. COD and N loads that do not end up in the effluent must be removed by oxidation processes, biomass growth, or methane production.

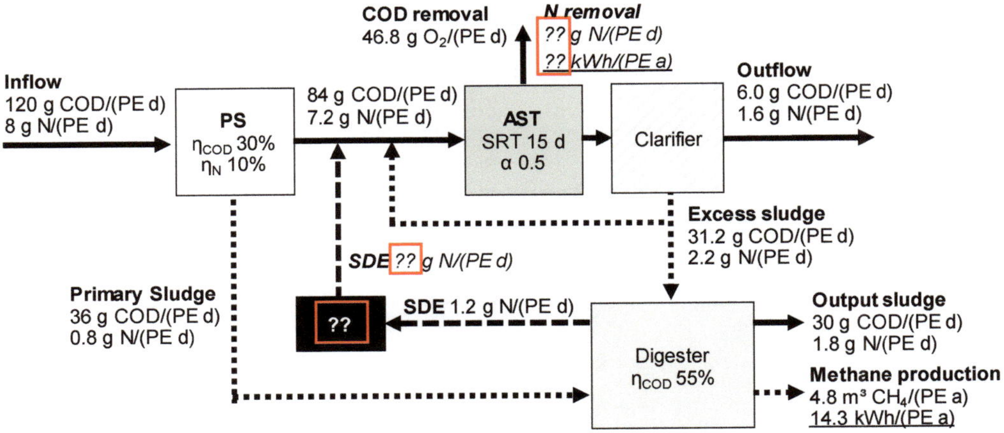

Figure 2. COD and N balances for single-stage activated sludge tanks.

The mechanical treatment step by primary sedimentation considers a COD and N removal of 30% and 10%, which end up in the primary sludge treated afterwards in the digester. Since the removal rate during primary sedimentation is independent of the SDE treatment concept, the COD and N loads to the digester and activated sludge tank (AST) are the same for all scenarios considered. The same applies to the excess sludge production during biological treatment, since all scenarios are based on a COD inflow load of 84 g

COD/(PE·d) and a SRT of 15 d. During biomass growth, approx. 7% of the removed COD is incorporated as nitrogen into the biomass [10].

The primary and excess sludge are treated in anaerobic digesters, whereby the output after digestion was assumed to be 30 g COD/(PE·d) and 1.8 g N/(PE·d), according to the experience at real plants (especially in Austria and Germany) and to modelling results [26]. The methane produced during digestion is converted by CHP with an electrical efficiency in the range of 30% to 40% (3 kWh_{el}/m^3 CH_4). For this study, the CHP efficiency was assumed to be 30%.

Incorporated nitrogen is released during digestion. Since the anaerobic degradation is the same, the nitrogen load in SDE is similar. Differences appear for the nitrogen load returned to the activated sludge tank, which depend on the applied SDE treatment concept. Thus, the energy demand for nitrification is driven by the nitrogen load after SDE treatment.

Figure 3 shows the COD and N mass balances for the two-stage WWTP concept. The same inflow loads were considered as for the single-stage design. However, there are differences with regard to the first-stage and the excess sludge production. COD removal in the first stage was calculated with 60%. Thus, the COD load to the second stage was 33.6 g COD/(PE·d) for all scenarios considered. In this context, the excess sludge produced in the activated sludge tank as well as the incorporated nitrogen resulted in the same amount. SRT in the first stage was defined to be 5 d. Since the biomass yield depends on the sludge retention time, the proportion of COD transferred into excess sludge accounts for 60% and 40% for oxidation (Figure 1). The amount of already oxidised nitrite or nitrate returned from SDE treatment to the first-stage reduces the COD load that has to be oxidised and in turn the aeration energy required. Thus, there appeared differences in the energy demand for aeration in the first and second stage.

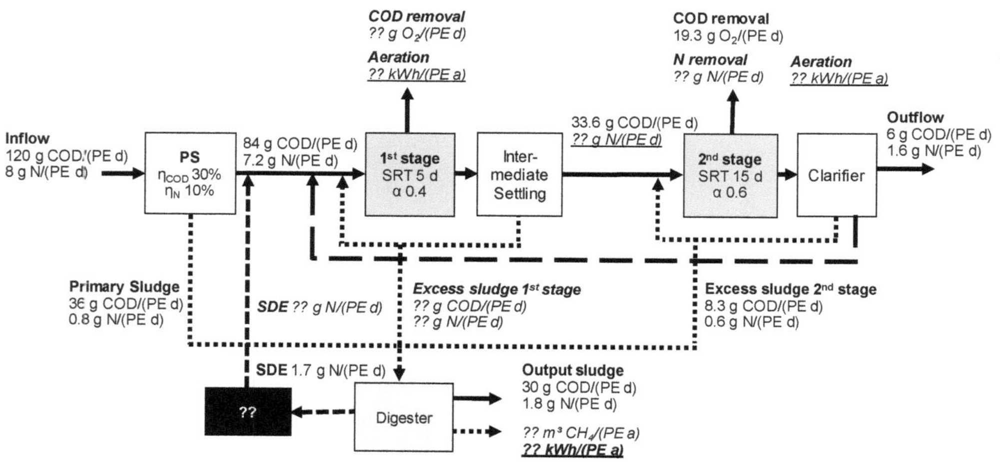

Figure 3. COD and N balances for two-stage activated sludge tanks.

2.3. Further Assumptions

The energy demand for the biological stage is linked to the efficiency of the aeration system. For tapwater, the standard aeration efficiency (SAE) of fine-bubble aeration systems is usually between 3.0 and 4.0 kg O_2/kWh [27]. The SAE in wastewater is well below this value and depends on different factors, e.g., aeration system, dissolved salts, detergents, reactor configuration, dry solid content, and SRT [28]. The α-factor is the ratio of the oxygen transfer into tapwater to the oxygen transfer into wastewater and a relevant design parameter for aeration systems. The α-factor should be considered carefully for the different treatment tanks, since it determines the energy demand for aeration. The main parameter affecting the α-factor seems to be the SRT, which is directly associated with the biomass concentration in the activated sludge tank [29]. Frey [27] reported a low SRT in the first stage of two-stage activated sludge plants with α-factors in the range of 0.3 to 0.4. For nitrifying and denitrifying plants (SRT from 12 to 14 d), the α-factor ranges from 0.5 to 0.7. Postulating an α-factor of 0.7 for separate SDE treatments, as indicated in practice, provides an additional advantage by reducing electricity demand for aeration aiming nitrification/nitritation. Nitrite produced in sidestream with a more favourable α-factor (lower electricity consumption) is used to remove a part of the COD load in mainstream through denitritation, avoiding that this COD is oxidisied aerobically by O_2 provided with less efficiency (lower α-factor). The energy saving is much higher at two-stage plants due to the lower α-factor in the first-stage. Table 3 summarises α-factors that were considered for the different treatment tanks and plant configurations.

Table 3. α-values considered for different treatment tanks.

Configuration	Treatment Tanks	α-Factors
Single-stage	activated sludge tank	0.5
Two-stage	first-stage	0.4
Two-stage	second-stage	0.6
SDE treatment	sidestream tank	0.7

3. Results

Table 4 summarises the results of the SDE treatment scenarios. The corresponding mass balances are shown in the Supplemental Information (File S1). The nitrogen load in the SDE of the single-stage WWTP was calculated with 1.2 g N/(PE·d), which corresponds to 15% of the incoming nitrogen load. The calculations presented are based on a TN removal of 80%. After SDE sidestream treatment, the NH_4-N load to the activated sludge tank clearly decreased, resulting in a lower energy demand for aeration. However, the total energy demand for aeration in the separate SDE tank and the activated sludge tank was in total approx. 12.2 kWh/(PE·a), nearly the same for all scenarios considered. Although some publications claim a lower oxygen and thus a lower electricity demand for the anammox process [30,31], the results of our study revealed that the lower electricity demand is not the case when considering the total aeration demand. The results illustrated that the overall energy consumption for the biological treatment at single-stage WWTPs can not be affected by the implementation of a SDE sidestream treatment. Similar results were found for the energy recovery from biogas utilisation, which was equal for all scenarios with approx. 14.3 kWh/(PE·a). However, a significant difference was found for the organic carbon demand for denitrification. The highest carbon demand accounted for scenario 1 (mainstream) with 4.8 kg COD/(PE·a), whereby the lowest carbon demand of 3.7 kg COD/(PE·a) was found for SDE nitritation and anammox. SDE pre-treatment with nitritation resulted in a 6% lower carbon demand. In the case of a limited COD availability or by focusing a target COD removal, SDE sidestream treatment offers suitable handling to ensure extensive nitrogen removal and to relieve the aeration system in the activated sludge tank. Besides the lower carbon demand, the results of the mass balances indicated that the energetic advantage of sidestream SDE treatment of single-stage WWTPs is less. Moreover, the operation of a sidestream treatment is linked to additional construction costs and the need of technical equipment (e.g., aeration system with blowers and stirrers). Control and handling of SDE sidestream treatments can be challenging when there arise foaming problems that can only be controlled by dosing anti-foaming agents. Operators and designers should carefully check the site-specific benefits for sidestream SDE treatments on single-stage WWTP design.

Table 4. Energy for aeration, energy from biogas utilisation, and OU_{DN} for TN removal of 80% in single-stage and two-stage WWTPs with different SDE treatment options.

Scenario		Energy from Biogas (kWh_{el}/(PE·a))	Energy for Aeration (kWh_{el}/(PE·a))	OU_{DN} (kg COD/(PE·a))
single-stage	1	14.3	12.3	4.80
	2	14.3	12.2	4.53
	3	14.3	12.2	3.70
two-stage	4	19.0	10.7	4.80
	5	19.9	9.8	4.41
	6	19.4	10.4	3.16

In the second part of the study, different SDE treatment concepts were considered for the two-stage plant design. The nitrogen load in the SDE of the two-stage WWTPs was calculated with 1.7 g N/(PE·d), which corresponds to 21% of the incoming nitrogen load. Sidestream concepts (scenario 5 and 6) consider the return of the pre-treated SDE to the first stage of a two-stage WWTP, which significantly reduces the energy demand for aeration in the first stage due to denitritation. Additionally, the higher oxygen transfer rate in the sidestream tank has a positive effect on the total energy balance compared with the single-stage plant design. Overall, the energy demand for aeration of the two-stage plant is considerably lower compared with the single-stage WWTP operation (−9.3%). The lowest energy demand for aeration was found with 9.8 kWh/(PE·a) for the scenario with sidestream nitritation of SDE. Moreover, energy savings are possible with SDE sidestream nitritation and anammox. Compared with SDE treatment in mainstream, the SDE treatment with nitritation allows energy savings of approx. 8%; 3% of energy savings are possible for nitritation and anammox. The results of the two-stage plant design indicate that differences in the aeration energy of the SDE treatment concepts considered can make up 0.3 to 0.9 kWh/(PE·a). Since the specific energy consumption is usually in the range of 20 to 30 kWh/(PE·a), the total energy savings can account for 4.5% or more than 90 MWh/a for WWTPs larger 100,000 PE.

A side-effect of the anoxic respiration and the low SRT in the first stage is an increased transfer of COD to the excess sludge with the benefit of higher biogas production in the digester. Compared with the single-stage WWTP operation (scenarios 1 to 3), the anaerobic stabilisation at two-stage WWTPs (scenarios 4 to 6) enables an overall higher gain of energy from anaerobic sludge stabilisation (+33%). The highest possible energy recovery from biogas was calculated with 19.9 kWh/(PE·a) for SDE sidestream nitritation, followed by nitritation and anammox with 19.4 kWh/(PE·a). The results indicate that sidestream SDE treatment allows an additional gain of energy from biogas up to 90 MWh/a considering a two-stage WWTP with a design capacity of 100,000 PE. Although SDE treatment in sidestream is attractive for saving energy and increasing the biogas yield, it was found in earlier studies that there can appear considerable N_2O emissions due to higher nitrite levels and loads. Especially sidestream nitritation was found to increase N_2O emissions of a WWTP [32]. Since N_2O is a climate relevant greenhouse gas, increased emissions affecting the CO_2 footprint of a WWTP and should also be considered.

Moreover, the calculations represented provide information about the carbon demand of the SDE treatment concepts. The lowest carbon demand for denitrification was found for the SDE treatment via nitriation and anammox with 3.16 kg COD/(PE·a), which is approx. 30% to 35% lower compared with the alternative scenarios. However, the carbon efficiency of the deammonification process presents no benefit, as the first stage needs to be operated under aerobic or anoxic conditions. Even though it is stated in many references that the deammonification of SDE reduces the energy demand of municipal WWTPs considerably, this is not the case for typical municipal wastewater where the oxygen demand for the carbon removal exceeds the oxygen demand for the nitrification. Moreover, there is an increased process control required in order to provide enough COD for denitrification in the second-stage. High N/COD ratios can restrict denitrification in the second-stage due to a lack of carbon and must be considered carefully.

The results confirmed that the concept of two-stage WWTPs is advantageous for an energy-efficient operation with an increased energy self-sufficiency. Nitritation in SDE pre-treatment tanks and the use of nitrite to substitute oxygen in the first stage of a two-stage WWTP appears to be the most advantageous SDE treatment concept for saving aeration energy. Nevertheless, it should be mentioned that different process settings and operational conditions can enable another SDE treatment option to the more favorable concept for an individual plant. So far, publications reporting full-scale operation that allow verifying the results presented are scarce. Full-scale experiences from the implementation of a SDE sidestream nitritation on the second-stage WWTP Kirchbichl were reported by Ref. [33]. The authors observed a stable nitritation after the startup of the process and temporary problems with foam. However, due to changes in the digester operation and internal return flows, it was not completely possible to evaluate the energetic effects out of the received data for the restricted experimental period. For the background of changeable operational conditions, mass balances have been proven to be an efficient tool to forecast energetic effects. The presented data serve to explain how mass balances can be applied with regard to assess SDE treatment.

4. Discussion

The goal of the study was to provide a comparison of different SDE treatment concepts by considering single- and two-stage plant design. The calculations performed by mass balances are based on specific data received from benchmark studies and full-scale experiences. Operators and engineers can use the described method to evaluate the best suitable SDE treatment concept for an individual WWTP by using plant-specific loads and data. The results presented demonstrated that the SDE sidestream treatment at single-stage WWTPs does not display an energetic advantage for one of the investigated treatment methods, since the energy demand for aeration was approx. 12.2 kWh/(PE a) for all scenarios considered. Moreover, the energy potential of the biogas was the same with 14.3 kWh/(PE·a). However, the deammonification process had a higher carbon efficiency with a 23% lower carbon demand compared with mainstream SDE treatment, which is a great benefit for single-stage activated sludge plants with high effluent requirements regarding nitrogen removal or a limited capacity of the aeration system. Overall, it was found that the two-stage WWTP operation allows an approx. 30% higher biogas production compared with single-stage WWTPs, since the anoxic respiration of heterotrophic bacteria and low SRT lead to a higher amount of COD transferred to sludge. This in turn reduces the energy demand for aeration considerably, whereby the energy demand for aeration is approx. 13% lower by considering the mainstream SDE treatment. SDE pre-treatment in sidestream via nitritation and the further treatment of nitrite-rich effluent in the first stage of a two-stage WWTP reduces the oxygen demand for COD respiration and is therefore the most favourable strategy to achieve an energy-efficient wastewater treatment with savings of approx. 8% of aeration energy.

Supplementary Materials: The following supporting information can be downloaded at: https://www.mdpi.com/article/10.3390/ijerph191912871/s1, File S1: SDE Treatment Concepts.

Author Contributions: Conceptualisation, T.B. and L.J.; project administration, V.P. and K.S.; supervision, J.K.; writing—original draft, T.B.; writing—review and editing, L.J. All authors have read and agreed to the published version of the manuscript.

Funding: This research was funded by the European Union Horizon 2020 as part of POWERSTEP (Grant Agreement 641661), a full-scale demonstration of energy-positive sewage treatment plant concepts towards market penetration. Open Access Funding by TU Wien.

Institutional Review Board Statement: Not applicable.

Informed Consent Statement: Not applicable.

Conflicts of Interest: The authors declare no conflict of interest.

References

1. Ganora, D.; Hospido, A.; Husemann, J.; Krampe, J.; Loderer, C.; Longo, S.; Bouyat, L.M.; Obermaier, N.; Piraccini, E.; Stanev, S.; et al. Opportunities to improve energy use in urban wastewater treatment: A European-scale analysis. *Environ. Res. Lett.* **2019**, *14*, 044028. [CrossRef]
2. Bodik, I.; Kubaska, M. Energy and sustainability of operation of a wastewater treatment plant. *Environ. Prot. Eng.* **2013**, *39*, 15–24. [CrossRef]
3. Ranieri, E.; Giuliano, S.; Ranieri, A.D. Energy consumption in anaerobic and aerobic based wastewater treatment plants in Italy. *Water Pract. Technol.* **2021**, *16*, 851–863. [CrossRef]
4. Svardal, K.; Kroiss, H. Energy requirements for wastewater treatment. *Water Sci. Technol.* **2011**, *64*, 1355–1361. [CrossRef] [PubMed]
5. Haslinger, J.; Lindtner, S.; Krampe, J. Operating costs and energy demand of wastewater treatment plants in Austria: Benchmarking results of the last 10 years. *Water Sci. Technol.* **2016**, *74*, 2620–2626. [CrossRef]
6. Zessner, M.; Lampert, C.; Kroiss, H.; Lindtner, S. Cost comparison of wastewater treatment in Danubian countries. *Water Sci. Technol.* **2010**, *62*, 223–230. [CrossRef] [PubMed]
7. Bell, K.Y.; Abel, S. Optimization of WWTP aeration process upgrades for energy efficiency. *Water Pract. Technol.* **2011**, *6*, 2. [CrossRef]
8. Gurung, K.; Tang, W.Z.; Sillanpää, M. Unit energy consumption as benchmark to select energy positive retrofitting strategies for Finnish Wastewater Treatment Plants (WWTPs): A case study of Mikkeli WWTP. *Environ. Process.* **2018**, *5*, 667–681. [CrossRef]
9. Henze, M.; Leslie Grady, C.P., Jr.; Gujer, W.; Marais, G.V.R.; Matsuo, T. A general model for single-sludge wastewater treatment systems. *Water Res.* **1987**, *21*, 505–515. [CrossRef]
10. DWA-A 131: Bemessung von einstufigen Belebungsanlagen. DWA-A 131: Design of single-stage activated sludge plant. Design Guideline. DWA Deutsche Vereinigung für Wasserwirtschaft, Abwasser und Abfall e.V., Hennef. 2016; ISBN 978-3-88721-331-2.
11. Gustavsson, D. Nitritation and Denitritation in Sludge Liquor Treatment. Ph.D. Thesis. Water and Environmental Engineering, Department of Chemical Engineering Lund University, Lund, Sweden, 2011.
12. Fux, C.; Velten, S.; Carozzi, V.; Solley, D.; Keller, J. Efficient and stable nitritation and denitritation of ammonium-rich sludge dewatering liquor using an SBR with continuous loading. *Water Res.* **2006**, *40*, 2765–2775. [CrossRef] [PubMed]
13. Magrí, A.; Béline, F.; Dabert, P. Feasibility and interest of the anammox process as treatment alternative for anaerobic digester supernatants in manure processing—An overview. *J. Environ. Manag.* **2013**, *131*, 170–184. [CrossRef]
14. Mouilleron, I.; Hyde, K.; Reid, K.; Keegan, A.; Rinck-Pfeiffer, R.; Krampe, J.; van den Akker, B. Start-up of a demonstration-scale deammonification reactor at Bolivar WWTP. Results of a demonstration trial at a wastewater treatment plant in South Australia. *Water J. Aust. Water Assoc.* **2014**, *41*, 119–123.
15. Buday, J.; Drtil, M.; Hutnan, M.; Derco, J. Substrate and Product Inhibition of Nitrification. *Chem. Pap.—Chem. Zvesti* **1999**, *53*, 379–383.
16. Wei, D.; Xue, X.; Yan, L.; Sun, M.; Zhang, G.; Shi, L.; Du, B. Effect of influent ammonium concentration on the shift of full nitritation to partial nitrification in a sequencing batch reactor at ambient temperature. *Chem. Eng. J.* **2014**, *235*, 19–26. [CrossRef]
17. DWA-M 349: Biologische Stickstoffelimination von Schlammwässern der anaeroben Schlammstabilisierung. DWA-M: 349 Biological nitrogen removal from sludge dewatered effluent from anaerobic sludge stabilization. Design Guideline. DWA Deutsche Vereinigung für Wasserwirtschaft, Abwasser und Abfall e.V., Hennef. 2019; ISBN 978-3-88721-824-9.
18. Lackner, S.; Gilbert, E.; Vlaeminck, S.; Joss, A.; Horn, H.; van Loosdrecht, M. Full-scale partial nitritation/anammox experiences—An application survey. *Water Res.* **2014**, *55*, 292–303. [CrossRef]
19. Jardin, N.; Hennerkes, J. Full-scale experience with the deammonification process to treat high strength sludge water—A case study. *Water Sci. Technol.* **2012**, *65*, 447–455. [CrossRef] [PubMed]
20. Strous, M.; Heijnen, J.J.; Kuenen, J.G.; Jetten, M.S.M. The sequencing batch reactor as a powerful tool for the study of slowly growing anaerobic ammonium-oxidizing microorganisms. *Appl. Microbiol. Biotechnol.* **1998**, *50*, 589–596. [CrossRef]
21. Ochs, P.; Martin, B.D.; Germain, E.; Wu, Z.; Lee, P.-H.; Stephenson, T.; van Loosdrecht, M.; Soares, A. Evaluation of a Full-Scale Suspended Sludge Deammonification Technology Coupled with an Hydrocyclone to Treat Thermal Hydrolysis Dewatering Liquors. *Processes* **2021**, *9*, 278. [CrossRef]
22. Han, X.; Zhang, S.; Yang, S.; Zhang, L.; Peng, Y. Full-scale partial nitritation/anammox (PN/A) process for treating sludge dewatering liquor from anaerobic digestion after thermal hydrolysis. *Bioresour. Technol.* **2020**, *297*, 122380. [CrossRef] [PubMed]
23. Böhnke, B. Das Adsorptions-Belebungsverfahren. The adsorption process for activated sludge. *Korresp. Abwasser* **1977**, *2*, 33–42.
24. Matsche, N.; Moser, D. Operation of a two-stage activated sludge package plant for high efficiency treatment. *Water Sci. Technol.* **1993**, *28*, 299–307. [CrossRef]
25. Lindtner, S.; Zessner, M. *Abschätzung von Schmutzfrachten in der Abwasserentsorgung bei Unvollständiger Datenlage. Estimation of Loads in Wastewater Disposal with Incomplete Data Basis*; TU Wien, Wiener Mitteilungen: Vienna, Austria, 2003; Volume 183, pp. 195–227. ISBN 3-85234-074-8.

26. Paravicini, V.; Schmidt, E.; Svardal, K.; Kroiss, H. Evaluating the stabilisation degree of digested sewage sludge: Investigations at four municipal wastewater treatment plants. *Water Sci. Technol.* **2006**, *53*, 81–90. [CrossRef]
27. Frey, W. *Planung und Gestaltung von Belüftungssystemen. Design of Aeration Systems*; TU Wien, Wiener Mitteilungen: Vienna, Austria, 1998; Volume 145, pp. 293–321.
28. Frey, W. *Messwerte und Kennzahlen der Maschinellen Ausrüstung auf Kläranlagen. Measured Values and Key Figures of the Mechanical Equipment on Sewage Treatment Plants*; TU Wien, Wiener Mitteilungen: Vienna, Austria, 2011; Volume 224, pp. 253–268.
29. Rosso, D.; Larson, L.; Stenstrom, M. Aeration of large-scale municipal wastewater treatment plants: State of the art. *Water Sci. Technol.* **2008**, *57*, 973–978. [CrossRef]
30. He, S.; Niu, Q.; Ma, H.; Zhang, Y.; Li, Y.Y. The treatment performance and the bacteria preservation of anammox: A review. *Water Air Soil Pollut.* **2015**, *226*. [CrossRef]
31. Aroraa, A.S.; Nawaza, A.; Qyyuma, M.A.; Ismail, S.; Aslam, M.; Tawfik, A.; Yun, C.M.; Lee, M. Energy saving anammox technology-based nitrogen removal and bioenergy recovery from wastewater: Inhibition mechanisms, state-of-the-art control strategies, and prospects. *Renew. Sustain. Energy Rev.* **2021**, *135*, 110126. [CrossRef]
32. Parravicini, V. Sidestream Nitritation at a two-stage WWTP as an energy efficient option—An Austrian case study. In Proceedings of the POWERSTEP Conference, Munich, Germany, 16–17 May 2018.
33. Baumgartner, T.; Parravicini, V. *WP4: Nitrogen Management in Side Stream, D4.4: Decision Support for Finding the Appropriate Resource and Energy Optimized SDE Treatment Technology*; Project Report POWERSTEP (641661); European Union: Brussel, Germany, 2018.

International Journal of
Environmental Research and Public Health

MDPI

Article

Risk Analysis of Heavy Metals Migration from Sewage Sludge of Wastewater Treatment Plants

Robert Kowalik [1,*], Jarosław Gawdzik [1], Paulina Bąk-Patyna [2], Piotr Ramiączek [2] and Nebojša Jurišević [3]

1 Faculty of Environmental, Geomatic and Energy Engineering, Kielce University of Technology, 25-314 Kielce, Poland
2 Faculty of Civil Engineering and Architecture, Kielce University of Technology, 25-314 Kielce, Poland
3 Department for Energy and Process Engineering, Faculty of Engineering, University of Kragujevac, 34000 Kragujevac, Serbia
* Correspondence: rkowalik@tu.kielce.pl

Abstract: More and more attention in sewage sludge management is being devoted to its environmental utilization. This approach is justified both from economic and environmental points of view. However, as with any method, there are certain possibilities and limitations. The goal of the natural utilization of sewage sludge is to recover the valuable agronomic properties and fertilizing potential of the sludge. The main aspect limiting the possibility of using sludge as a fertilizer is the heavy metal content. In this paper, an analysis of the risk of environmental contamination in the case of application of sewage sludge with different forms of sludge treatment was carried out. Risk indices such as Igeo and PERI, based on the comparison of total metal content in sludge and soil, as well as RAC and ERD indices, which take into account the mobility of metals in soil, were calculated. It was shown that high levels of potential risk and geoaccumulation indicators do not necessarily disqualify the use of sewage sludge, the key aspect is the form of mobility in which the heavy metals are found in the sludge, and this should be the only aspect taken into account for the possibility of their environmental use.

Keywords: heavy metals; sewage sludge; mobility; environmental pollution

Citation: Kowalik, R.; Gawdzik, J.; Bąk-Patyna, P.; Ramiączek, P.; Jurišević, N. Risk Analysis of Heavy Metals Migration from Sewage Sludge of Wastewater Treatment Plants. *Int. J. Environ. Res. Public Health* **2022**, *19*, 11829. https://doi.org/10.3390/ijerph191811829

Academic Editors: Yung-Tse Hung, Hamidi Abdul Aziz and Issam A. Al-Khatib

Received: 18 August 2022
Accepted: 15 September 2022
Published: 19 September 2022

1. Introduction

The progress of civilization in the last century has contributed significantly to the improvement of human life quality. As a result of these changes, the water demand and pollutant load of discharged wastewaters have increased [1–3]. It is, therefore, necessary to continuously expand and modernize the infrastructure for wastewater collection and treatment. However, this contributes to an increase in the amount of sewage sludge generated as a byproduct of sewage treatment plant processes. So far, no waste-free method or effective solution has been developed to completely eliminate sewage sludge from the environment [4,5]. However, there are many methods to utilize the sludge generated this way. Agricultural use of sludge fertilizers is particularly beneficial due to its high soil-forming and fertilizing properties [6–8]. To this end, the method of sewage sludge management is primarily determined by the amount and properties of the sludge [9,10]. Sludge with high reclamation fertilizer values can be used as an organic fertilizer as long as the micropollutant content does not have a negative impact on the soil environment [11–13]. The permissible levels of heavy metals in the application of sewage sludge in Poland and the world are listed in Table 1.

Table 1. Heavy metal limit values in sewage sludge intended for natural use (mg/kg d.m.).

Metal	Limit Values for Heavy Metals in Sewage Sludge Intended for Natural Use					
	Poland Regulation [14]	EU Directive 86/278/EEC [15]	Chinese Regulation GB 18918-2002 [16]		USA Regulation 40 CFR Part 503, 503.13 [17]	South African Guideline (Pollutant Class a) [18]
			pH < 6.5	pH > 6.5		
Cd	20	20–40	5	20	39	40
Ni	300	300–400	100	200	420	420
Zn	2500	2500–4000	500	1000	2800	2800
Cu	1000	1000–1750	250	500	1500	1500
Cr	500	-	600	1000	-	1200
Pb	750	750–1200	300	1000	300	300

The chemical forms of metals present in sewage sludge can be identified by sequential extraction or speciation based on the fractionation of compounds. The use of this analytical procedure ensures the separation of the test material into fractions characterized by different degrees of mobility [19].

Soluble metals, which are highly mobile and readily available, pose the greatest threat to soil inhabitants as micronutrients that enter ground and surface waters move up the trophic chain. Heavy metals in the soil are not immediately absorbed by plants; however, they can slowly form hazardous solutions over time [20]. Some essential elements, such as Fe, Co, Cu, Cr, Mo, Mn, Se, Ni, and Zn, are required for organisms in trace amounts; however, they become toxic at higher levels. Nonessential elements such as Sb, Pb, Hg, Ag, and As are toxic and not needed by living organisms [21]. However, most wastewaters and wastes contain heavy metals in amounts sufficient to cause toxicity to crop plants [22].

Sewage sludge, a byproduct of wastewater treatment, can be managed in several ways. However, the most favorable variant from an ecological standpoint and in terms of a circular economy is its use for agricultural purposes.

Municipal sewage sludge can be used as a substrate for the production of organic fertilizers or plant growth aids, but the most important criterion it must meet is the total content of heavy metals. In that regard, the goal of this study is to confirm that a high concentration of heavy metals in the sludge does not always rule out the possibility of sludge agricultural use. The key, therefore, is the content of metals in fractions that tend to migrate deep into the environment and, thus, can easily enter the food chain.

This study investigates the sewage sludge content of four wastewater treatment plants in Poland using different wastewater treatment technologies. The investigation considers heavy metal concentrations, mobility, and the risk of contamination of the environment. Based on the results, the geoaccumulation index (Igeo), potential environmental risk index (PERI), risk assessment code (RAC), and environmental risk determinant (ERD) were calculated. All indicators were then compared to sewage sludge use regulations in Poland and Europe. It was also determined whether or not the treatment technology is critical in terms of the content of heavy metals in mobile forms. The importance of analyzing the form in which heavy metals are present became apparent when considering their use for agricultural purposes.

2. Materials and Methods

2.1. Characteristics of Collection Points and Potential Uses of Sludge

Sludge samples were collected from four different wastewater treatment plants located in the Swietokrzyskie province in Poland (Figure 1). The characteristics of the treatment plants are presented in Table 2. The plants differ in the type of sludge treatment, which are: oxygen stabilization, dewatering on belt press, Imhoff fermentation, and internal digester fermentation. Reference points for the content of heavy metals in soil were measuring stations prepared within the *Framework of Monitoring of Chemistry of Arable Soils in Poland* [23] located not far from the sampling points (Figure 1). Test Point 361 in Wola Kopcowa was selected as a point for potential use of sewage sludge. The soil was characterized by complex 2z (medium grassland), type: A (podzolic soil), and valuation class: IVb. The soil type was sandy loam. The "pH" in H_2O suspension was 5.5, while in KCL, it was 4.5. The humus content of the soil was 3.24%, organic carbon 1.88%, total nitrogen 0.09%, while the C/N ratio was 20.89 [23].

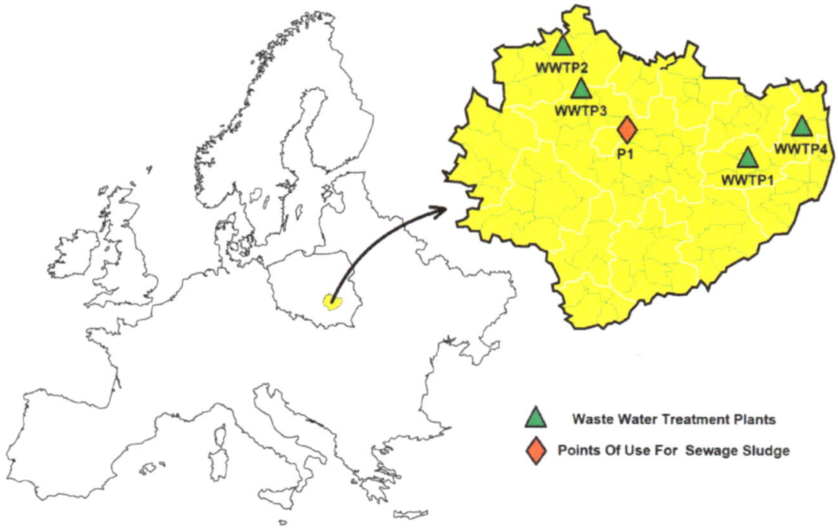

Figure 1. Location of WWTPs and potential sites of agricultural use of sewage sludge (own research).

Table 2. Characteristics of WWTPs (own research).

	Wastewater Treatment Plant			
	WWTP1	**WWTP2**	**WWTP3**	**WWTP4**
Location of WWTP	Opatow	Kornica	Mniow	Ozarow
Type of WWTP	Mech.-biol.	Mech.-biol.	Mech.-biol.	Mech.-biol.
Equivalent Number of Residents	15,240	21,594	9550	9660
SS treatment	Internal digester fermentation	Imhoff fermentation	Oxygen stab.	Dewatering on belt press
Distance of the WWTP from the point use of SS (km)	56	61	32	80

2.2. Heavy Metal Speciation

Heavy metals can be classified into four mobility fractions based on their migration capacity [24]. These are:

FI fraction—associated with carbonates, the most mobile;
FII fraction—associated with amorphous iron and manganese oxides;
FIII fraction—associated with organic and sulfide matter;
FIV fraction—associated with silicates—a completely chemically stable fraction.

The study used a four-step procedure developed by the European Community Reference Bureau, or BCR for short [25,26]:

- Step I: CH_3COOH extraction–(FI—exchangeable fraction);
- Step II: extraction $NH_2OH{\cdot}HCl$–(FII—reducible fraction);

- Step III: extraction H_2O_2/CH_3COONH_4–(FIII—oxidizable fraction);
- Step IV: mineralization of the residual fraction with a mixture of concentrated acids (HCl, HF, HNO_3)–(FIV—residual fraction).

2.3. *Heavy Metal Accumulation Risk Indicators*

2.3.1. Geoaccumulation Index of Heavy Metal in Soil (Igeo)

This specific method was proposed by Muller [27] to determine and classify the state of sludge/soil contamination at five levels, ranging from uncontaminated to highly contaminated. The geoaccumulation index (Igeo) measures the level of sediment or soil contamination by inorganic or organic trace substances of environmental concern and bioelements. It compares current concentrations to precivilization concentrations or, in the case of synthetic substances that do not occur in nature but have recently been produced, to reference values derived from an assumed uniform global distribution of these substances. Igeo is defined by the equation [27,28]:

$$\text{Igeo} = log_2 \frac{C_n}{1.5 \cdot B_n} \tag{1}$$

where:

C_n—the concentration of a specific heavy metal element in sewage sludge, mg·kg^{-1} d.m.; B_n—content of a given element from the group of heavy metals present in the soil, mg·kg^{-1} d.m.

Table 3 presents the classification of the heavy metals geoaccumulation index and risk assessment code.

Table 3. Classification of Igeo [28,29].

Igeo	Pollution Value
<0	No pollution
0–1	No pollution, moderate pollution
1–2	Moderate pollution
2–3	moderate pollution or high
3–4	High pollution

2.3.2. Risk Assessment Code (RAC)

The risk assessment code (RAC) is a quantitative method for determining the mobility and bioavailability of heavy metals based on total metal concentration and chemical fraction. Because the acid-extractive fraction (F1), which consists of exchange fraction, has higher bioavailability, its mass fraction is used to evaluate metals in soils or sediments [28]. The RAC index introduced by Perin et al. [29], was classified into five risk categories (Table 4). It is calculated in accordance with [29,30]:

$$\text{RAC} = \frac{\text{F1}}{\text{HM}} \cdot 100\% \tag{2}$$

where:

F1—acid heavy metal concentration—soluble/free fraction; mg·kg^{-1}; HM—total heavy metal concentration, mg·kg^{-1}.

Table 4. Classification of RAC [29–31].

RAC	Risk Value
<1	No risk
1–10	Low risk
11–30	Medium risk
31–50	High risk
>50	Very high risk

2.3.3. Potential Environmental Risk Index (PERI)

The potential ecological risk index (PERI), developed by Hakanson (1980) [28], is based on the principles of sedimentology. Scientists use it extensively to evaluate the pollution and potential ecological risk associated with heavy metals in sewage sludge. This index takes into account not only the heavy metal content of sewage sludge, but also the ecological and environmental effects of heavy metals [30]. It is calculated by the following formulas [30–32]:

$$C_f^i = \frac{C_D^i}{C_R^i} \quad (3)$$

where:
C_f^i —pollution factor;
C_D^i—concentration of the i-th element from the HM's group present in sludge, mg·kg^{-1} d.m.;
C_R^i—concentration of the i-th element from the HM's group in the soil, mg·kg^{-1}.

$$E_r^i = T_r^i \cdot C_f^i \quad (4)$$

where:
E_r^i– index of the potential ecological risk of the i-th element from the HM group;
T_r^i–toxicity factor of the i-th element from the HM group;
The degree of heavy metal toxicity varies according to the toxicity factor. (T_r^i): Cd-30, Cu, Ni and Pb-5, and Zn-1 [30].

The sum of potential ecological risk from sludge in the ground is defined by the equation [31]:

$$\text{PERI} = \sum_{i=1}^{n} E_r^i \quad (5)$$

The risk level is classified into 5 categories as shown in Table 5.

Table 5. PERI indicator classification [30–32].

E_r^i	PERI	Risk Value
<40	<150	Low
40–80	150–300	Medium
80–320	300–600	High
>320	>600	Very high

2.3.4. Environmental Risk Determinant (ERD)

Considering the mobility of heavy metals, it can be seen that only Fraction IV does not migrate into the soil-water environment under any conditions. The mobile fractions (FI, FII) are considered to be the most mobile, while Fraction FIII can be mobile under certain conditions, i.e., when the organic matter in the soil is fully processed by microorganisms and when there is an ozone storm. Metals bound to iron and manganese oxides are released into the environment relatively slowly. Under certain conditions of pH and oxidation-reduction potential, metals bound to FII can exhibit significant bioavailability [33]. Environmental risk assessment is carried out based on the first three fractions, taking into account the level of individual predisposition of each fraction to release heavy metals into the soil environment.

The ERD calculates the content of heavy metal elements based on their distribution in the four fractions. Each fraction is assigned a weight ranging from 0 to 1. The authors proposed using the ERD index because none of the indicators using the mobility issue take into account the weight of each fraction. Consider that the FI, FII, and FIII fractions are mobile, but the FI fraction is much more mobile than FII and FIII, which takes into account the formula for the ERD index. The adopted weight ranges were proposed based on the scale analysis of the other indicators. The content of metals found in Fraction I is taken into account in its entirety to determine the risk of ecological contamination, while Fractions FII and FIII, no longer completely mobile, were reduced by a procedure of potentiating the values to the second and third power, respectively. The applied scale ranges were proposed based on the evaluation of the scales of the other indicators. Its determinant is defined by the equation [33,34]:

$$ERD = F_{p1} + F_{p2} + F_{p3} \quad (6)$$

where:

$Fp_1 = F_1$; F_1—metal content in Fraction FI on a scale of 0–1; $F_{p2} = {F_2}^2$; F_2—metal content in Fraction FII on a scale of 0–1; $F_{p3} = {F_3}^3$; F_3—metal content in Fraction FIII on a scale of 0–1.

The risk level is classified into 4 categories as shown in Table 6.

Table 6. ERD indicator classification [33,34].

ERD	Risk Value
$0 < ERD \leq 0.35$	Low risk
$0.35 < ERD \leq 0.6$	Medium risk
$0.6 < ERD \leq 0.8$	High risk
$0.8 < ERD$	Very high risk

3. Results

This section is divided into subsections that provide a concise description of the experimental results, their interpretation, and experimental conclusions. The results of chemical speciation of heavy metals in sewage sludge are shown in Table 7.

Table 7. Chemical speciation of heavy metal in sewage sludge, for sludge from all four treatment plants, the results are the statistical average of four separate measurements for each sludge, excluding coarse errors, mg·kg^{-1}.

	Heavy Metal (mg/kg s.m.)					
Fraction	Cu	Cr	Cd	Ni	Pb	Zn
			Sewage sludge—S1			
Fraction I	3.3 ± 0.2	2.0 ± 0.3	0.3 ± 0.1	3.5 ± 0.1	5.2 ± 0.1	79.4 ± 0.7
Fraction II	1.8 ± 0.1	1.1 ± 0.1	0.3 ± 0.1	1.4 ± 0.1	0.5 ± 0.2	122.8 ± 2.6
Fraction III	57.1 ± 1.1	16.1 ± 0.7	1.9 ± 0.1	5.9 ± 0.1	7.8 ± 0.1	323.8 ± 1.5
Fraction IV	22.8 ± 0.7	22.0 ± 0.7	1.1 ± 0.8	9.2 ± 0.3	54.7 ± 9.4	170.8 ± 1.3
ΣFI … IV	85.0 ± 2.3	41.2 ± 0.9	3.6 ± 0.9	20.0 ± 0.5	68.2 ± 11.3	696.8 ± 2.6
			Sewage sludge—S2			
Fraction I	0.2 ± 0.1	5.2 ± 0.2	0.1 ± 0.1	0.0 ± 0.1	0.1 ± 0.1	161.3 ± 2.0
Fraction II	1.1 ± 0.1	0.8 ± 0.2	0.6 ± 0.1	0.4 ± 0.1	0.6 ± 0.2	71.7 ± 0.7
Fraction III	47.4 ± 0.8	12.1 ± 0.3	0.5 ± 0.1	0.3 ± 0.1	7.4 ± 0.8	356.0 ± 3.5
Fraction IV	18.3 ± 0.4	18.9 ± 0.4	0.6 ± 0.1	2.1 ± 0.3	9.1 ± 0.8	899.0 ± 9.2
ΣFI … IV	67.0 ± 0.9	37.0 ± 2.4	1.8 ± 0.5	2.8 ± 0.5	17.2 ± 3.1	1488 ± 8.0

Table 7. *Cont.*

	Heavy Metal (mg/kg s.m.)					
Fraction	**Cu**	**Cr**	**Cd**	**Ni**	**Pb**	**Zn**
			Sewage sludge—S3			
Fraction I	3.3 ± 0.2	1.6 ± 0.1	0.4 ± 0.1	2.6 ± 0.2	9.4 ± 0.9	99.2 ± 2.2
Fraction II	1.6 ± 0.1	1.5 ± 0.1	0.3 ± 0.1	6.1 ± 0.5	11.1 ± 1.3	123.2 ± 2.9
Fraction III	36.1 ± 0.3	14.3 ± 0.3	1.7 ± 0.1	9.1 ± 0.7	9.9 ± 1.1	499.9 ± 8.2
Fraction IV	22.1 ± 0.3	19.1 ± 0.4	1.4 ± 0.1	4.3 ± 0.3	98.5 ± 9.9	324.5 ± 8.1
ΣFI … IV	63.1 ± 0.9	36.5 ± 0.5	3.8 ± 0.2	22.1 ± 0.7	128.9 ± 3.7	1047 ± 19.6
			Sewage sludge—S4			
Fraction I	2.1 ± 0.2	4.9 ± 0.3	1.1 ± 0.1	1.6 ± 0.2	4.2 ± 0.5	121.2 ± 1.1
Fraction II	3.7 ± 0.1	0.1 ± 0.05	1.6 ± 0.1	3.8 ± 0.5	3.7 ± 0.4	111.2 ± 1.4
Fraction III	33.2 ± 0.2	16.1 ± 0.2	0.8 ± 0.1	0.2 ± 0.1	4.6 ± 0.5	329.0 ± 3.1
Fraction IV	2.3 ± 0.3	17.1 ± 0.5	0.9 ± 0.1	8.1 ± 0.6	58.0 ± 6.1	350.5 ± 2.1
ΣFI … IV	41.3 ± 1.2	38.2 ± 1.3	4.4 ± 0.6	13.7 ± 0.7	70.5 ± 11.7	911.9 ± 7.7

The Igeo index is largely dependent on the heavy metal content of the soil at the site of potential use. Because of this fact, a common site for sludge from all four WWTPs was chosen as a potential sludge use point. Figure 2 shows the Igeo value for all samples analyzed

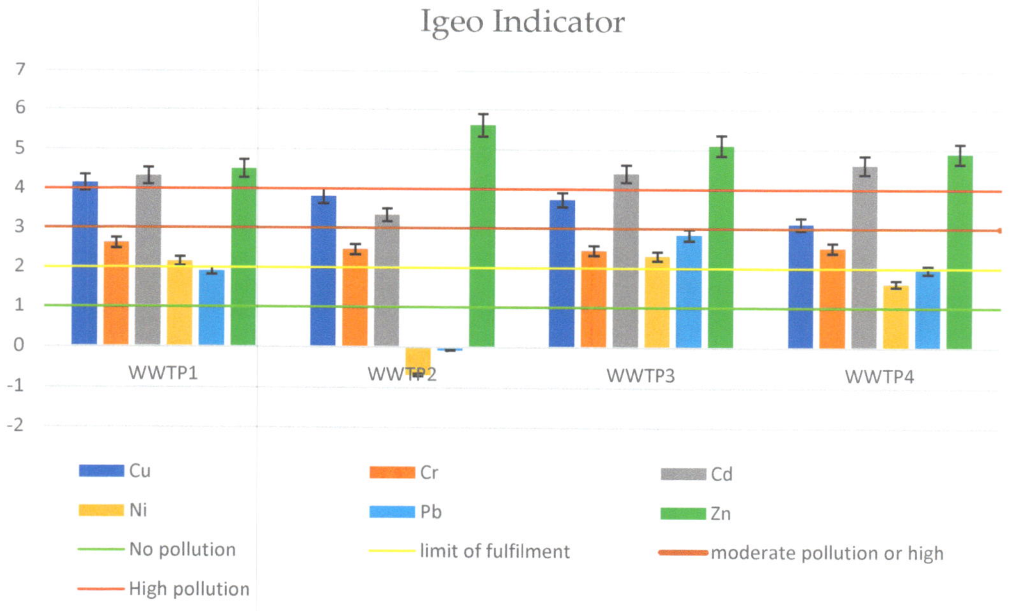

Figure 2. The geoaccumulation index (Igeo) of heavy metals in sewage sludge.

For all analyzed cases of sewage sludge, the RAC showed no high risk of environmental contamination. This is due to the low content of heavy metals in the FI fraction. Statistically, sediments collected from WWTP4 have the highest percentage of metals in the FI fraction. The highest percentage was recorded for cadmium from WWTP4, which was 25%, but it did not reach the high-risk level. The outcomes can be considered satisfactory; however, it should be noted that the RAC index only considers the FI fraction, ignoring the heavy metals in the FII and FIII fractions. Figure 3 shows the RAC value for all sewage sludge samples.

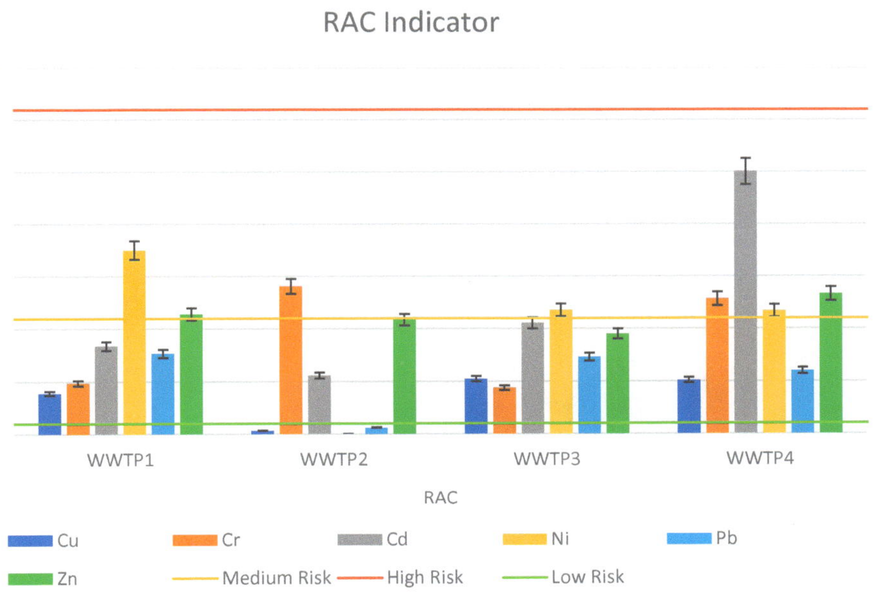

Figure 3. Risk assessment code (RAC) of HMs in sewage sludge.

Analyzing the PERI indicator for the studied sewage sludge samples, it can be concluded that copper, cadmium, and zinc are the main heavy metals that pose a threat to the environment. Other heavy metals showed low levels of potential environmental contamination. Figure 4 shows the PERI values.

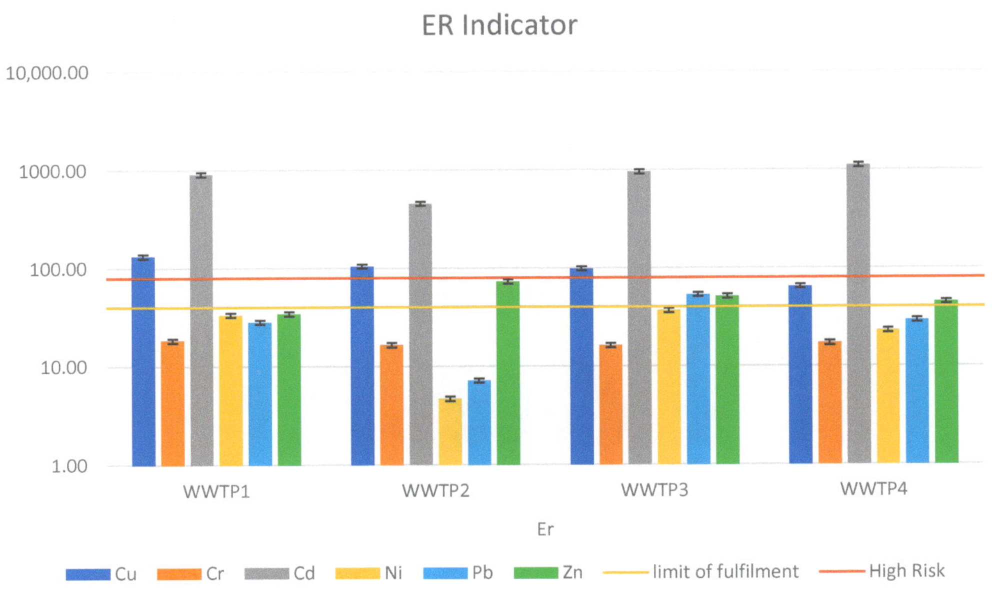

Figure 4. PERI indicator of heavy metals in sewage sludge.

The ERD index revealed similar risk levels to the RAC, but it was more accurate due to the inclusion of metals from fractions II and III. As with the RAC, sewage sludge collected from Wastewater Treatment Plant 4 proved to be the most hazardous in terms of potential metal migration. Figure 5 depicts the ERD values.

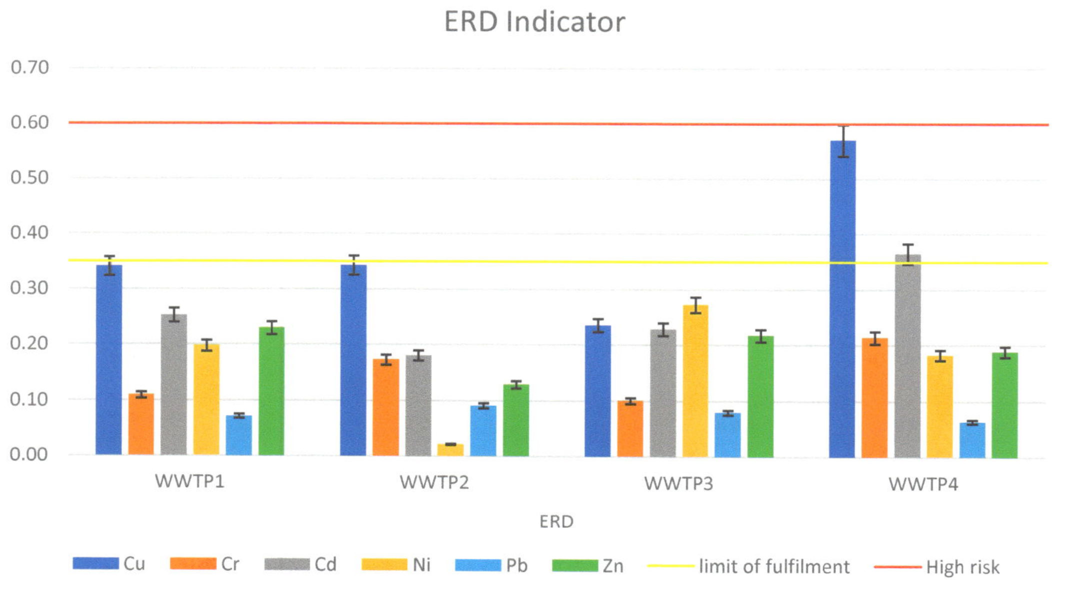

Figure 5. The environmental risk determinant (ERD) indicator of heavy metals in sewage sludge.

For the results of potential risk, for all indices and wastewater treatment plants, a noncompliance table was prepared (Table 8). The heavy metals listed in the table did not meet the criterion that would qualify the sludge for potential environmental use. As can be seen in Figures 2 and 4, indicators comparing total sediment content to content at the point of use were far more critical than indicators considering mobility. Although the metal content of sediment is high, it may be in stable fractions that cannot migrate in soil or vegetation. As a result, it appears appropriate to consider the form in which the metal occurs when evaluating the possibility of natural sewage sludge use.

Table 8. Schedule of failure to meet heavy metal toxicity criterion from analyzed sites for four pollutant indicators.

WWTP *	Igeo	RAC	PERI	ERD
WWTP1	Ni, Zn, Cu, Cr, Cd	Ni, Zn	Cu, Cd	-
WWTP2	Zn, Cu, Cr, Cd,	Cr	Cu, Cd, Zn	-
WWTP3	Ni, Zn, Pb, Cu, Cr, Cd,	Ni	Cu, Cd, Pb, Zn	-
WWTP4	Zn, Cu, Cd, Cr,	Cr, Cd, Ni, Zn	Cu, Cd, Zn	Cd, Cu

* WWTP1 (internal digester fermentation), WWTP2 (Imhoff fermentation), WWTP3 (oxygen stab.) WWTP4 (dewatering on belt press).

4. Discussion

This paper presents the analysis of sewage sludge taken from four treatment plants that use various water treatment technologies. The heavy metal content in all samples did not exceed the permissible metal content limit for agricultural use. However, the results of the analyzed indicators of contamination risk were not so encouraging. The most stringent proved to be the Igeo index, which compares the metal content of the sludge with the content in the geological substrate at the site of potential use. According to Igeo, all sediments posed a very high risk of ecological contamination, with zinc and cadmium proving particularly toxic. Another indicator considering metal mobility was RAC. According to the indicator value, the sediments showed a medium or low risk of contamination. This is related to their exchangeable fraction content. With respect to RAC, the most toxic sludge sample was the one taken from Treatment Plant 4: Heavy Metal. In all samples, cadmium and copper proved to be the most toxic and risky. The other metals showed medium or low levels of toxicity. The index proposed by the authors (ERD), which is based primarily on the issue of heavy metal mobility, identified WWTP 4 as having the highest risk of ecological contamination. The other three WWTPs, on the other hand, do not pose a high contamination potential risk. Copper is the heavy metal most likely to penetrate deep into the soil from the sludge samples of all analyzed wastewater treatment plants.

5. Conclusions

This paper examines the risk of environmental contamination caused by the agricultural use of sewage sludge from four wastewater treatment plants in Poland. Each of the analyzed facilities uses different wastewater treatment technology. Sewage sludge from all wastewater treatment plants met the applicable heavy metals limits imposed by legal acts, which is the primary criterion for using sewage sludge as a fertilizer. Conventional indices based solely on total heavy metal content, such as Igeo and PERI, were significantly more critical in assessing the feasibility of sludge use than the ERD and RAC, which also look at heavy metal chemical forms. As a result, most of the metals in the sludge were in a completely stable form, and despite their high concentration, there was no possibility of migration and entry into the crop. The study found that wastewater treatment technology has no significant impact on the total content of heavy metals in the sludge; however, sludge from a treatment plant that uses press dewatering has the highest content in the mobile sections. In all sewage sludge samples, copper was the most mobile heavy metal.

Author Contributions: Conceptualization, R.K. and J.G.; methodology, R.K. and J.G.; software, P.B.-P. and P.R.; validation, R.K., P.R. and P.B.-P.; formal analysis, J.G.; investigation, P.B.-P.; resources, P.R.; data curation, N.J.; writing—original draft preparation, R.K.; writing—review and editing, R.K., N.J. and J.G.; visualization, P.B.-P. and N.J.; supervision, J.G.; project administration, J.G.; funding acquisition, P.B.-P. All authors have read and agreed to the published version of the manuscript.

Funding: The APC was funded by program of the Minister of Science and Higher Education under the name: Regional Initiative of Excellence in 2019–2022 project number 025/RID/2018/19, financing amount PLN 12.000.000.

Institutional Review Board Statement: Not applicable.

Informed Consent Statement: Not applicable.

Data Availability Statement: Not applicable.

Conflicts of Interest: The authors declare no conflict of interest.

References

1. Miksch, K.; Sikora, J. *Biotechnologia Ścieków;* WN PWN: Warszawa, Poland, 2012.
2. Bień, J.; Neczaj, E.; Worwąg, M.; Grosser, A.; Nowak, D.; Milczarek, M.; Janik, M. Kierunki zagospodarowania osadów w Polsce po roku 2013. *Inżynieria Ochr. Śr.* **2011**, *14*, 375–384.
3. Ministry of Environment. *Municipal Sewage Sludge Treatment Strategy for the Years 2019–2022;* Ministry of Environment: Warszawa, Poland, 2018. Available online: https://www.gov.pl/attachment/2846e2b3-68c7-46eb-b36e-7643e81efd9a (accessed on 19 May 2022).
4. McBride, M.B. Toxic elements in sewage suldge-amended soils: Has promotion of beneficial use discounted the risks? *Adv. Environ. Res.* **2003**, *8*, 5–19. [CrossRef]
5. Gawdzik, J. *Mobility of Selected Heavy Metals in Sewage Sludge;* Kielce University of Technology: Kielce, Poland, 2013.

6. National Waste Management Plan. 2016. Available online: https://bip.mos.gov.pl/fileadmin/user_upload/bip/strategie_plany_programy/DGO/Krajowy_plan_gospodarki_odpadami_2022_____M.P._poz._784_.pdf (accessed on 30 May 2021).
7. Eurostat. Available online: https://ec.europa.eu/eurostat/data/database (accessed on 30 May 2021).
8. Durdević, D.; Trstenjak, M.; Hulenić, I. Sewage sludge thermal treatment technology selection by utilizing the analytical hierarchy process. *Water* **2020**, *12*, 1255–1271. [CrossRef]
9. Fytili, D.; Zabaniotou, A. Utilization of sewage sludge in EU application of old and new methods—A review. *Renew. Sustain. Energy Rev.* **2008**, *12*, 116–140. [CrossRef]
10. Latosińska, J. Influence of temperature and time of sewage sludge incineration on the mobility of heavy metals. *Environ. Prot. Eng.* **2017**, *43*, 105–122. [CrossRef]
11. Urra, J.; Alkorta, I.; Mijangos, I.; Epelde, L.; Garbisu, C. Application of sewage sludge to agricultural soil increases the abundance of antibiotic resistance genes without altering the composition of prokaryotic communities. *Sci. Total Environ.* **2019**, *10*, 1410–1420. [CrossRef]
12. Latosińska, J. Risk assessment of soil contamination with heavy metals from sewage sludge and ash after its incineration. *Desalin. Water Treat* **2020**, *199*, 297–306. [CrossRef]
13. Waste Act (J.L. 2013, No. 0, Item. 21). Available online: https://isap.sejm.gov.pl/isap.nsf/download.xsp/WDU20130000021/T/D20130021L.pdf (accessed on 19 March 2022).
14. Minister of the Environment. *Regulation of the Minister of the Environment of 6 February 2015 on the Municipal Sewage Sludge (J. L. 2015, No. 0, Item. 257)*; Minister of the Environment: Warsaw, Poland, 2015.
15. Minister of the Economy. *Regulation of the Minister of the Economy of 16 of 16 July 2015 on the Acceptance of Waste for Landfilling (J.L. 2015, No.0, Item. 1277)*; Minister of the Economy: Warsaw, Poland, 2015.
16. Minister of Agriculture and Rural Development. *Regulation of the Minister of Agriculture and Rural Development of 18 June 2008 on the Implementation of Certain Provisions of the Act on Fertilizers and Fertilization (J.L. 2008, No. 119, Item.765)*; Minister of Agriculture and Rural Development: Warsaw, Poland, 2008.
17. Code of Federal Regulations. *Protection of Environment. Chapter I—Environmental Protection Agency (Continued), Subchapter O—Sewage Sludge, Part 503—Standards for The Use Or Disposal Of Sewage Sludge, Subpart B—Land Application, Section 503.13—Pollutant Limits*; Office of the Federal Register (United States): Washington, DC, USA, 2010.
18. Water Research Commission. *Guidelines for the Utilisation and Disposal of Wastewater Sludge. Vol. 2. Requirements for the Agricultural Use of Wastewater Sludge*; WRC Report No.: TT 262/06; WRC: Pretoria, South Africa, 2009.
19. Gawdzik, J. Mobility of heavy metals in sewage sludge on the example of a selected sewage treatment plan. *Eng. Environ. Prot.* **2012**, *15*, 5–15.
20. Mizerna, K.; Król, A. Sequential extraction of heavy metals in mineral-organic composite. *Ecol. Eng. Environ. Technol.* **2018**, *19*, 23–29. [CrossRef]
21. Bąk, Ł.; Szeląg, B.; Górski, J.; Górska, K. The Impact of Catchment Characteristics and Weather Conditions on Heavy Metal Concentrations in Stormwater—Data Mining Approach. *Appl. Sci.* **2019**, *9*, 2210. [CrossRef]
22. Latosińska, J.; Kowalik, R.; Gawdzik, J. Risk assessment of soil contamination with heavy metals from municipal sewage sludge. *Appl. Sci.* **2021**, *11*, 548–561. [CrossRef]
23. Monitoring of Soil and Land Quality in Poland. Monitoring of Soil and Land Quality in Poland. Chief Inspectorate of the Environmental Protection. Poland. Available online: http://www.gios.gov.pl/pl/stan-srodowiska/monitoring-jakosci-gleby-i-ziemi (accessed on 19 May 2022).
24. Jasińska, A. The Importance of Heavy Metal Speciation from the Standpoint of the Use of Sewage Sludge in Nature. *Eng. Prot. Environ.* **2018**, *21*, 239–250. [CrossRef]
25. Larner, B.L.; Sean, A.J.; Townsend, A.T. Comparative study of optimised BCR sequential extraction scheme and acid leaching of elements in the certified reference material NIST 271. *Anal. Chim. Acta* **2006**, *556*, 444–449. [CrossRef]
26. Ure, A.M.; Quevauviller, P.; Muntau, H.; Gripink, B. Speciation of heavy metals in soils and sediments. An account of the improvement and harmonization of extraction techniques undertaken under auspicies of the BCR of the Commision of the European Communities. *Int. J. Environ. Anal. Chem.* **1993**, *51*, 135–151. [CrossRef]
27. Muller, G. Index of geoaccumulation in sediments of the Rhine River. *Geol. J.* **1969**, *2*, 109–118.
28. Hakanson, L. An ecological risk index for aquatic pollution control. A sedimentological approach. *Water Res.* **1980**, *14*, 975–1101. [CrossRef]
29. Perin, G.; Craboledda, L.; Lucchese, M.; Cirillo, R.; Dotta, L.; Zanetta, M.L.; Oro, A.A. Heavy metal speciation in the sediments of northern Adriatic Sea. A new approach for environmental toxicity determination. In *Heavy Metals in the Environment*; Lakkas, T.D., Ed.; CEP Consultants: Edinburgh, Scotland, 1985; Volume 2, pp. 454–456.
30. Zhang, J.; Tian, Y.; Zhang, J.; Li, N.; Kong, L.; Yu, M.; Zuo, W. Distribution and risk assessment of heavy metals in sewage sludge after ozonation. *Environ. Sci. Pollut. Res.* **2017**, *24*, 5118–5125. [CrossRef]
31. Xiao, Z.; Yuan, X.; Leng, L.; Jiang, L.; Chen, X.; Zhibin, W.; Xin, P.; Jiachao, Z.; Zeng, G. Risk assessment of heavy metals from combustion of pelletized municipal sewage sludge. *Environ. Sci. Pollut. Res.* **2016**, *23*, 3934–3942. [CrossRef]
32. Tytła, M.; Widzewicz, K.; Zielewicz, E. Heavy metals and its chemical speciation in sewage sludge at different stages of processing. *Environ. Technol.* **2016**, *37*, 899–908. [CrossRef]

33. Kowalik, R.; Latosińska, J.; Gawdzik, J. Risk Analysis of Heavy Metal Accumulation from Sewage Sludge of Selected Wastewater Treatment Plants in Poland. *Water* **2021**, *13*, 2070. [CrossRef]
34. Kowalik, R.; Latosińska, J.; Metryka-Telka, M.; Porowski, R.; Gawdzik, J. Comparison of the Possibilities of Environmental Usage of Sewage Sludge from Treatment Plants Operating with MBR and SBR Technology. *Membranes* **2021**, *11*, 722. [CrossRef]

International Journal of *Environmental Research and Public Health*

MDPI

Case Report

Municipal Wastewater Connection for Water Crisis and Jaundice Outbreaks in Shimla City: Present Findings and Future Solutions

Ranju Kumari Rathour, Deepak Sakhuja, Arvind Kumar Bhatt and Ravi Kant Bhatia *

Department of Biotechnology, Himachal Pradesh University Shimla, Shimla 171005, India
* Correspondence: ravibiotech07@hpuniv.ac.in

Abstract: The felicitous tourist destination "Hills Queen" and the capital city of Himachal Pradesh, an enticing state in the Himalayan region, are met with water crisis every year and jaundice outbreaks occasionally. In 2016, there was a severe jaundice outbreak in Shimla city. In a contemporaneous investigation, we attempted to trace out the possible reason for these crises in Shimla. Samples were collected month wise from different water-supply sources and their physicochemical and microbial loads were analyzed. The microbiological examination found a totally excessive microbial load (1.064×10^9 cfu/mL on common) throughout the year with a maximum ($>1.98 \times 10^{10}$ cfu/mL) in the wet season and minimum ($>3.00 \times 10^7$ cfu/mL) in the winter. Biochemical and morphological evaluation confirmed that most of the water resources reported a high number of coliforms and Gram-negative microorganisms due to sewage-water infiltration. These microorganisms in the water are responsible for the liver infection that ultimately causes jaundice. For safe and potable water, infiltration of municipal wastewater must be prevented at any cost. Scientific disposal of wastewater and purification of uncooked water have to be conducted earlier than consumption or use for different domestic functions, to avoid water crises and fetal ailment outbreaks in the near future.

Keywords: Shimla; municipal wastewater; jaundice; pathogens; water crisis

Citation: Rathour, R.K.; Sakhuja, D.; Bhatt, A.K.; Bhatia, R.K. Municipal Wastewater Connection for Water Crisis and Jaundice Outbreaks in Shimla City: Present Findings and Future Solutions. *Int. J. Environ. Res. Public Health* **2022**, *19*, 11266. https://doi.org/10.3390/ijerph191811266

Academic Editors: Yung-Tse Hung, Hamidi Abdul Aziz and Issam A. Al-Khatib

Received: 27 July 2022
Accepted: 3 September 2022
Published: 7 September 2022

1. Introduction

Water is the primary component of every living organism, with fluids derived from numerous sources on Earth. It is necessary for all known forms of life on Earth and covers 71% of the planet's surface [1–3]. India is endowed with a vast range of snow-capped Himalayas and a network of rivers that contribute to the country's water supply. Rivers play a vital role in the lives of the Indian people by providing water for drinking and various other anthropogenic activities [4,5]. There is a definite link between water availability and the prevention of numerous water-borne illnesses. Although access to clean drinking water has increased in nearly every corner of the world over the last several decades, however one billion people (globally) still lack access to drinking and sanitation [6]. The infiltration of home and industrial effluent into the water sources has degraded the available pure water. Drinking water polluted by various chemical or physical sources has the biggest impact on human health, particularly in underdeveloped nations [7]. However, it is anticipated that by 2025, more than half of the world's population will be vulnerable to water-related hazards.

According to one of WHO's reports, by 2030, water shortage in some developing countries may surpass 50%, resulting in 80% of water-borne or water-linked human diseases owing to biological pollution of drinking water [8]. Drinking water should ideally not include any harmful microorganisms or bacteria suggestive of fecal contaminations (increased amount of *E. coli* and other coliforms) [9,10]. Many cities and other rural/urban regions have had outbreaks of water-borne infections as a result of inadequate management of water supplies and poor waste disposal. Consumption of drinkable water with sewage water in the Municipal Corporation of Shimla in 2016 resulted in a jaundice epidemic, which infected more than 1600 people and reported 16 casualties [11].

Authorities halted the supply of drinking water from polluted sources (the city's primary water sources), exacerbating Shimla's water crisis. The epidemic was caused by a rise in fecal coliform concentrations in drinking water. Aside from jaundice, other disorders, such as ear infections, dysentery, typhoid fever, viral and bacterial gastroenteritis, and so on, are also associated with the presence of fecal coliforms such as *Enterobacter*, *Klebsiella*, *Citrobacter*, *E.coli*, etc. [12]. Regular extensive treatment and analysis of drinking water is required to keep an eye on and to minimize the microbial burden. Controlling microbiological contamination at all checkpoints might be a solution to water-borne illnesses and its link to water crises in all mega cities and significant tourist locations, such as Shimla, to prevent against any future epidemic disasters and water crises.

2. Materials and Methods

2.1. Area under Investigation

Shimla, a picturesque hill station and popular tourist destination in the Himalayas (Figure 1) faced water crisis in 2016. The city is spread out across an uneven altitude of 2100 m above sea level. This magnificent hill station is located between 31.06° N and 77.13° E. The city has a total size of 25 km^2 [13,14]. Shimla's water-supply system dates back to 1875, and additional water sources have been located to supplement the supply in order to fulfill the city's ever-increasing demand for water. Dhalli Catchment area, Cherot Nallah, Jagroti Nallah, Chair Nallah, Gumma Khad, and Ashwani Khad pump and provide surface water from rivulets at various altitudes. The total installed capacity is 47.54 million liters. Four water treatment plants exist in Gumma Khad, Ashwani Khad, Cherot Nallah and Dhalli Catchment area.

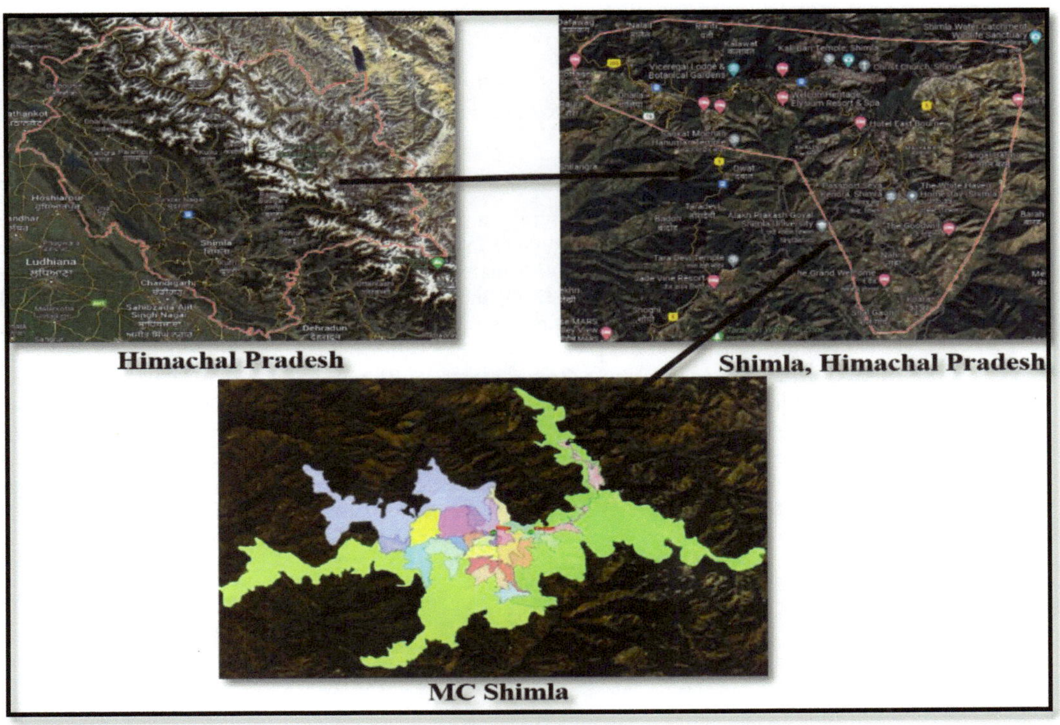

Figure 1. Geographical location of study area.

2.2. Sample Collection

Water samples were collected as per WHO guidelines from different water-supply sources every month of one year in sterile bottles and brought to laboratory for further processing.

2.3. Analysis of Physiochemical Parameters

Various physio-chemical parameters such as temperature, pH, TDS, salt concentration, color and conductivity were recorded.

2.4. Analysis of Microbiological Load

The microbial load was analyzed by standard plate-count technique using a serial dilution agar-plate method up to 10^{-6}-fold dilution of samples. Further, 100 μL of each dilution was spread on different basic as well as selective media (Nutrient, MacConkey, EMB, Endo) plates and incubated at 37 °C for 24 h [4]. After 24 h, the plates were analyzed for different morphological appearances on different media and then further sub-cultured to obtain pure culture by repeated streaking.

2.5. Identification of Microbes

Selected isolates were examined for various morphological as well as biochemical features such as colony morphology, Gram's staining, methyl red, oxidase test, catalase test, glucose, lactose, maltose, sorbitol fermentation test, Vogues–Proskaure test, hydrolysis of casein, etc., following the standard protocols.

2.6. DO, BOD and COD of Collected Samples

Standard protocols were used for analysis of dissolved, biological and chemical oxygen demands [14,15].

3. Results

3.1. Collection of Samples

Samples were chronologically collected from source, i.e., raw water, to finale filtration tanks and finally to distribution points, as shown in Figure 2 of the three main water-supply sources.

3.2. Analysis of Physicochemical Parameters

The physical and chemical parameters of water were analyzed at the time of sampling and data generated are shown in Tables 1–3. All the water sources, i.e., Dhalli Catchment area, Cherot Nallah/Jagroti Nallah and Ashwani Khad, showed average pH (8.5), TDS (108 ppm), conductivity (131 μS) and salt (70 ppm), which were under the permitted limit of WHO norms for drinking water [10,16,17]. The above-mentioned range of parameters might be due to the presence of rocks and alkaline soil in the catchment area of water sources, but these are within the permitted limits and water is safe to use for domestic purposes. Average water temperature of all the sources was found in the range of 15–25 °C and it provides an ambient temperature for the survival of mesophilic microorganisms. This low range of temperature is due to the fact that all these water sources are fed by melting snow caps and dense forest streams. However, all these parameters are within the permitted range of WHO guidelines (Table 4), but at the same time, these factors act as add-on factors during infiltration of litter and sewage to the water sources and help in the proliferation of opportunistic pathogens.

Figure 2. Different sampling sites investigated during course of study. (**A**) Churat Nallah, (**B**) Ashwani Khad hard workers (**C**) Water treatment plant Dhalli.

Table 1. Analysis of samples collected from source Churat/Jagroti Nallah.

Parameters	WHO Standards	Raw Water Churat Nallah						Raw Water Jagroti Nallah					
		Jan–Feb	March–April	May–June	July–August	September–October	November–December	January–February	March–April	May–June	July–August	September–October	November–December
Color	Colorless	Colorless	Colorless	Colorless	Slightly Turbid	Colorless	Colorless	Colorless	Colorless	Colorless	Slightly turbid	Colorless	Colorless
pH	6.5–8.5	8.4 ± 0.28	8.2 ± 0.27	8.5 ± 0.15	8.2 ± 0.11	8.9 ± 0.44	8.2 ± 0.33	8.6 ± 0.27	8.4 ± 0.56	8.4 ± 0.20	8.0 ± 0.15	8.5 ± 0.57	8.2 ± 0.44
Temp (°C)		15.7 ± 0.56	17.3 ± 0.33	18.2 ± 0.11	16.2 ± 0.25	18.7 ± 0.57	12.7 ± 0.24	16.2 ± 0.20	16.8 ± 0.33	24.7 ± 0.57	25.1 ± 0.44	19.2 ± 0.56	20.1 ± 0.11
TDS (ppm)	300	86.9 ± 0.25	93.5 ± 0.11	105.0 ± 0.27	71.8 ± 0.24	103.0 ± 0.28	132.9 ± 0.65	89.5 ± 0.19	95.3 ± 0.25	131.0 ± 0.28	120.0 ± 0.33	101.0 ± 0.27	72.3 ± 0.19
Conductivity (µS/cm)	250	123.5 ± 0.19	107.5 ± 0.19	148.8 ± 0.25	100.2 ± 0.20	141.9 ± 0.19	98.0 ± 0.25	113.5 ± 0.57	117.2 ± 0.28	141.5 ± 0.24	136.2 ± 0.56	181.3 ± 0.28	101.8 ± 0.33
Salt conc. (mg/L)	30–60	59.8 ± 0.15	76.3 ± 0.25	69.4 ± 0.56	48.9 ± 0.33	54.2 ± 0.11	67.5 ± 0.56	63.8 ± 0.44	83.3 ± 0.19	69.4 ± 0.60	55.6 ± 0.25	87.3 ± 0.20	50.6 ± 0.44

Table 2. Analysis of samples collected from source Ashwani Khad.

Parameters	Raw Water						Filtration Tank					
	Jan–Feb	March–April	May–June	July–August	September–October	November–December	January–February	March–April	May–June	July–August	September–October	November–December
Color	Colorless	Colorless	Colorless	Turbid	Colorless	Colorless	Colorless	Colorless	Colorless	Turbid	Colorless	Colorless
pH	8.4 ± 0.25	8.2 ± 0.15	8.6 ± 0.29	8.3 ± 0.23	8.5 ± 0.56	8.3 ± 0.28	8.3 ± 0.24	8.4 ± 0.27	8.5 ± 0.22	8.2 ± 0.17	8.2 ± 0.24	8.6 ± 0.29
Temp (°C)	15.6 ± 0.33	17.2 ± 0.27	21.4 ± 0.43	22.0 ± 0.53	18.5 ± 0.25	16.7 ± 0.20	15.8 ± 0.44	16.4 ± 0.53	24.5 ± 0.53	20.1 ± 0.27	19.2 ± 0.25	15.8 ± 0.48
TDS (ppm)	118.0 ± 0.43	116.5 ± 0.53	128.0 ± 0.24	123.0 ± 0.56	125.0 ± 0.22	137.0 ± 0.25	123.0 ± 0.27	123.0 ± 0.25	126.0 ± 0.11	101.8 ± 0.20	110.0 ± 0.29	123.4 ± 0.27
Conductivity (µS/m)	122.3 ± 0.13	128.3 ± 0.33	180.3 ± 0.13	86.6 ± 0.19	98.3 ± 0.45	107.0 ± 0.26	100.8 ± 0.32	118.3 ± 0.26	190.0 ± 0.25	72.3 ± 0.29	96.2 ± 0.19	127 ± 0.57
Salt conc. (ppm)	85.0 ± 0.34	83.0 ± 0.44	87.0 ± 0.23	60.7 ± 0.26	74.2 ± 0.34	82.0 ± 0.29	88.8 ± 0.45	83.7 ± 0.24	91.9 ± 0.18	50.6 ± 0.25	83.4 ± 0.54	89.0 ± 0.20

Table 3. Analysis of samples collected from source Dhalli WTP.

Parameters	January–February			March–April			May–June			
	Raw Water	**Filtration Tank**	**Storage Tank**	**Raw Water**	**Filtration Tank**	**Storage Tank**	**Raw Water**	**Settling Tank**	**Public Tap 1**	**Public Tap 2**
Color	Colorless	Colorless	Colorless	Colorless	Colorless	Colorless	Colorless	Colorless	Colorless	Colorless
pH	8.3 ± 0.19	8.5 ± 0.26	8.4 ± 0.28	8.7 ± 0.25	8.5 ± 0.27	8.8 ± 0.28	8.30 ± 0.25	8.45 ± 0.19	8.67 ± 0.29	8.54 ± 0.23
Temp. (°C)	12.7 ± 0.28	12.4 ± 0.27	13.0 ± 0.25	16.3 ± 0.57	17.5 ± 0.28	20.5 ± 2	15.7 ± 0.19	16.4 ± 0.32	17.3 ± 0.57	16.8 ± 0.28
TDS (ppm)	132.2 ± 0.34	118.5 ± 0.54	129 ± 0.23	125 ± 0.43	145.3 ± 0.43	106.8 ± 0.19	102.2 ± 0.28	118.8 ± 0.23	145 ± 0.54	149.8 ± 0.35
Conductivity (μS/m)	74.3 ± 0.45	108.6 ± 0.32	115 ± 0.36	114 ± 0.28	127.6 ± 0.34	148.3 ± 0.32	84.8 ± 0.21	94.5 ± 0.36	102 ± 0.43	97.6 ± 0.38
Salt conc. (ppm)	85.3 ± 0.21	78.1 ± 0.17	94.2 ± 0.43	88.3 ± 0.36	78.2 ± 0.29	74.5 ± 0.44	87.4 ± 0.57	78.6 ± 0.27	78.5 ± 0.29	75.8 ± 0.27
Parameters	July–August			September–October			November–December			
	Raw Water	**Settling Tank**	**Storage Tank**	**Filtrated Water**	**Public Tap 1**	**Public Tap 2**	**Raw Water**	**Settling Tank**	**Public Tap 1**	**Public Tap 2**
Color	Colorless	Turbid	Slightly turbid	Slightly turbid	Colorless	Colorless	Colorless	Colorless	Colorless	Colorless
pH	8.9 ± 0.21	8.7 ± 0.36	8.9 ± 0.11	9.7 ± 0.52	8.7 ± 0.38	8.6 ± 0.17	8.23 ± 0.33	8.15 ± 0.23	8.79 ± 0.20	8.33 ± 0.19
Temp. (°C)	18.0 ± 2	22.8 ± 2	18.7 ± 2	18.0 ± 2	20.3 ± 2	24.0 ± 2	12.70 ± 2	15.4 ± 2	16.3 ± 2	16.80 ± 0.20
Cond (μS/m)	89.9 ± 0.18	83.5 ± 0.19	103 ± 0.43	145 ± 0.16	107 ± 0.17	94.6 ± 0.28	132.9 ± 0.42	118.8 ± 0.37	135 ± 0.42	139.8 ± 0.23
TDS (ppm)	115.0 ± 0.54	122.8 ± 0.39	141.9 ± 0.15	212.0 ± 0.37	150.0 ± 0.26	132.8 ± 0.27	98 ± 0.31	100.8 ± 0.18	107 ± 0.28	97.6 ± 0.29
Salt Conc. (ppm)	58.8 ± 0.24	61.7 ± 0.33	69.5 ± 0.26	98.0 ± 0.53	72.8 ± 0.39	66.1 ± 0.36	67.5 ± 0.28	68.5 ± 0.43	72.8 ± 0.43	65.1 ± 0.54

Table 4. Standard limits of different parameters in potable water.

Sr. No.	Parameter	Permissible Limit	Instrument Used for Analysis
1	**Color**	Colorless	-
2	**pH**	6.5–8.5	pH meter
3	**Temp. (°C)**	Varies with environmental conditions	Thermometer
4	**TDS (ppm)**	1000	TDS meter
5	**Conductivity (µS/m)**	400	Conductivity meter
6	**Salt conc. (ppm)**	250	Conductivity meter
7	**BOD (ppm)**	1–2	Titration
8	**COD (ppm)**	10	Titration

3.3. Analysis of Microbiological Load

The standard agar-plate method was used to analyze the microbial load in the water samples. *E. coli* and *Klebsiella* sp. were most-commonly present lactose-fermenting *Enterobacter*, while *Pseudomonas* sp. and *Proteus* sp. were non-lactose fermenting isolates commonly present in all samples (Figure 3). More than 8.97×10^9 cfu/mL of bacteria were observed in different samples based of their morphology on different media and biochemical analysis. Svanevik and Lunestad [18] also used this method to study microbial contaminations in water. During the microbial load analysis, it was found that all the water samples were heavily contaminated with microorganisms, especially opportunistic pathogens, i.e., *E. coli* and *Pseudomonas* sp., which might be responsible for the jaundice, and this microbial load was very high as compared to that of the standard limits of WHO.

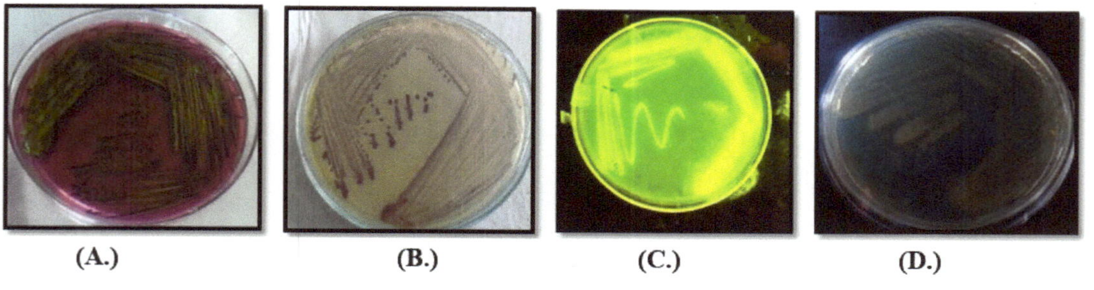

Figure 3. Colony morphology of different microbes on different media (**A.**) *E. coli* on EMB (**B.**) *Klebsiella* on *Klebsiella* selective agar and (**C.**) *Pseudomonas* sp. on Kings B medium under UV transilluminator and (**D.**) *Proteus* sp. on nutrient agar.

Further, it has been also found that microbiological load was $>2.25 \times 10^7$ cfu/mL in January–February, $>4.36 \times 10^9$ cfu/mL March–April, 4.72×10^9 cfu/mL May–June, $>4.064 \times 10^{10}$ cfu/mL July–August, September–October 5.54×10^9 cfu/mL and $>4.01 \times 10^9$ cfu/mL in November–December. Microbiological contamination was almost similar in all the samples. *E. coli* and *Pseudomonas* sp. were prominently observed throughout the year. Presence of *E. coli* and *Pseudomonas* sp. was also even recorded in tap water. *Klebsiella* sp. and *Proteus* sp. were also reported in the Dhalli water treatment plant during rainy session, which might be due to infiltration of sewage from nearby villages/urban localities to the water streams. The standard limit is 2.7×10^2 cfu/mL [19], while the microbiological load was found to be very high as compared to standard limits. Month-wise microbiological load of all the collected samples is shown in Table 5.

Table 5. Microbial load of different water samples.

Months	Microbial Load									
	Jagroti Nallah	Churat Nallah	Ashwani Khad		Dhalli WTP					
	Raw Water	Raw Water	Raw Water	Filtration Tank	Raw Water	Settling Tank	Storage Tank	Filtration Tank	Public Tap 1	Public Tap 2
January–February	1.81×10^6	2.31×10^6	1.82×10^7	1.82×10^5	1.93×10^3	2.43×10^3	4.76×10^4	1.20×10^4	3.26×10^2	3.54×10^2
March–April	1.31×10^6	1.82×10^7	2.02×10^7	2.02×10^7	5.42×10^2	3.12×10^3	5.43×10^3	3.80×10^7	6.80×10^2	6.38×10^2
May–June	2.11×10^7	5.43×10^7	2.82×10^7	2.13×10^7	5.76×10^4	5.42×10^4	4.80×10^7	3.76×10^6	5.93×10^3	4.43×10^2
July–August	8.42×10^7	1.18×10^7	6.82×10^7	5.80×10^7	3.80×10^7	6.43×10^5	4.93×10^7	3.82×10^7	8.82×10^2	5.42×10^3
September–October	1.98×10^7	6.21×10^7	4.32×10^7	3.76×10^6	6.20×10^6	6.20×10^6	8.82×10^7	4.20×10^6	2.93×10^3	1.13×10^3
November–December	4.52×10^6	3.11×0^3	2.75×10^7	2.76×10^6	4.52×10^7	6.12×10^3	6.20×10^6	4.27×10^6	3.43×10^2	2.42×10^2

3.4. DO, BOD and COD Analysis

DO, BOD and COD are measured by the amount of organic compounds in water. Biochemical/ biological oxygen demand is the amount of dissolved oxygen needed by a biological system in water to break down organic materials present in a given water sample at certain temperature over a specific time period. It is most commonly expressed in milligrams of oxygen consumed per liter of sample during 5 days of incubation at 20 °C and used as an index to determine organic pollution. The maximum BOD was observed in raw water of Churat Nallah, i.e., 16.8 ± 0.053 mg/L, followed by the raw water of Ashwini Khad water treatment plant (14.4 ± 0.065 mg/L). BOD value of public-tap water was in range from 0.2 to 1.0 mg/L. BOD were calculated on the basis of the difference between the DO value of the water sample on first ady and the DO value after a specific time interval (5 days).

Chemical oxygen demand (COD) is an indicative measure of the amount of oxygen that can be consumed by chemical reactions in the solution under test. It is commonly expressed in mass of oxygen consumed over volume of solution, which in SI units is milligrams per liter (mg/L). Maximum COD was observed with the raw water of Churat Nallah (0.384 ± 0.005 mg/L). Kadam and Agrawal [20] also analyzed the BOD and COD values to ensure potability of water. The DO, BOD and COD of samples are shown in Table 6. BOD and COD are measures of water contamination. The higher value of BOD indicates the presence of more biological contaminations in water [21]. The total bacterial count and the coliform density were directly related to the biochemical oxygen demand, but inversely related to the dissolved oxygen. Thus, tests for dissolved oxygen are of the utmost importance for controlling water pollution [22].

Table 6. BOD and COD for different water samples.

Sr. No	Sample	DO (0 Day) mg/L	DO (5 Days) mg/L	BOD (mg/L)	COD (mg/L)
01.	Blank	8 ± 0.078	7.2 ± 0.054	4.8 ± 0.049	07.00 ± 0.022
02.	Ashwani Khad raw water	4.8 ± 0.065	2.4 ± 0.022	14.4 ± 0.065	0.0228 ± 0.005
03.	Ashwani Khad Filtration tank	4.6 ± 0.028	2.4 ± 0.029	13.2 ± 0.080	0.0126 ± 0.002
04.	Jagroti Nallah raw water	4.8 ± 0.022	3.9 ± 0.077	5.4 ± 0.76	0.0128 ± 0.001
05.	Churat Nallah raw water	3.6 ± 0.053	0.8 ± 0.032	16.8 ± 0.053	0.384 ± 0.005
06.	Churat Nallah filtration tank	3.6 ± 0.049	2 ± 0.078	9.6 ± 0.069	0.348 ± 0.005
07.	Dhali WTP Sand filtration	3.2 ± 0.054	1.6 ± 0.029	9.6 ± 0.062	0.0164 ± 0.002

Table 6. *Cont.*

Sr. No	Sample	DO (0 Day) mg/L	DO (5 Days) mg/L	BOD (mg/L)	COD (mg/L)
08.	Dhali WTP Settling tank	3.4 ± 0.065	1.60 ± 0.056	10.8 ± 0.045	0.0125 ± 0.003
09.	Dhali WTP filtration tank	3 ± 0.052	1.2 ± 0.055	10.8 ± 0.055	0.0064 ± 0.002
10.	Blank	2 ± 0.039	1.6 ± 0.062	0.4 ± 0.059	06.80 ± 0.065
11.	Churat Nallah Raw water	1.2 ± 0.054	0.5 ± 0.067	0.7 ± 0.063	0.0232 ± 0.003
12.	Dahli WTP	0.9 ± 0.022	0.3 ± 0.060	0.6 ± 0.068	0.0141 ± 0.001
13.	Chlorinated water	0.8 ± 0.065	0.5 ± 0.074	0.3 ± 0.065	0.0122 ± 0.003
14.	Public tap 1 Dhalli	1.6 ± 0.067	1 ± 0.075	0.6 ± 0.047	0.254 ± 0.007
15.	Public tap 2 Dhalli	1.2 ± 0.056	0.9 ± 0.025	0.3 ± 0.071	0.282 ± 0.006
16.	Public tap 1 Sanjauli	1.0 ± 0.046	0.9 ± 0.072	0.3 ± 0.067	0.276 ± 0.007
17.	Public tap 2 Sanjauli	1.3 ± 0.022	1.0 ± 0.069	0.2 ± 0.054	0.287 ± 0.003
18.	Public tap 1 Summer hill	1.3 ± 0.049	0.8 ± 0.058	0.5 ± 0.041	0.256 ± 0.004
19.	Public tap 2 Summer hill	1.4 ± 0.054	0.9 ± 0.055	0.4 ± 0.065	0.285 ± 0.006
20.	Public tap 3 Summer hill	1.2 ± 0.076	1.0 ± 0.065	0.6 ± 0.065	0.263 ± 0.005

4. Discussion

Different water samples were collected month-wise from different sites of the main water-supply sources of Shimla and almost all physiochemical parameters were found to be suitable for potable water according to WHO standards, except microbial load. Kistemann et al. [23] also analyzed physiochemical parameters to ensure potability of drinking water. Behailu et al. [24] also analyzes the potability of drinking water and found that the average pH of all samples was in the range 7.6–8.2. The average pH and temperature of all samples were found to be 8.5 and 19 °C, respectively. The pH and temperature of all water pumping sources were ambient and conducive for the growth of opportunistic pathogens such as *E. coli, Pseudomonas* sp. *Klebsiella* sp. and *Proteus* sp. etc. and resulted in higher microbial growth, which supports our findings that a microbiological load higher than that of the permissible range of water potability standards might be a possible reason for jaundice outbreak in Shimla. Further, shortage in the rainfall intensified the issue and increases the concentration of microbes in the water sources, resulting in water shortage and crisis in the hill region. The observed TDS in the present study was varied from 75 to 225 ppm. According to the WHO, the TDS standard for drinking water is <1000 ppm. The collective amount of all dissolved cation and anions in the water is known as TDS and it is mainly associated with conductivity [25]. High TDS increases the density of water, decreases solubility of gases such as oxygen, and ultimately makes the water unsuitable for drinking [26]. The measure of competence of water to pass electrical flow is its conductivity, which is directly related to the concentration of ions in water coming from dissolved salts and inorganic materials such as alkalis, chlorides, sulfides and carbonate compounds. Conductivity of drinking water is in the range of 5–50 mS/m, (5,000,000 μS/m) [27]. Conductivity was found to be high in the month of July to September when microbial load was high. Increase in conductivity leads to increase in microbial load as minerals and matter provide necessary metabolites and factors for the rapid growth of microorganisms [28]. The lower TDS and conductivity ensure the potability of water.

BOD and COD of water samples were also analyzed using standard protocol and were found to be higher than the permissible limit. *E. coli* and *Pseudomonas* sp. were prominently observed in all the samples throughout the year. Both of these bacteria are opportunistic pathogens for humans. *E. coli* is an indicator organism of water pollution. Water was highly contaminated with *Klebsiella* sp. in the months of June, July, August and September. Onyango et al. [29] also reported presence of *E. coli* and other coliforms in drinking water during their study. Presence of *E. coli* or fecal contaminants in water is the main cause for the outbreak of jaundice [30,31]. According to a report [32], the DO of drinking water must be in the range 2–5 mg/L, BOD of drinking water 2–10 mg/L and COD 5–10 mg/L. According to a report [33], the jaundice epidemic in 2016 was caused by contaminated water from Ashwami Khad, from where Shimla used to acquire its largest share of water. After the report of jaundice cases, government officials stopped the water supply from these contaminated water sources, which increased the water shortage in the city and ultimately deepened the water crisis.

5. Conclusions and Future Perspectives

All the water distribution channels in Shimla were badly contaminated with various opportunistic pathogens, as revealed through our investigation. Fortunately, much less contamination appeared in the distribution line after treatment, but microbial load was still beyond the permitted limit. Results of this research could help the MC/IPH officials to take remediate measures in good time. Authorities should also look for alternative water sources to prevent any water crisis in the future that may also arise due to a shortage of rainfall in the catchment areas. The efforts will also help to establish a cleaning schedule for the MC water-supply sources and create proper wastewater treatment facilities for the villages/towns situated nearby the catchment areas, to prevent the infiltration of domestic and industrial sewage, which will help to provide safe drinking water to the consumers in MC Shimla. Moreover, based upon our findings, the MC/IPH department may install a heavy-duty UV system to control the heavy microbial load. Using these recommended corrective measures, such future disasters could be avoided, and the public of such areas will have potable drinking water free from all types of contaminants and opportunistic pathogens.

Author Contributions: R.K.R.: Conceptualization, Data curation, Formal analysis, Validation, Visualization, and Writing-original draft, Writing- review & editing, D.S.: Data curation, Formal analysis, R.K.B.: Conceptualization, Methodology, Formal analysis, Writing- review & editing, A.K.B.: Formal analysis, Funding acquisition, Investigation. All authors have read and agreed to the published version of the manuscript.

Funding: This research was funded by SCSTE Govt. of H.P. This research was funded by HIMCOSTE, Govt. of Himachal Pradesh, grant number [SCST/E (8)-1/2016-Vol-1, 5517].

Institutional Review Board Statement: Not applicable.

Informed Consent Statement: Not applicable.

Data Availability Statement: Not applicable.

Acknowledgments: Authors are highly thankful to for funding this study and providing financial assistance in the form of project fellow to Ranju Kumari Rathour. The facilities availed from Department of biotechnology are duly acknowledged.

Conflicts of Interest: The authors declare no conflict of interest.

References

1. Gleick:, P.H. *Water in Crisis: A Guide to the World's Freshwater Resources*; Oxford University Press: Oxford, UK, 1993; p. 13.
2. Liu, J.; Zhao, M.; Yao, L.; Liu, W.; Fan, G. Evaluating the value of ecological water considering water quality and quantity simultaneously. *Water Environ. J.* **2020**, *34*, 635–647. [CrossRef]
3. Hasan, H.A.; Muhammad, M.H. A review of biological drinking water treatment technologies for contaminants removal from polluted water resources. *J. Water Process Eng.* **2020**, *33*, 101035. [CrossRef]

4. Antony, R.M.; Renuga, F.B. Microbiological analysis of drinking water quality of Ananthanar channel of Kanyakumari district, Tamil Nadu, India. *Rev. Ambiente Agua* **2012**, *7*, 42–48. [CrossRef]
5. Kaur, J.; Kaur, H. Water potability test of drinking water collected from sub regions of district Jalandhar, Punjab, India: A Review. *Int. Res. J. Biol. Sci.* **2015**, *4*, 62–65.
6. Akoto, O.; Adiyiah, J. Chemical analysis of drinking water from some communities in the Brong Ahafo region. *Int. J. Environ. Sci. Technol.* **2007**, *4*, 211–214. [CrossRef]
7. Palamuleni, L.; Akoth, M. Physico-chemical and microbial analysis of selected borehole water in Mahikeng, South Africa. *Int. J. Environ. Res. Public Health* **2015**, *12*, 8619–8630. [CrossRef] [PubMed]
8. World Health Organization (W.H.O.). *Guidelines for Drinking Water Quality Geneva 2006*; Report No: WHO/SDE/WSH 06.07; WHO: Geneva, Switzerland, 2006.
9. Estim, A.; Saufie, S.; Mustafa, S. Water quality remediation using aquaponics sub-systems as biological and mechanical filters in aquaculture. *J. Water Process Eng.* **2019**, *30*, 100566. [CrossRef]
10. World Health Organization (W.H.O.). *Guidelines for Drinking Water Quality, Geneva, 2004*; Report No: WHO/SDE/WSH; WHO: Geneva, Switzerland, 2004. Available online: http://apps.who.int/iris/bitstream/handle/10665/42852/9241546387.pdf;jsessionid=30E7F64DE4C472C1F29E09A2034DDD15?sequence=1 (accessed on 26 July 2022).
11. Bodh, A. Jaundice Outbreak Reaching Epidemic Proportions in Shimla. A Report, The Times of India. 2016. Available online: https://timesofindia.indiatimes.com/city/shimla/Jaundice-outbreak-reaching-epidemic-proportions-in-Shimla/articleshow/50517708.cms (accessed on 26 July 2022).
12. Fresno County Department of Public Health (F.C.A.). *E. coli* or Fecal Coliform Bacteria Contamination in Your Water Supply. 2009. Available online: https://www.co.fresno.ca.us/home/showdocument?id=4747 (accessed on 26 July 2022).
13. Shimla Municipal Corporation. (S.M.C.). 2007. Available online: https://web.archive.org/web/20070403005805/http://www.shimlamc.org/mcshimla.htm (accessed on 26 July 2022).
14. Shimla District Census (S.D.C.). 2011 Handbook (PDF). Census of India. (Urban Section). 2016. Available online: https://censusindia.gov.in/nada/index.php/catalog/483/download/1580/DH_2011_0211_PART_B_DCHB_SHIMLA.pdf (accessed on 26 July 2022).
15. Delzer, G.C.; McKenzie, S.W. Five-day biochemical oxygen demand. In *USGS TWRI Book 9-a7*, 3rd ed.; U.S. Geological Survey: Reston, VA, USA, 2003; pp. 11–20.
16. Belkin, S.; Brenner, A.; Abeliovich, A. Effect of inorganic constituents on chemical oxygen demand-I. Bromides are unneutralizable by mercuric sulfate complexation. *Water Res.* **1992**, *26*, 1577–1581. [CrossRef]
17. *IS 10500-1991*; Indian Standard for Drinking Water as per BIS Specifications. A Report: 010-03-21. Bureau of Indian Standards (B.I.S.): Old Delhi, India, 1991; Volume 23, p. 25.
18. Svanevik, C.S.; Lunestad, B.T. Microbiological water examination during laboratory courses generates new knowledge for students, scientists and the government. *FEMS Microbiol. Lett.* **2015**, *362*, 151. [CrossRef] [PubMed]
19. European Commission Council Directive (E.C.C.D.). 2006/7/EC of the European Parliament and of the Council of 15 February 2006 Concerning the Management of Bathing Water Quality and Repealing Directive 76/160/EECOJ. 2006. Available online: https://eur-lex.europa.eu/legal-content/EN/TXT/PDF/?uri=CELEX:32006L0007&from=EN (accessed on 26 July 2022).
20. Kadam, S.S.; Agrawal, B.A. Microbiology and physico-chemical analysis of different sources of drinking water in Dahanu Taluka of Thane district. *Eur. J. Exp. Biol.* **2015**, *5*, 13–22.
21. Kedra, M. Sensitivity of mountain catchments to global warming: A case study of the San Basin, Poland. *Water Environ. J.* **2020**. [CrossRef]
22. Olutiola, P.O.; Awojobi, K.O.; Oyedeji, O.; Ayansina, A.D.; Cole, O.O. Relationship between bacterial density and chemical composition of a tropical sewage oxidation pond. *Afr. J. Environ. Sci. Technol.* **2010**, *4*, 595–602.
23. Kistemann, T.; Claben, T.; Koch, C.; Dangendorf, F.; Fischeder, R.; Gebel, J.; Vacata, V.; Exner, M. Microbial load of drinking water reservoir tributaries during extreme rainfall and runoff. *Appl. Environ. Microbiol.* **2002**, *68*, 2188–2197. [CrossRef] [PubMed]
24. Behailu, T.W.; Badessa, T.S.; Tewodros, B.A. Analysis of physical and chemical parameters in ground water used for drinking around Konso Area, Southwestern Ethiopia. *J. Anal. Bioanal. Tech.* **2017**, *8*, 379.
25. Guimaraes, J.T.; Souza, A.L.; Brígida, A.I.S.; Furtado, A.A.; Chicrala, P.C.; Santos, V.R.; Mesquita, E.F. Quantification and characterization of effluents from the seafood processing industry aiming at water reuse: A pilot study. *J. Water Process Eng.* **2018**, *6*, 138–145. [CrossRef]
26. Asadullah, A.; Nisa, K.; Khan, S.I. Physico-chemical properties of drinking water available in educational institutes of Karachi city. *Sci. Tech. Dev.* **2013**, *32*, 28–33.
27. Gray, R.J. Conductivity analyzers and their application. In *Environ Instrumentation and Analysis Handbook*; Wiley: Hoboken, NJ, USA, 2004; pp. 491–510.
28. Kaptan, B.; Kayisoglu, S.; Demirci, M. The relationship between some physico-chemical, microbiological characteristics and electrical conductivity of milk stored at different temperature. *J. Tekirdag Agri. Faculty* **2011**, *8*, 13–20.
29. Onyango, A.E.; Okoth, M.W.; Kunyanga, C.N.; Aliwa, B.O. Microbiological Quality and Contamination Level of Water Sources in Isiolo County in Kenya. *J. Environ. Public Health* **2018**, *10*, 2139867. [CrossRef] [PubMed]
30. Jaiswal, B.E. *Coli* Cause of Jaundice Outbreak in Cuttack. TNN. 2017. Available online: https://timesofindia.indiatimes.com/city/cuttack/e-coli-cause-of-jaundice-outbreak-in-cuttack/articleshow/56523999.cms (accessed on 26 July 2022).

31. Sharma, A. Jaundice outbreak in Shimla: 10 Dead. The Indian Express. 2016. Available online: https://indianexpress.com/article/india/india-news-india/jaundice-outbreak-in-shimla-10-dead (accessed on 26 July 2022).
32. UNESCO/WHO/UNE. Water Quality Assessments—A guide to Use of Biota, Sediments and Water in Environmental Monitoring -2nd Edition. Selection of Water Quality Variables. 1996. Available online: http://www.earthprint.com (accessed on 26 July 2022).
33. Chatterji, R. Shimla Water Crisis Explained: Sewage Contaminated a Stream, a Jaundice Outbreak Followed, then a High Court Order Left the City High and Dry 2018. Available online: https://www.huffpost.com/archive/in/entry/shimla-water-crisis-explained-sewage-contaminated-a-stream-a-jaundice-outbreak-followed-then-a-high-court-order-left-the-city-high-and-dry_a_23451060 (accessed on 26 July 2022).

International Journal of
Environmental Research and Public Health

MDPI

Article

Assessment of Magnetic Nanomaterials for Municipality Wastewater Treatment Using Biochemical Methane Potential (BMP) Tests

Gloria Amo-Duodu *, Emmanuel Kweinor Tetteh, Sudesh Rathilal and Martha Noro Chollom

Green Engineering Research Group, Department of Chemical Engineering, Faculty of Engineering and the Built Environment, Durban University of Technology, Durban 4001, South Africa
* Correspondence: gamoduodu04@gmail.com; Tel.: +27-8499-92841 or +233-2489-49519

Abstract: Wastewater as a substrate potential for producing renewable energy in the form of biogas is gaining global attention. Herein, nanomaterials can be utilised as a nutrient source for microorganisms for anaerobic digestion activity. Therefore, this study explored the impact of seven different magnetic nanomaterials (MNMs) on the anaerobic digestion of wastewater via biochemical methane potential (BMP) tests for biogas production. The BMP assay was carried out with eight bioreactors, where each was charged with 50% wastewater and 30% activated sludge, leaving a headspace of 20%. Aside the control bioreactor, the other seven (7) bioreactors were dosed with 1.5 g of MNMs. This was operated under anaerobic conditions at a mesophilic temperature of 35 °C for 31 days. At the degree of 80% degradation of contaminants, the results that showed bioreactors charged with 1.5 g MNMs of TiO_2 photocatalyst composites were more effective than those constituting metallic composites, whereas the control achieved 65% degradation. Additionally, the bioreactor with magnetite (Fe_3O_4) produced the highest cumulative biogas of 1172 mL/day. Kinetically, the modified Gompertz model favoured the cumulative biogas data obtained with a significant regression coefficient (R^2) close to one.

Keywords: anaerobic digestion; biochemical methane potential; chemical oxygen demand; nanomaterials; kinetics

Citation: Amo-Duodu, G.; Tetteh, E.K.; Rathilal, S.; Chollom, M.N. Assessment of Magnetic Nanomaterials for Municipality Wastewater Treatment Using Biochemical Methane Potential (BMP) Tests. *Int. J. Environ. Res. Public Health* **2022**, *19*, 9805. https://doi.org/10.3390/ijerph19169805

Academic Editors: Paul B. Tchounwou, Yung-Tse Hung, Hamidi Abdul Aziz and Issam A. Al-Khatib

Received: 22 June 2022
Accepted: 26 July 2022
Published: 9 August 2022

1. Introduction

The development of energy-efficient centralised wastewater treatment systems to mitigate emerging pollutants and environmental challenges associated with the water sector is gaining global attention [1,2]. Meanwhile, there is a significant risk of freshwater and energy resource depletion because of increasing population growth and industrialization, as well as anthropogenic CO_2 emissions [1–3].

South Africa, a country known for its scarcity of water, is also faced with high-strength organic chemical pollutants emitted by industries such as pharmaceuticals, petrochemicals, agrochemicals, mining, textiles, pulp and paper, and so on, posing threats to water bodies [4–6]. These recalcitrant pollutants typically enter the aquatic medium via industrial effluent that does not meet the discharge standards due to ineffective municipal Wastewater Treatment Plants (WWTPs) [5,7]. Water-soluble substances, in general, are easier to distribute and transport in the water cycle, and their direct impact on the ecosystem can be seen in a short period of time [6,8].

Furthermore, advancements in scientific environmental assessments reveal that recalcitrant contaminants (such as antibiotics) can still be detected in wastewater streams after a long period of time [9]. As a result, conventional WWTPs are incapable of dealing with high-concentration organics and emerging contaminants (ECs) such as pharmaceuticals (antibiotics), biomolecules (COVID-19 RNA), personal care products, food additives, and customized nanomaterials [6,8]. This has piqued the interest of many water sector stakeholders in improving the efficacy of WWTPs.

To comply with stringent bylaws, the energy required for the operation of conventional wastewater treatment plants (WWTPs) in conjunction with disposal and distribution systems is costly (South African-German Energy Programme—GIZ-SAGEN, 2015) [4,10,11]. As a result, using wastewater treated residue as an energy resource for biogas production to offset the WWTP's energy consumption becomes interesting area for researchers [10,12]. Herein, anaerobic digestion (AD) has been one of the global technologies used for the degradation of high organic content of wastewater into biogas [12].

AD, on the other hand, involves the hydrolysis of complex organics to soluble and degradable molecules, as well as acidogenesis, acetogenesis, and methanogenesis [13,14]. Consequentially, the AD processes, which are predominantly employed as biological processes for industrial and municipal wastewater treatment, have their own challenges [15,16]. This is because most of them either operate at a very low capacity due to the numerous challenges posed by emerging and biorecalcitrant compounds and the rest have completely collapsed [17–19]. Thus, this issue must be addressed to improve water quality and biogas production. Several studies, which include reactor adjustment, addition of nanomaterials, co-digestion, etc., have been adopted over the years to curb this challenge [18,20].

Nanomaterials have exceptional size-dependent properties (10–150 nm), making them indispensable and superior for a wide range of applications. Examples include chitosan, titanium dioxides (TiO_2), iron oxides, zinc oxides, zeolites, carbon nanotubes, copper oxides, and so on [21–23]. Furthermore, the presence of most of these chemical additives (iron- and aluminium-based coagulants) alters the chemistry of the sludge, resulting in complex organic contents. As a result, reducing sludge production while increasing caloric value via biological treatment in conjunction with NPs is a possibility [22,24–26]. A study by Ajay, et al. [20] reports that these metal NPs (iron, cobalt, nickel etc.) are inorganic additives that serve as micronutrients to the microorganisms in the AD to enhance methane and biogas production.

In this vein, the current study was conducted, where the application of magnetic nanomaterials for wastewater treatment using the biochemical methane potential test was evaluated to ascertain the effects on contaminant removals, the methane and biogas yield.

2. Materials and Methods

The wastewater and sludge were obtained from the eThekwini municipal wastewater treatment plant (Umbilo) in South Africa's KwaZulu-Natal province of which the samplings of the wastewater and sludge were performed at the biofiltration system (BS) of the plant. The wastewater and sludge were characterized in accordance with water and wastewater examination standard methods [27]. The results obtained are shown in Table 1. The nanomaterials used in this study was synthesized using the co-precipitation synthesis method, which has been detailed in studies by Tetteh, et al. [28], Amo-Duodu, et al. [29], and Amo-Duodu, et al. [30]. The characteristics of the nanomaterials have been reported in these studies. The selection of these nanomaterials was conducted based on a study by Amo-Duodu, et al. [31].

Table 1. Characterization of wastewater and activated sludge samples.

Wastewater	
Parameters	**Biofiltration System (BS)**
Chemical oxygen demand (COD) (mg/L)	2380 ± 32
Colour (465 nm, Pt.Co)	570 ± 7.6
Turbidity (NTU)	73.2 ± 12.5
pH	7.42 ± 3.6
Activated sludge	
Total Solids (TS) (mg TS/L)	304.5 ± 23.6
Volatile Solid (VS) (mg VS/L)	229.5 ± 2.65
VS/TS (%)	75.37 ± 3.5

2.1. Biochemical Methane Potential (BMP) Test

The BMP test was performed in accordance with the protocol reported by Jingura and Kamusoko [13] and Hülsemann, et al. [14] to attest to the effectiveness of the synthesized magnetic nanomaterials (MNMs) used for the study. This was completed using 8 Duran Schott bottles (1 L bioreactors) with air-tight caps and three outlets on each cap, which were placed in a thermostatic water bath (Figure 1). Table 2 presents the wastewater, activated sludge, and MNPs load distribution for each bioreactor. After charging the bioreactors (A–H), they were purged with nitrogen gas for 2 min and allowed to stand for two days to create an anaerobic environment. The bioreactor systems (A–H) were then run at a temperature of 35 °C for 30 days. The downward displacement technique was used to monitor the daily amount of biogas produced.

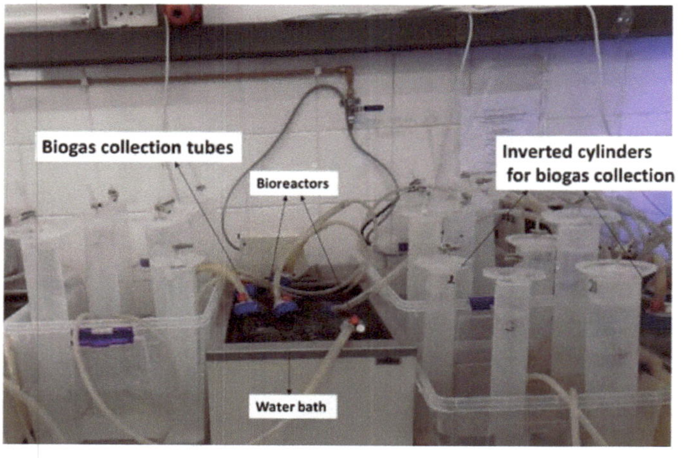

Figure 1. The schematic diagram of the biochemical methane potential (BMP) test setup.

Table 2. Experimental matrix for BMP test.

Setup	MNPs Loading (g)	Symbol (s)	Wastewater (L)	Sludge (L)
A	1.5 Fe_3O_4	mF	0.5	0.3
B	1.5 $NiFe_2O_4$	NmF	0.5	0.3
C	1.5 $CuFe_2O_4$	CmF	0.5	0.3
D	1.5 $TiO_2Fe_2O_4$	TmF	0.5	0.3
E	1.5 Chitosan$TiO_2Fe_2O_4$	ChTmF	0.5	0.3
F	1.5 $CuTiO_2Fe_2O_4$	CTmF	0.5	0.3
G	1.5 $ALTiO_2Fe_2O_4$	ATmF	0.5	0.3
H	No MNPs (Control)	n/a	0.5	0.3

2.2. Water Quality Analysis

At the end of the 30-day digestion period, samples of the supernatant were taken and analysed from each setup. The remaining content was decanted from each setup, leaving the sludge behind for analysis. Before analysing, 5 mL of supernatant liquid was measured and diluted with distilled water using a dilution factor of 10. The reactor efficiency was calculated by estimating the contaminants removal (Equation (1)). The first-order and modified Gompertz models were used to determine the degree of degradation and stability of the biological system as a function of the cumulative biogas data obtained for a given run.

$$Reactor\ efficiency = \left(\frac{C_i - C_f}{C_i}\right) \times 100 \quad (1)$$

where, C_i = Substrate influent and C_f = Substrate effluent.

Additionally, the cumulative biogas data obtained was fitted on a modified Gompertz model and first-order kinetics model as expressed in (2) and (3), respectively. This was used to estimate the biogas yield.

$$Y(t) = Ym.\exp\left(-\exp\left[\frac{2.7183R_{max.}}{Ym}[\lambda - t]\right] + 1\right) \quad (2)$$

$$Y(t) = Ym\,[1 - \exp(-kt)] \quad (3)$$

where $Y(t)$ = Cumulative methane yield (mL/g COD), Ym = maximum methane yield (mL/g COD), k = rate constant (1/day), R_{max} = maximum methane production rate (mL/g COD.day), $k = (R_{max}.e/Cm)$ = maximum specific substrate uptake rate per the maximum biogas production (1/day), λ = Lag phase (day), and t = time (day).

3. Results and Discussion

The biostimulation effect and treatability efficiency of MNMs (Table 2) by each bioreactor (setup A–G), compared with the control system (setup H), is presented in this section. This was based on the BMP results obtained.

3.1. Effect of MNMs on Contaminants Removal from BS Wastewater

The BMP test's treatability performance was evaluated, and the contaminants' removal (%COD, %colour, %turbidity) from the BS wastewater sample was evaluated for each bioreactor. From the findings of the study, magnetised photocatalyst (TmF) had a high significance in the removal of specific contaminants in wastewater samples (Figure 2) of a removal efficiency of COD, colour, and turbidity, which was found to be 91.60%, 78.95%, and 97.81%, respectively.

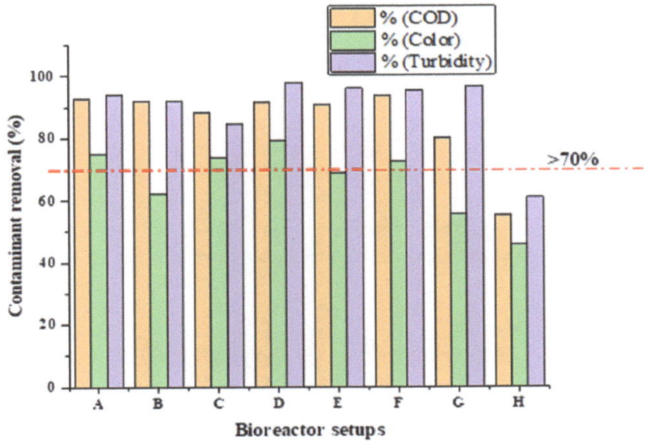

Figure 2. Contaminant removal for bioreactors A (mF), B (NmF), C (CmF), D (TmF), E (ChTmF), F (CTmF), G (ATmF), and H (no MNM) of MNM loading of 1.5 g at 35 °C for 30 HRT.

Figure 2 depicts the outcome of evaluating MNMs for the treatment of BS wastewater, where setup D with MNM (TmF) performed admirably with approximately 79% colour removal. This could be because the additive MNM (TmF) is composed of TiO_2 and Fe, which are trivalent ions with high oxidation-reduction properties capable of oxidizing a wide range of organic pollutants [32–34]. In this case, the overall performance of the MNM-containing bioreactors was found to be preferable and superior to that of the control bioreactor H (no MNM). The order of COD degradation was found to be F (93.70%) > A (92.59%) > B (91.90%) > D (91.60%) > E (90.76%) > C (88.07%) > G (79.83%) > H (54.96%).

The colour and turbidity removal performance of the bioreactors with MNMs was compared to the control (setup H). It was observed that the control has a removal of 45.61% and 60.79% for colour and turbidity removal, respectively, as shown in Table 3 and Figure 2. The removal efficiency of colour for the bioreactors with MNMs dosed had above 60% and a turbidity removal above 80%, as presented in Table 3. The findings also suggested that bioreactors with MNMs composed of TiO_2 photocatalyst (ChTmF, ATmF, TmF, and CTmF) had good biodegradability of the contaminant, which could be attributed to their high sorption ability for high-strength organic contaminants [35]

Table 3. The water quality analysis for BS wastewater.

Setup	COD Removal (%)	Colour Removal (%)	Turbidity Removal (%)
A	92.59	74.86	94.13
B	91.90	61.98	91.94
C	88.07	73.68	84.56
D	91.60	78.95	97.81
E	90.76	68.77	95.90
F	93.70	72.63	95.36
G	79.83	55.61	96.45
H	54.96	45.61	60.79

A (mF), B (NmF), C (CmF), D (TmF), E (ChTmF), F (CTmF), G (ATmF), and H (no MNM).

3.2. *Biogas and Methane Yield of BMP System for BS Wastewater*

Figure 3 depicts the cumulative biogas production of reactors A-H. Bioreactor A, dosed with magnetite (mF) additives, produced 1172 mL/day of biogas, which was almost double that compared to control reactor H (525 mL/day). This result was in contrast to reactor D with MNM (TmF), which was very effective for water quality improvement efficiency for colour and turbidity removal, as reported by other studies [22,36,37]. TmF is a well-known photocatalyst whose surface is more active and gets excited when exposed to UV-light [38]. This excitation causes the release of electrons and holes radicals, which help in the adsorption of contaminants and reduction of CO_2 to methane in the presence of hydron ions [38–40]. However, in this study the TmF was not exposed to any UV-light; hence, it was not as effective as it could have been, and this could be the reason for the above observation of its low biogas yield. Similarly, the increase in biogas production validates previous reports because MNMs produce radical ions that act as a reducing agent during methanogenesis activity, as shown in Equation (4) [15,18,41].

$$CO_2 + 4H_2 \rightarrow CH_4 + 2H_2O \quad (4)$$

Figure 4 depicts the methane yield for the BMP setups A-H. The findings of this study support previous research on the use of MNMs and other trace metal solutions such as Fe, Cu, Ni, Zn, Ti, and Mg for biogas generation and methane enhancement [42–44]. Liang, et al. [43] reported on the usage of iron-based nanomaterials to enhance methane and biogas generation. Aside from biogas production, the MNM additions improved the methanation mechanism. This resulted in a significant percentage (>80%) of methane composition as compared to the control setup H (Figure 4). Thus, setups A, B, C, and D recorded >90% methane composition, but the control (setup H) had a methane output of 65%. Importantly, the increase in methane content is very efficient in terms of heat and power consumption.

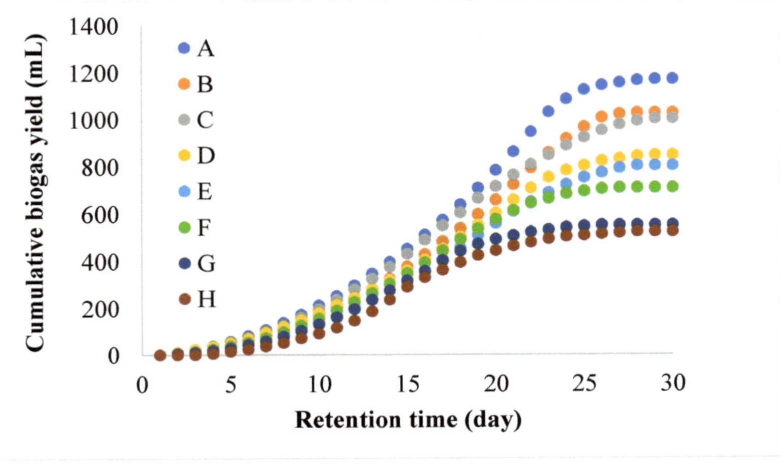

Figure 3. Cumulative biogas yield for bioreactors A (mF), B (NmF), C (CmF), D (TmF), E (ChTmF), F (CTmF), G (ATmF), and H (no MNM) of MNM loading of 1.5 g at 35 °C for 30 HRT.

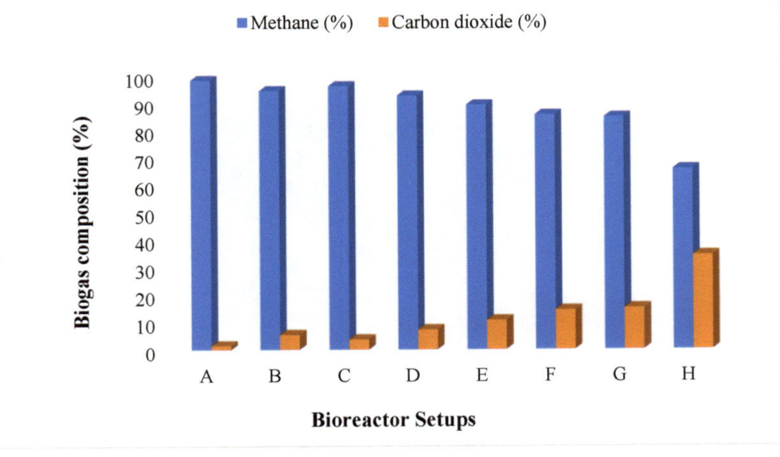

Figure 4. Biogas composition for bioreactors A (mF), B (NmF), C (CmF), D (TmF), E (ChTmF), F (CTmF), G (ATmF), and H (no MNM) of MNM loading of 1.5 g at 35 °C for 30 HRT.

3.3. Kinetic Study of the BMP System

Kinetic modelling was used as an acceptable method to determine the kinetic conditions of the bioreactors as a function of the biogas produced. This was carried out to obtain information about the reactor's kinetic degradation to avoid impending AD reactor failure due to poor operation [45]. The obtained cumulative biogas data was fitted using the modified Gompertz and first-order models to ascertain the substrate–microbe utilization for the biogas production [18,45]. Table 4 and Figure 5 show that the systems fitted better on the modified Gompertz models with R^2 values greater than 0.98. Clearly, the results (Figure 3) indicate that the presence of MNM additives accelerated the kinetics degradation activity, which increased the biogas production [45]. As a result, the models' predicted biogas values were relatively higher than the measured biogas values (Yt). In essence, the BS wastewater stream data from the bioreactors with MNM additives favoured the modified Gompertz kinetic model, with their lag of phase (λ) being within 4–9 days, attesting to the reactor's rapid response [45]. Moreover, the minimum sum of squares errors (SSE) denotes the models' statistical significance ($p > 0.05$) and predictability. Likewise, it may provide knowledge on how to design an industrial-scale reactor operating under similar conditions to be viable with MNM additives.

Table 4. Summary of the kinetic study for bioreactors: A (mF), B (NmF), C (CmF), D (TmF), E (ChTmF), F (CTmF), G (ATmF), and H (no MNM) fitted on first-order and modified Gompertz models.

		Modified Gompertz Model			First-Order Model		
Setup	Measured Yield, (mL/day)	Predicted Yield (mL/day), Y_2	Y1–Y_2 (mL/day)	R^2	Predicted Yield (mL/day), Y_3	Y_1–Y_3 (mL/day)	R^2
A	1172	1460	288	0.9931	1872	700	0.9688
B	1028	1316	288	0.9943	1956	928	0.9689
C	1004	1174	170	0.9986	1372	368	0.9786
D	848	986	138	0.9952	3476	2658	0.9758
E	804	899	95	0.9960	3387	2583	0.9716
F	710	729	19	0.9854	1162	452	0.9618
G	553	557	4	0.9813	584	31	0.9404

A (mF), B (NmF), C (CmF), D (TmF), E (ChTmF), F (CTmF), G (ATmF), and H (no MNM).

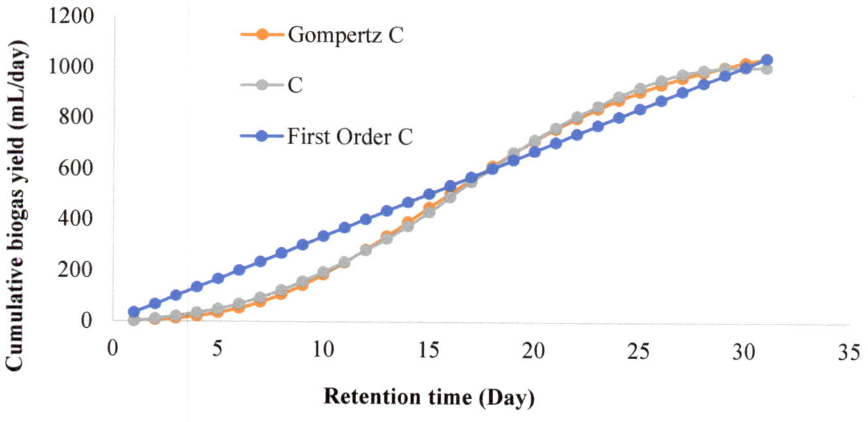

Figure 5. Fitting of cumulative biogas yield of bioreactor C (CmF) with highest R^2 (0.9986) on first-order and modified Gompertz kinetic models.

4. Conclusions

The magnetic nanomaterials (MNMs) investigated demonstrated having potential to enhance the AD process methanogens' rapid response for biogas production, reducing sludge production and improving the wastewater treatment quality. A biochemical methane (BMP) test was used in this study to evaluate the MNMs' biostimulation effect on anaerobic digestion of wastewater for biogas production and treated wastewater for reuse. This also enhanced the kinetic stability of the AD system and improvement of the biogas produced. MNMs composed of TiO_2 photocatalyst composites (ATmF, TmF, CTmF, and ChTmF) were found to be more effective than those composed of metallic composites. Furthermore, the degree of degradation with BMP setups charged with MNMs demonstrated 70–80 percent removal of the COD, colour, and turbidity when compared to the control system, which achieved 50–65 percent efficiency without any MNM additives. It is found that bioreactor with TmF additives demonstrated a critical pathway for converting wastewater into circular-economy resources (energy).

Author Contributions: Conceptualization was of E.K.T. and S.R.; methodology, investigation, and writing of the original draft preparation was performed by G.A.-D.; whereas supervision, review, and editing were performed by E.K.T., M.N.C. and S.R. All authors have read and agreed to the published version of the manuscript.

Funding: This research was funded by the Water Research Commission of South Africa under project identification WRC Project: C2019/2020-00212.

Institutional Review Board Statement: Not applicable.

Informed Consent Statement: Not applicable.

Data Availability Statement: All data are available in the manuscript.

Acknowledgments: The authors would like to express their gratitude to the Durban University of Technology, the Green Engineering Research Group, and the Water Research Commission (WRC) for their assistance with WRC Project C2019/2020-00212 as well as to the National Research Foundation (NRF) for their scholarship grant number 130143.

Conflicts of Interest: We, the authors, do declare that there is no conflict of interest in this work and that its publication has been approved by all co-authors as well as the relevant authorities at the institute where the study was performed. In addition, the manuscript includes all relevant data.

References

1. Isiaka, A.A.; Olaniran, A.O. Treatment of industrial oily wastewater by advanced technologies: A review. *Appl. Water Sci.* **2021**, *11*, 98.
2. Divyapriya, G.; Singh, S.; Martínez-Huitle, C.A.; Scaria, J.; Karim, A.V.; Nidheesh, P. Treatment of real wastewater by photoelectrochemical methods: An overview. *Chemosphere* **2021**, *276*, 130188. [CrossRef]
3. Ángeles, R.; Vega-Quiel, M.J.; Batista, A.; Fernández-Ramos, O.; Lebrero, R.; Muñoz, R. Influence of biogas supply regime on photosynthetic biogas upgrading performance in an enclosed algal-bacterial photobioreactor. *Algal Res.* **2021**, *57*, 102350. [CrossRef]
4. Adeogun, A.I.; Bhagawati, P.; Shivayogimath, C. Pollutants removals and energy consumption in electrochemical cell for pulping processes wastewater treatment: Artificial neural network, response surface methodology and kinetic studies. *J. Environ. Manag.* **2021**, *281*, 111897. [CrossRef]
5. Stafford, W.; Cohen, B.; Pather-Elias, S.; Von Blottnitz, H.; Van Hille, R.; Harrison, S.T.; Burton, S.G. Technologies for recovery of energy from wastewaters: Applicability and potential in South Africa. *J. Energy S. Afr.* **2013**, *24*, 15–26. [CrossRef]
6. Kweinor Tetteh, E.; Opoku Amankwa, M.; Armah, E.K.; Rathilal, S. Fate of covid-19 occurrences in wastewater systems: Emerging detection and treatment technologies—a review. *Water* **2020**, *12*, 2680. [CrossRef]
7. Ponce-Robles, L.; Masdemont-Hernández, B.; Munuera-Pérez, T.; Pagán-Muñoz, A.; Lara-Guillén, A.J.; García-García, A.J.; Pedrero-Salcedo, F.; Nortes-Tortosa, P.A.; Alarcón-Cabañero, J.J. WWTP effluent quality improvement for agricultural reuse using an autonomous prototype. *Water* **2020**, *12*, 2240. [CrossRef]
8. Riaz, S.; Park, S.-J. An overview of TiO2-based photocatalytic membrane reactors for water and wastewater treatments. *J. Ind. Eng. Chem.* **2020**, *84*, 23–41. [CrossRef]
9. Chollom, M.N.; Rathilal, S.; Swalaha, F.M.; Bakare, B.F.; Tetteh, E.K. Removal of antibiotics during the anaerobic digestion of slaughterhouse wastewater. *Planning* **2020**, *15*, 335–343. [CrossRef]
10. Nethengwe, N.S.; Uhunamure, S.E.; Tinarwo, D. Potentials of biogas as a source of renewable energy: A case study of South Africa. *Int. J. Renew. Energy Res.* **2018**, *8*, 1112–1123.
11. Jordaan, G.P. Evaluating the Sustainable Potential of Biogas Generation in South Africa. Master's Thesis, Stellenbosch University, Stellenbosch, South Africa, 2018.
12. McCabe, B.K.; Schmidt, T. *Integrated Biogas Systems: Local Applications of Anaerobic Digestion towards Integrated Sustainable Solutions*; IEA Bioenergy: Paris, France, 2018.
13. Jingura, R.M.; Kamusoko, R. Methods for determination of biomethane potential of feedstocks: A review. *Biofuel Res. J.* **2017**, *4*, 573–586. [CrossRef]
14. Hülsemann, B.; Zhou, L.; Merkle, W.; Hassa, J.; Müller, J.; Oechsner, H. Biomethane potential test: Influence of inoculum and the digestion system. *Appl. Sci.* **2020**, *10*, 2589. [CrossRef]
15. Abdelwahab, T.A.M.; Mohanty, M.K.; Sahoo, P.K.; Behera, D. Application of nanoparticles for biogas production: Current status and perspectives. *Energy Sources Part A Recovery Util. Environ. Eff.* **2020**, 1–13. [CrossRef]
16. Abdelsalam, E.M.; Samer, M. Biostimulation of anaerobic digestion using nanomaterials for increasing biogas production. *Rev. Environ. Sci. Bio/Technol.* **2019**, *18*, 525–541. [CrossRef]
17. Kweinor Tetteh, E.; Rathilal, S. Biogas production from wastewater treatment-evaluating anaerobic and biomagnetic systems. *Water-Energy Nexus* **2021**, *4*, 165–173. [CrossRef]

18. Zaidi, A.A.; RuiZhe, F.; Shi, Y.; Khan, S.Z.; Mushtaq, K. Nanoparticles augmentation on biogas yield from microalgal biomass anaerobic digestion. *Int. J. Hydrogen Energy* **2018**, *43*, 14202–14213. [CrossRef]
19. Hassanein, A.; Lansing, S.; Tikekar, R. Impact of metal nanoparticles on biogas production from poultry litter. *Bioresour. Technol.* **2019**, *275*, 200–206. [CrossRef] [PubMed]
20. Ajay, C.; Mohan, S.; Dinesha, P.; Rosen, M.A. Review of impact of nanoparticle additives on anaerobic digestion and methane generation. *Fuel* **2020**, *277*, 118234. [CrossRef]
21. Abou Kana, M.T.; Radi, M.; Elsabee, M.Z. Wastewater treatment with chitosan nano-particles. *Int. J. Nanotechnol. Appl.* **2013**, *3*, 39–50.
22. Brar, S.K.; Verma, M.; Tyagi, R.; Surampalli, R. Engineered nanoparticles in wastewater and wastewater sludge–Evidence and impacts. *Waste Manag.* **2010**, *30*, 504–520. [CrossRef]
23. Esakkimuthu, T.; Sivakumar, D.; Akila, S. Application of nanoparticles in wastewater treatment. *Pollut. Res* **2014**, *33*, 567–571.
24. Lombi, E.; Donner, E.; Tavakkoli, E.; Turney, T.W.; Naidu, R.; Miller, B.W.; Scheckel, K.G. Fate of zinc oxide nanoparticles during anaerobic digestion of wastewater and post-treatment processing of sewage sludge. *Environ. Sci. Technol.* **2012**, *46*, 9089–9096. [CrossRef] [PubMed]
25. Inozemtseva, O.A.; German, S.V.; Navolokin, N.A.; Bucharskaya, A.B.; Maslyakova, G.N.; Gorin, D.A. Encapsulated magnetite nanoparticles: Preparation and application as multifunctional tool for drug delivery systems. In *Nanotechnology and Biosensors*; Elsevier: Amsterdam, The Netherlands, 2018; pp. 175–192.
26. Majidi, S.; Zeinali Sehrig, F.; Samiei, M.; Milani, M.; Abbasi, E.; Dadashzadeh, K.; Akbarzadeh, A. Magnetic nanoparticles: Applications in gene delivery and gene therapy. *Artif. Cells Nanomed. Biotechnol.* **2016**, *44*, 1186–1193. [CrossRef] [PubMed]
27. American Public Health Association. *Standard Methods for the Examination of Water and Wastewater*; American Public Health Association: Washington, DC, USA, 2012.
28. Tetteh, E.K.; Amo-Duodu, G.; Rathilal, S. Synergistic Effects of Magnetic Nanomaterials on Post-Digestate for Biogas Production. *Molecules* **2021**, *26*, 6434. [CrossRef]
29. Amo-Duodu, G.; Kweinor Tetteh, E.; Rathilal, S.; Chollom, M.N. Synthesis and characterization of magnetic nanoparticles: Biocatalytic effects on wastewater treatment. *Mater. Today Proc.* **2022**, *62*, S79–S84. [CrossRef]
30. Amo-Duodu, G.; Tetteh, E.K.; Rathilal, S.; Armah, E.K.; Adedeji, J.; Chollom, M.N.; Chetty, M. Effect of Engineered Biomaterials and Magnetite on Wastewater Treatment: Biogas and Kinetic Evaluation. *Polymers* **2021**, *13*, 4323. [CrossRef]
31. Amo-Duodu, G.; Rathilal, S.; Chollom, M.N.; Kweinor Tetteh, E. Application of metallic nanoparticles for biogas enhancement using the biomethane potential test. *Sci. Afr.* **2021**, *12*, e00728. [CrossRef]
32. Enayati Ahangar, L.; Movassaghi, K.; Emadi, M.; Yaghoobi, F. Photocatalytic application of TiO_2/SiO_2-based magnetic nanocomposite (Fe_3O_4@ SiO_2/TiO_2) for reusing of textile wastewater. *Nanochemistry Res.* **2016**, *1*, 33–39.
33. Esfandiari, N.; Kashefi, M.; Afsharnezhad, S.; Mirjalili, M. Insight into enhanced visible light photocatalytic activity of Fe_3O_4–SiO_2–TiO_2 core-multishell nanoparticles on the elimination of Escherichia coli. *Mater. Chem. Phys.* **2020**, *244*, 122633. [CrossRef]
34. Yang, C.; Wang, P.; Li, J.; Wang, Q.; Xu, P.; You, S.; Zheng, Q.; Zhang, G. Photocatalytic PVDF ultrafiltration membrane blended with visible-light responsive Fe (III)-TiO_2 catalyst: Degradation kinetics, catalytic performance and reusability. *Chem. Eng. J.* **2021**, *417*, 129340. [CrossRef]
35. Vaiano, V.; Sacco, O.; Sannino, D.; Stoller, M.; Ciambelli, P.; Chianese, A. Photocatalytic removal of phenol by ferromagnetic N-$TiO_2/SiO_2/Fe_3O_4$ nanoparticles in presence of visible light irradiation. *Chem. Eng. Trans.* **2016**, *47*, 235–240.
36. Baniamerian, H.; Isfahani, P.G.; Tsapekos, P.; Alvarado-Morales, M.; Shahrokhi, M.; Vossoughi, M.; Angelidaki, I. Application of nano-structured materials in anaerobic digestion: Current status and perspectives. *Chemosphere* **2019**, *229*, 188–199. [CrossRef] [PubMed]
37. Zaidi, N.S.; Sohaili, J.; Muda, K.; Sillanpää, M. Magnetic field application and its potential in water and wastewater treatment systems. *Sep. Purif. Rev.* **2014**, *43*, 206–240. [CrossRef]
38. Amorós-Pérez, A.; Cano-Casanova, L.; Lillo-Rodenas, M.A.; Román-Martínez, M.C. Cu/TiO_2 photocatalysts for the conversion of acetic acid into biogas and hydrogen. *Catal. Today* **2017**, *287*, 78–84. [CrossRef]
39. Alvarado-Morales, M.; Tsapekos, P.; Awais, M.; Gulfraz, M.; Angelidaki, I. TiO2/UV based photocatalytic pretreatment of wheat straw for biogas production. *Anaerobe* **2017**, *46*, 155–161. [CrossRef] [PubMed]
40. Wu, J.; Lin, H.-M. Photo reduction of CO_2 to methanol via TiO_2 photocatalyst. *Int. J. Photoenergy* **2005**, *7*, 115–119. [CrossRef]
41. Abdelsalam, E.; Samer, M.; Attia, Y.; Abdel-Hadi, M.; Hassan, H.; Badr, Y. Influence of zero valent iron nanoparticles and magnetic iron oxide nanoparticles on biogas and methane production from anaerobic digestion of manure. *Energy* **2017**, *120*, 842–853. [CrossRef]
42. Ren, G.; Chen, P.; Yu, J.; Liu, J.; Ye, J.; Zhou, S. Recyclable magnetite-enhanced electromethanogenesis for biomethane production from wastewater. *Water Res.* **2019**, *166*, 115095. [CrossRef] [PubMed]
43. Liang, Y.-G.; Li, X.-J.; Zhang, J.; Zhang, L.-G.; Cheng, B. Effect of microscale ZVI/magnetite on methane production and bioavailability of heavy metals during anaerobic digestion of diluted pig manure. *Environ. Sci. Pollut. Res.* **2017**, *24*, 12328–12337. [CrossRef]

44. Liu, L.; Zhang, T.; Wan, H.; Chen, Y.; Wang, X.; Yang, G.; Ren, G. Anaerobic co-digestion of animal manure and wheat straw for optimized biogas production by the addition of magnetite and zeolite. *Energy Convers. Manag.* **2015**, *97*, 132–139. [CrossRef]
45. Mu, Y.; Wang, G.; Yu, H.-Q. Kinetic modeling of batch hydrogen production process by mixed anaerobic cultures. *Bioresour. Technol.* **2006**, *97*, 1302–1307. [CrossRef] [PubMed]

International Journal of
Environmental Research and Public Health

Review

Application of Advanced Oxidation Technology in Sludge Conditioning and Dewatering: A Critical Review

Jiahua Xia [1], Juan Ji [1], Zhiqiang Hu [1], Ting Rao [1], Ankang Liu [2], Jingqian Ma [3] and Yongjun Sun [3,*]

[1] Nanjing Jiangbei New Area Public Utilities Holding Group Co., Ltd., Nanjing 210044, China; xiayu_610@sina.com (J.X.); jj13814186775@163.com (J.J.); airhzq@126.com (Z.H.); raot@mail.ustc.edu.cn (T.R.)
[2] Nanjing Water Purification Environmental Research Institute Co., Ltd., Nanjing 211100, China; lakyx2@126.com
[3] College of Urban Construction, Nanjing Tech University, Nanjing 211800, China; majingqian123123@126.com
* Correspondence: sunyongjun@njtech.edu.cn

Abstract: Sludge dewatering is an important link in sludge treatment. In practical engineering, the dewatering effect of unconditioned sludge is very poor. The use of advanced oxidation technology can improve sludge dewatering performance, reduce sludge capacity, and remove micro-pollutants, which is beneficial for sludge post-treatment and disposal. Based on the current status of sludge conditioning and dehydration, the characteristics of the advanced oxidation method for sludge dehydration were systematically explained using various free radical reaction mechanisms and dehydration conditions. The effects of various advanced oxidation technologies on sludge conditioning and dewatering has been extensively discussed. Finally, the application prospects of the advanced oxidation technology in sludge conditioning and dewatering are presented.

Keywords: advanced oxidation technology; sludge reduction; sludge conditioning; sludge dewatering; oxidation

check for updates

Citation: Xia, J.; Ji, J.; Hu, Z.; Rao, T.; Liu, A.; Ma, J.; Sun, Y. Application of Advanced Oxidation Technology in Sludge Conditioning and Dewatering: A Critical Review. *Int. J. Environ. Res. Public Health* **2022**, *19*, 9287. https://doi.org/10.3390/ijerph19159287

Academic Editors: Yung-Tse Hung, Hamidi Abdul Aziz and Issam A. Al-Khatib

Received: 9 June 2022
Accepted: 26 July 2022
Published: 29 July 2022

1. Introduction

Considerable achievements have been made in China's sewage treatment industry through the completion and operation of a large number of urban sewage treatment plants. The sharp increase in the amount of sewage treatment plants has led to the production of a large amount of activated sludge, which has led to a gradual increase in the load on sewage treatment plants [1]. Sludge treatment has become a key challenge for the Chinese water industry [2]. Sludge has the characteristics of high-water content, large volume, and high treatment cost, showing the performance of the fluid or semi-fluid state, causing great inconvenience for subsequent transportation and treatment [3].

Research focus. The main methods of sludge treatment are sanitary landfills and incineration. The moisture content of the sludge be less than 60% and 50%, respectively, based on the requirements of sanitary landfills and incineration [4]. In most cases, the moisture content of excess sludge is more than 95%, and there are four main types of sludge water, namely free water, interstitial water, adsorbed water, and bound water. After the excess sludge is concentrated, the moisture content is slightly reduced but still cannot meet the requirements of sanitary landfill and incineration [5]. Hence, further deep dewatering is required. Deep dewatering helps to protect the environment, strengthen land use, reasonably degrade a large amount of sludge pollutants, and ensure the safe integration of sludge resources into the circular ecosystem [6].

The methods of sludge conditioning include physical conditioning, chemical conditioning, and biological conditioning [7]. In physical conditioning, the microbial cells in the sludge are destroyed by physical methods, the structure of the sludge is changed, the binding effect of the sludge and water is reduced, and the inner water is released [8]. Among them, thermal conditioning has obvious effects on improving the dewatering rate of most sludges, but the treatment effect on sludges with low organic content is poor. Although microwave conditioning time is short, requires low heating temperature, has high thermal efficiency, and uses simple equipment, it requires a certain amount of sludge [9]. Freeze conditioning can be achieved spontaneously without adding the chemical additives, and the cost is low, but the scope of use is limited [10]. Ultrasonic conditioning has a high level of efficiency and environmental protection, and has application prospects, but its principle of action and complex effects are still to be discussed further [11]. Overall, the high energy consumption and high equipment cost of physical conditioning limit its large-scale use [12]. Microbial conditioning promotes mutual coagulation and precipitation of colloidal suspensions in water and improves sludge dewatering performance [13]. Microbial conditioning has strong advantages, such as high efficiency, non-toxicity, no secondary pollution, biodegradation, compact sludge floc, and a wide range of application [14]. However, the current research on such flocculants is not comprehensive, the research level is low, the mechanism of flocculants is not fully understood, and there are problems, such as large dosage, high production costs and low relevance [15].

Chemical conditioning is widely used as the optimal choice in the context of the comprehensive comparison. Advanced oxidation (AOP) is a typical chemical regulation method, and the free radicals it produces have strong oxidative effects [16]. Free radicals are generated by different oxidants, and different oxidants generate different free radicals with different reaction mechanisms [17]. Extracellular polymer (EPS) is the main component of sludge, and its components can promote the flocculation of sludge. However, too high an EPS content would lead to deterioration of the dehydration results due to hydrophilicity and strong intermolecular forces [18]. Therefore, reducing the EPS content and increasing its hydrophobicity is beneficial to the removal of water. Under the oxidation of AOP, EPS is effectively hydrolyzed, which weakens the strong binding ability of EPS and release the bound water in the sludge, thereby improving the dewatering performance of the sludge [19]. Advanced oxidation is an efficient sludge reduction treatment technology with characteristics of strong oxidizing ability, enhanced sludge-water separation effect, and it is not easy to generate secondary pollution [20]. Its application in the field of sludge treatment has gradually developed into a research focus in recent years.

This paper focuses on the application of different oxidation methods in sludge dewatering improvement and analyzes their applications in sludge dewatering and micropollutant removal. We have found that the advanced oxidation method destroys the EPS by free radicals, breaks the complex bond between water molecules and organic matter, reduces the hydration energy, dissolves the microbial cell wall, and helps the degradation of sludge and the removal of micro-pollutants in the sludge system. Hydroxyl and sulfate ions are more beneficial for improving the dewatering performance of sludge. However, before choosing the oxidizing agent and the AOP method, geographical conditions, purpose of use, level of economic development, and climatic conditions should be considered. This review could provide researchers with new ideas to employ multiple approaches for advanced oxidation by generating different or identical free radicals, thereby making it possible to develop new methods of dehydration that are more efficient, economical, and environmentally friendly.

2. Application of Hydrogen Peroxide in Sludge Dewatering

As a common oxidizing agent, hydrogen peroxide is widely used in advanced oxidation processes in Figure 1. The strongly oxidizing hydroxyl radicals formed in hydrogen peroxide play an important role in water treatment. Hydroxyl radicals inactivate bacteria during disinfection and release water through oxidation of EPS during sludge dewatering. Hydroxyl free radicals can also damage the cell wall of microorganisms or bacteria, increase the permeability of the cell membrane, destroy the cell wall, and release the intracellular water from the cell [21]. The Fenton system refers to a mixed system composed of Fe^{2+} and H_2O_2 [22]. Under acidic conditions, Fe^{2+} acts as a catalyst to promote the decomposition of H_2O_2 into hydroxyl groups with high reactivity. If Fe^{3+} is present, a Fenton-like process takes place at the same time, in which Fe^{2+} ions are formed in the Fenton process [23]. Fe^{3+}

also reacts with water in solution and generating Fe^{2+} and hydroxyl radicals [24]. In the generation of ·OH by Fenton or Fenton-like methods, Fe^{2+} plays a key role in the initiation and progression of the reaction.

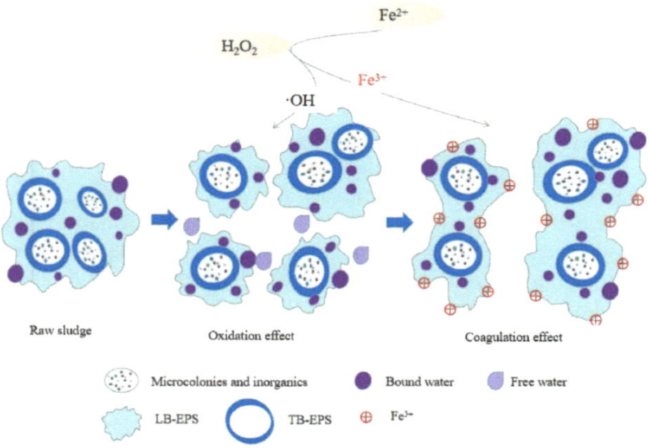

Figure 1. Mechanism of Enhanced Sludge Dewatering by Hydrogen Peroxide Pretreatment.

H_2SO_4 was used to adjust the pH of the sludge. Under the optimal dosage of Fe^{2+} and H_2O_2, the SRF and CST of the sludge were reduced to 6.149×10^9 m/kg and 15.7 s, which were 93% and 48.5% lower than the original sludge, respectively, indicating that the sludge dewatering effect is significantly improved [25]. LIU is also in the optimal dose. The sludge specific resistance SRF decreases by 95% below pH = 5, w(Fe^{2+}) = 40 mg/g DS, w(H_2O_2) = 32 mg/g DS, lime = 400 mg/g DS [26]. Fenton's reagent has been shown to be an effective strong oxidant in several experiments, but it also has several shortcomings and limitations. The pH of most Fenton reactions is below 4, which increases the burden on subsequent sludge treatment. In order to overcome this difficulty, the dewatering performance of sludge was tested under five systems, namely, Fenton-lime, Fenton, Fe^{3+}, Fe^{2+}, and H_2O_2, and it was found that the Fenton-lime, Fenton, Fe^{3+} system could significantly reduce the sludge SRF and CST [27]. Among them, Fenton and lime together adjusted the sludge, which can increase the pH while maintaining a good dehydration effect. Tao et al. further studied the effect of combining iron-rich biochar with H_2O_2 on sludge dewatering ability [28]. The iron contained in biochar combines with H_2O_2 to accelerate the Fenton reaction, and red mud is added as a framework material during the reaction. The WC of the mud cake decreased from 96.49% to 46.38%, and the CST and SRF decreased by 90.24% and 98.01%, respectively. The addition of biomass charcoal increases the drainage channel, which promotes the outflow of more sludge water under the same pressure during the dehydration process, thereby improving the dehydration efficiency.

Under the coupling action of Fenton, light, and ultrasonic waves, the polysaccharides in the sludge filtrate increased significantly, the particle size of the sludge became smaller than that under the action of Fenton oxidation alone, and the dewatering performance was significantly enhanced. Fenton oxidation coupled bioleaching with can improve sludge dewatering performance and remove trace metals in sludge [27–29]. For example, when the dosage of bioleaching 5d and H_2O_2 was 2 g/L, the removal effect of each micro-metal was significantly improved, and the removal rate of Cd increased from 90% to 99.5%, the removal rate of Cu and Zn increased from 60% to 70%, and Pb removal rate increased from 20% to 39% [29].

As shown in Table 1, the use of Fenton-type Fenton oxidation is fast, economical, and environmentally friendly, and the sludge dewatering performance and stability are significantly improved. The coexistence of Fe^{2+} and H_2O_2 can have an oxidizing effect. In addition, Fe^{3+} can also improve the dehydration performance. However, Fe^{3+} reacts with hydroxyl to form hydroxide precipitate, which increases the alkalinity of the solution and has a certain influence on the precipitation of Fe^{3+}. Meanwhile, the generation of ·OH will be hindered if the H_2O_2 is excessive [30]. Therefore, researchers further strengthen sludge dewatering through Fenton coupling technology, which aids in the complete mineralization of organic matter, removes harmful substances such as trace metals in sludge, and reduces the environmental hazard of sludge. The Fenton reagent is coupled with the skeleton construct composite conditioner, which can form a rigid grid structure with high permeability in the mud cake, reduce the compressibility of the mud cake, and further enhance the dehydration effect. The Fenton reaction conditions are also acidic, and a neutralizing agent must *be* added to make the filtrate close to neutral. Yu used lime as the neutralizer, and through the corresponding surface design, the pH of the filtrate was increased under the premise of ensuring the dehydration effect [31]. Lime was used as the sludge floc skeleton, which can transfer part of the pressure to the inside of the floc under high pressure and provide a channel for water release [32]. However, lime production generates a considerable amount of CO_2, and therefore red mud has been used as a neutralizer to cause the red mud to form a new mineralization stage and a hard network structure to help free water flow out [33]. In addition, slaked lime, pulverized coal, lignite, and sulfuric acid modified fly ash are also commonly used in the sludge conditioning process. Therefore, the combined use of Fenton reagent and physical conditioner can effectively improve the sludge drying rate and reduce the pH value of the filtrate, but issues, such as high treatment cost and low utilization rate of hydroxyl radicals, persist. Through the coupling of Fenton technology with light and ultrasonic waves, the cost of conditioners can be reduced effectively, the scope of application of sludge can be increased, and is more environment friendly as it ensures the dewatering effect.

Table 1. Study results of enhanced sludge dewatering by hydrogen peroxide pretreatment.

Preprocessing Method	Optimal Pretreatment Conditions	Conditioning Dehydration Effect	Ref.
Fenton	Fe^{2+}: 5000 mg/L; H_2O_2: 6000 mg/L	CST reduction: 48.5%; SRF reduction: 93.3%	[25]
Fenton	Fe^{2+}: 25 mg/g DS; H_2O_2: 50 mg/g DS	CST reduction: 95.0%	[34]
Lime + Fenton	Fe^{2+}: 40 mg/g DS; H_2O_2: 32 mg/g DS; pH: 5; Lime: 400 mg/g DS	SRF reduction: 95.0%	[26]
Lime + Fenton	Fe^{2+}: 47.9 mg/g DS; H_2O_2: 34.3 mg/g DS; pH: 6.6 ± 0.2; Lime: 43.2 mg/g DS	Water content: 55.8 ± 0.6%	[27]
Lime + Fenton	Fe^{2+}/H_2O_2: 50/30 mg/g DS; pH: 3; Lime: 50 mg/g DS; *t*: 10 min	SRF reduction: 90.0%	[27,32]
Red mud + Fenton	Fe^{2+}: 31.9 mg/g DS; H_2O_2: 33.7 mg/g DS; Red mud: 275.1 mg/g DS	Water content: 59.8 ± 0.4%	[33]
Bioleaching + Fenton	H_2O_2: 2.0 g/L; *t*: 5 d	CST reduction: 89.8%; SRF reduction: 83.8%	[29]
Iron rich biochar + Fenton	Biochar: 0.792 g/g VS; H_2O_2: 0.072 g/g VS	CST reduction: 90.2%; SRF reduction: 98.0%; Water content: 46.0%	[28]

3. Application of Persulfate in Sludge Dewatering

As shown in Figure 2, compared with the advanced oxidation process of hydrogen peroxide, one of the advantages of the sulfate oxidation method is that the raw material is solid, which makes it easy to transport and store. Persulfate (PS) and peroxymonosulfate (PMS) are commonly used as sulfate-generating oxidants [35]. Persulfate can ionize persulfate ion ($S_2O_8{}^{2-}$) in water. Its oxidizing ability is comparable to that of ozone, and its oxidation potential is 2.01 V. Under the catalytic activation of light, heat, transition metals, etc., a large amount of $\cdot SO_4{}^-$ can be generated, and the oxidation potential is up to 2.60 V.

The advanced sulfate oxidation process also has other excellent advantages, such as longer half-life and wider operating pH range (2–10).

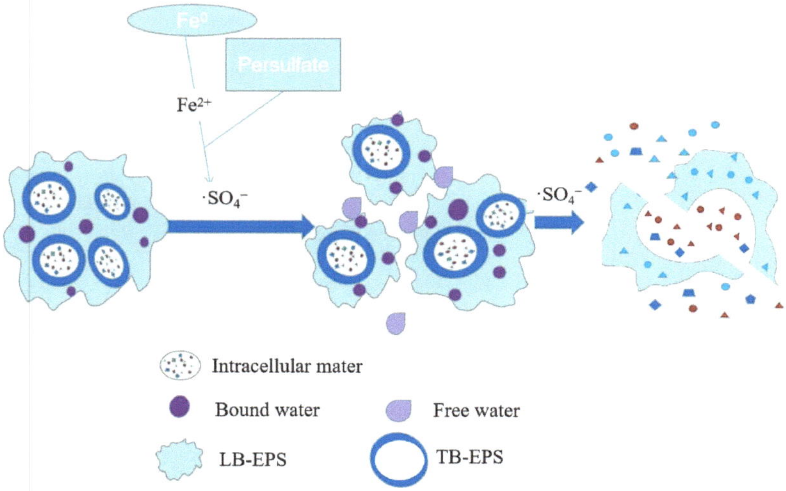

Figure 2. Mechanism of enhanced sludge dewatering by persulfate pretreatment.

Zhen et al. used activated persulfate to enhance sludge dewatering for the first time and reported that at room temperature, persulfate in the pH range of 3.0–8.5 was catalytically activated by transition metal ions (Fe^{2+}) and acted on excess sludge, which could significantly improve the sludge dewatering performance [36]. In addition, when the $S_2O_8{}^{2-}$ and Fe^{2+} were 1.2 mmol/g (calculated by the dosage in 1 g of volatile suspended matter VSS, the same below) and 1.5 mmol/g, the CST decreased rapidly, which could be reduced by 88.8% within 1 min. The $\cdot SO_4{}^-$ produced by Fe^{2+} catalytic activation has strong oxidizing property. $\cdot SO_4{}^-$ can destroy the floc structure of sludge, release EPS bound water and intracellular bound water, and greatly improve the dewatering performance of sludge. Zhou et al. used zero-valent iron instead of Fe^{2+} as $S_2O_8{}^{2-}$ inducer and found that when $K_2S_2O_8$ was 4 g/L and Fe^0 was 15 g/L, the sludge CST was reduced by more than 50%. Heat treatment was introduced into the $Fe^{2+}/S_2O_8{}^{2-}$ pretreatment system and the iron-sulfur oxidation-heat coupling integrated sludge enhanced dewatering technology was constructed. Moreover, the disintegration of EPS and separation of sludge and water were strengthened through a synergistic mechanism, which effectively reduced the dosage of chemicals and saved costs. When the temperature was 80 °C, the Fe^{2+} dosage was 1.2 mmol/g, and the $S_2O_8{}^{2-}$ dosage was 1.5 mmol/g, the CST decreased by 96.5% (88.8% at normal temperature), the structure of the sludge floc was destroyed, and the bound water of EPS was released. The sludge dewatering efficiency also improved greatly [37].

As shown in Table 2, different from Fenton oxidation, when activated persulfate is used to enhance sludge dewatering, pH adjustment is not required, and thus, the operation is simple, fast and efficient. It is an efficient and promising sludge pretreatment enhanced dewatering technology. The $\cdot SO_4{}^-$ produced by $Fe^{2+}/S_2O_8{}^{2-}$ can realize high-efficiency degradation of EPS, and under the coupling action of low-pressure electrolysis $Fe^{2+}/S_2O_8{}^{2-}$, the micellar structure of sludge bacteria and the microbial cell wall are dissolved and destroyed, resulting in a large amount of sludge and the releasing of floc interstitial water and intracellular bound water. However, the oxidation of $S_2O_8{}^{2-}$ and Fe^{2+} and its derivative processes are still in the exploratory stage. The effect of using zero-valent iron to replace Fe^{2+} as a spontaneous catalyst is not ideal because Fe^0 needs to dissolve and decompose Fe^{2+} first, and then induce $S_2O_8{}^{2-}$ to form $\cdot SO_4{}^-$, and the surface of Fe^0 is easily passivated, resulting in the failure of Fe^0 in the inner core, and thus, the dehydration efficiency cannot be compared with that of Fe^0. Fe^{2+} is comparable to [38]. In addition, the oxidation potential of persulfate is 2.01 V when applied to sludge dewatering, the reaction rate is too low, and the effect is not ideal. However, under the catalytic activation of heat, light, transition metals, etc., persulfate can generate a large amount of $\cdot SO_4^-$, and the oxidation potential is as high as 2.60 V. High heat and high-pressure pretreatment destroys the sludge floc structure and improves the sludge dewatering performance, but the heat treatment consumes considerable energy, and thus, the operating cost is high. Electrolysis is also an effective method for sludge conditioning and enhanced dewatering, and the sludge treatment by electrolysis has little effect on the environment. However, the results showed that although low-pressure electrolysis alone could improve the sludge dewatering efficiency, it had limited effect on the closely attached EPS. In the future, the sludge enhanced dewatering process and theoretical system based on $Fe^{2+}/S_2O_8^{2-}$ oxidation should be further enriched and improved to provide reserve technologies for the front-end safety treatment and end-end ecological disposal of sludge.

Table 2. Study results of enhanced sludge dewatering by persulfate pretreatment.

Preprocessing Method	Optimal Pretreatment Conditions	Conditioning Dehydration Effect	Ref.
$Fe^{2+}/S_2O_8^{2-}$	Fe^{2+}: 1.5 mmol/g; $S_2O_8^{2-}$: 1.2 mmol/g	CST reduction: 88.8%	[36]
$Fe^{2+}/S_2O_8^{2-}$	Fe^{2+}: 1.5 mmol/g; $S_2O_8^{2-}$: 1.2 mmol/g; T: 80 °C	CST reduction: 97.0%	[37]
$Fe^0/S_2O_8^{2-}$	Fe^0: 15.0 g/L; $S_2O_8^{2-}$: 4.0 g/L	CST reduction: 50.2%	[39]
$Fe^0/S_2O_8^{2-}$	Fe^0: 2.0 g/L; $S_2O_8^{2-}$: 0.5 g/L	CST reduction: 90.0%	[40]
Electrolysis + $Fe^{2+}/S_2O_8^{2-}$	Fe^{2+}: 0.5 mmol/g; $S_2O_8^{2-}$: 0.4 mmol/g;	Water content: 82.5%	[41]
Acid-ZVI/PMS	ZVI: 0.15 g/g DS; Oxone: 0.07 g/g DS; pH: 3	CST reduction: 19.6%; Water content: 76.3%	[42]
Fe^{2+}/PMS	PMS 0.1 g/g TSS; Fe^{2+} 0.5 g/g TSS	CST reduction: 66.0%; SRF reduction: 88.0%	[43]

4. Application of Ozone in Sludge Dewatering

As shown in Figure 3, the mechanism of action of O_3 includes direct and indirect reactions. Ozone first destroys the sludge floc structure by oxidizing EPS and bridges and breaking up the colloidal structure. Since ozone has a strong oxidizing power, it damages cell membranes, releasing intracellular and extracellular substances into the liquid phase. Ozone can react with EPS and soluble substances in cells, including proteins, polysaccharides, and intracellular substances containing C, N, and P. Intracellular substances containing C can generate volatile fatty acids, resulting in a drop in pH. Ozone can also mineralize dissolved EPS and organic matter released by cells to generate CO_2 and H_2O, resulting in a reduction in the amount of ozone that reacts with the sludge. After ozonation, the mixed liquor suspension and mixed liquor volatile sludge suspended solid were reduced due to cell lysis. Authors should discuss the results and how they can be interpreted from the perspective of previous studies and of the working hypotheses. The findings and their implications should be discussed in as broad a context as possible. Future research directions can also be highlighted.

Chacana et al., treated the primary and digested sludge with ozone, and found that the SCOD and biodegradable COD of the primary sludge did not increase under the action of ozone oxidation; however, part of the organic matter was mineralized and converted to CO_2 [44,45]. When the ozone mass concentration was 37.8 mg/g SS, the oxidation efficiency was the best, but the sludge CST increased significantly from (25.5 ± 2.4) s to (289.0 ± 25.9) s, and the filtration performance decreased significantly [46] because the size of the sludge flocs becomes smaller, and part of the free water is converted into adsorbed water. The polysaccharides and proteins in the filtrate increased from (4.46 ± 0.21) mg/L to (220.90 ± 24.87) mg/L and from (6.26 ± 0.28) mg/L to (386.54 ± 32.15) mg/L, respectively. The polysaccharides and proteins in EPS also move into the sludge when the intracellular polysaccharides and proteins are released. Therefore, polysaccharide and protein content increased while EPS decreased. Demir et al. also confirmed that after ozone treatment, the water content of the sludge cake was reduced and the sedimentation rate was accelerated, which could improve the mud-water separation efficiency to a certain extent, but the sludge SRF increased and the filtration performance decreased [45]. The degree of mineralization of sludge increased after O_3 treatment, VSS/SS decreased from 0.86 to below 0.50, and 35% to 95% of solid carbon was mineralized. Weemaes reported that with an O_3 dosage of 0.2 kg/kg COD, approx. 67% of the solid organic components were converted, 29% of which were dissolved, and the remaining 38% were directly mineralized [47]. After sludge ozonation, both inorganic and organic matter was dissolved into the supernatant, and the reduced value of MLVSS was slightly smaller than that of MLSS. The decrease in the MLVSS is the main reason for the decrease in the MLSS. With the increase in O_3 dosage, the organic matter entering the supernatant becomes much larger than the inorganic matter, and thus, the sludge VSS/SS decreases with the increase in the degree of ozonation, thereby promoting the invisible growth of the sludge and reducing the sludge dry base production. The higher the soluble organic matter (SCOD), the more favorable for the subsequent aerobic/anaerobic digestion of the sludge. Thermal-chemical technology was used to treat sludge and it was found that SCOD increased significantly during O_3 treatment [48]. At an O_3 dose of 0.1 g/g DS, the SCOD increased from 240 mg/L to 960 mg/L. The increased SCOD can be attributed to O_3 disrupting the microbial cell wall and releasing the cytoplasm into solution [49]. At the same time, the SCOD increase in the O_3 strong oxidizing treatment solution of is used to ensure that the quality of the effluent during the O_3 treatment process is up to the standard. The recirculation of ozonated sludge meant that the O_3 treatment process did not affect the quality of the effluent. In this experiment, the SS and SCOD values of the effluent remained at 10 and 15 mg/L, respectively, while the dry sludge yield was greatly improved.

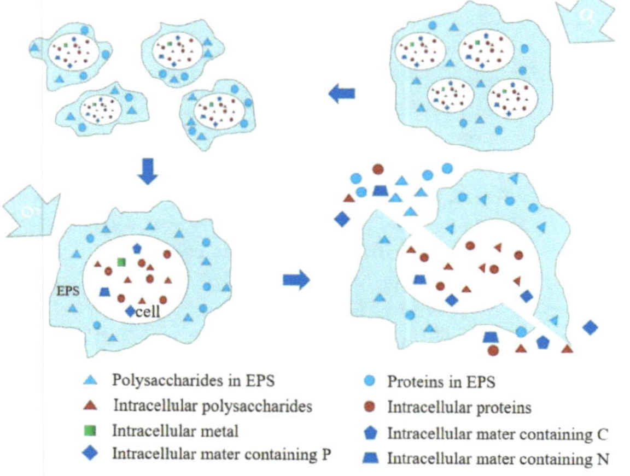

Figure 3. Enhanced sludge dewatering mechanism by ozone pretreatment.

As shown in Table 3, the dewatering performance of sludge was significantly reduced after ozonation because as the size of the sludge flocs becomes smaller, part of the free water is converted into adsorbed water, the cells are ruptured, the intracellular substances and extracellular polymers enter the liquid phase, and the interstitial water and bound water content increased. When ozone and chitosan (CT) were combined for sludge conditions, effective sludge dewatering results were observed [50]. In the ozone/CT system, EP and sludge cells were destroyed by ozone, and CT as a coagulant could increase the size of sludge flocs and enhance the dewatering ability. Accordingly, rational control of the amount of O_3 can stimulate its catalytic oxidation performance. At present, ozone is generally used as a catalytic material to catalyze the formation of ·OH from H_2O_2, which is used to strengthen sludge dissolution and methane fermentation. After ozonation, the hydrophobicity of sludge decreased, and the hydrophilicity increased. The effect of ozone on enhancing sludge dewatering is very limited, but ozone treatment promotes the anaerobic digestion process of sludge and improves methane yield. The anaerobic fermentation treatment of the excess sludge after ozone treatment can strengthen the anaerobic fermentation speed and increase biogas production, which has a significant effect on the energy conversion and reduction of sludge [51]. However, high ozone dose or long pretreatment time will lead to mineralization of dissolved organic matter and inactivation of key microorganisms, which is not conducive to methane production and energy recovery [52]. High cost is the main obstacle restricting the engineering application of ozone sludge treatment. In practice, factors such as sludge reduction and input cost should be considered comprehensively [53].

Table 3. Study results of enhanced sludge dewatering by ozone pretreatment.

Preprocessing Method	Optimal Pretreatment Conditions	Conditioning Dehydration Effect	Ref.
O_3	O_3: 37.8 mg/g SS	SS reduction: 13.5%;VSS reduction 11.9%	[46]
O_3	O_3: 0.5 g/g TS	SS reduction 77.8%; VSS reduction 71.6%	[45]
O_3	O_3: 0.1 g/g COD	Methane growth rate: 180%	[47]
O_3	O_3: 0.1 g/g DS	Methane growth rate: 25%	[49]
O_3	O_3: 50 mg/g DS	SS reduction: 49.1%; VSS reduction: 45.7%	[52]
O_3 + CT	O_3: 60 g/g TS; CT 20 mg/g TS	Water content: 56.5%	[29,50]

5. Application of Permanganate and Ferrate in Sludge Dewatering

Recent research and applications have promoted the addition of H_2O_2 and PS/PMS as substrates that serve as catalysts for the production of hydroxyl and sulfate ions in Figure 4. In addition, potassium permanganate oxidation and ferrate oxidation can also be used as reliable oxidation methods in AOP. Under the catalysis of ferrous iron, permanganate can generate manganese dioxide with strong oxidation. Different forms of manganese dioxide can generate strong oxidizing$\cdot O_2^-$ in the dark. Iron in ferrate is in an oxidized state and has a strong oxidizing property. At the same time, ferrate act as a flocculation when reduced to iron ions or ferric hydroxide.

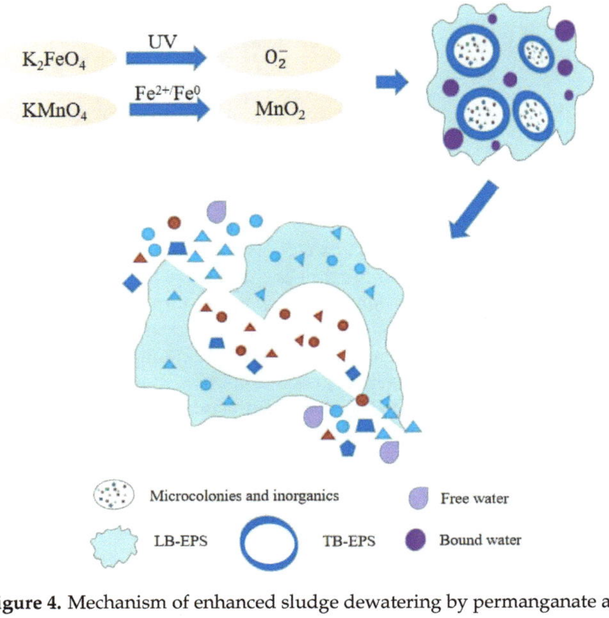

Figure 4. Mechanism of enhanced sludge dewatering by permanganate and ferrate pretreatment.

Fe^{2+} and ZVI have catalytic effects on potassium permanganate, and under the catalysis of Fe^{2+}, $KMnO_4$ can react with it to generate strong oxidizing manganese dioxide (MnO_2) [54]. After the sludge treatment, the WC, CST, and SRF of the sludge can be reduced effectively, and the effect is stronger than that of $KMnO_4$ alone. The use of Fe(II)-$KMnO_4$ has a highly effective effect on water purification, which includes the oxidation and flocculation ability of organic matter, and has been widely used in the treatment of high algae and arsenic-containing water. Likewise, heat or acid treatment prior to oxidation has also been proven effective, reducing the moisture content to 59.32%. It can be seen that permanganate itself has a certain oxidizing ability, and the catalytic system can improve the oxidizing ability of the permanganate. Under acidic and neutral conditions, the addition of ferrate can reduce WC in sludge without the aid of catalytic reagents. When using biochar as a framework material, the WC reduction is less than 60%. Therefore, the use of ferrate requires control of the appropriate pH for ferrate addition to work [55].

As shown in Table 4, the combined application of the potassium permanganate oxidation method can effectively improve sludge dewatering. The Fe(II)-$KMnO_4$ system can be combined with pyrolysis sludge-based carbon (SBC) to construct a Fe-Mn-SBC system which aims to effectively adsorb AS(III) in groundwater and remediate groundwater pollution. Ferrate (VI) is also used in water treatment as a strong oxidant and disinfectant because iron is in the +6-oxidation state [56,57]. The oxidation mechanism is the strong oxidation of ferrate in the reduction process. The oxidizing power of ferrate (VI) is highly dependent on pH, water composition, type, and concentration of contaminants. Ferrate (VI) has a flocculation effect when reduced to iron ions or ferric hydroxide, and is an economical, energy-saving, and environmentally friendly chemical reagent, but maintaining a certain pH value will increase the use of acid/base and generate additional costs. In the latest study, Wu et al. found that when the ferrate (VI) system was formed using a combination of UV and ferrate (VI), the oxidative power increases and superoxide radicals could be generated. This result is more significant for the degradation of phenolic pollutants [58]. Through the coupling of ferrate and permanganate with other technologies, the cost of conditioners can be effectively reduced, the scope of application of sludge can be increased, and on the premise of ensuring the dewatering effect, is more environmentally friendly.

Table 4. Study results of enhanced sludge dewatering by permanganate and ferrate pretreatment.

Preprocessing Method	Optimal Pretreatment Conditions	Conditioning Dehydration Effect	Ref.
TA + $Fe^0/KMnO_4$	$KMnO_4$ 0.4 mM/g DS; Fe^0 2.5 mMg/g DS; T: 45 °C	Water Content: 60.1%	[54]
Ferrate (VI)	VI: 25.3 mmol/L; T: 25 °C; pH: 7.0	tetrabromobisphenol A reduction: 99.1%	[57]
K_2FeO_4	K_2FeO_4: 1200 mg/L; pH: 3	SRF reduction: 30.5%; Water content: 73.4%	[55]
UV + Ferrate (VI)	VI: 0.1 g/L; pH: 7.0; UV: 0.198 mW/cm^2	2,4-DCP degradation growth rate: 1020%	[58]
$KMnO_4$ + Fe^{2+}	$KMnO_4/Fe^{2+}$ = 3/1 (mass ration)	SRF reduction: 83.8%	[59]

6. Application of Other Technologies in Sludge Dewatering

As shown in Figure 5, wet air oxidation (WAO) is a high-temperature and high-pressure oxidation treatment technology for organic or inorganic substances using oxygen as an oxidizing agent in a solution. Organic or inorganic substances can be converted into carbon dioxide, water, and biodegradable short-chain organic acids, with acetic acid being the most important short-chain organic acid. The influencing factors include reaction temperature, reaction pressure, pH value of the solution, reaction time, properties of the object to be treated, and the catalyst dosage. Heavy metals such as Cu and Ni, can be used as catalysts for wet air oxidation, namely catalytic wet air oxidation (CWAO). CWAO is developed based on WAO, with fast response speed and low running cost.

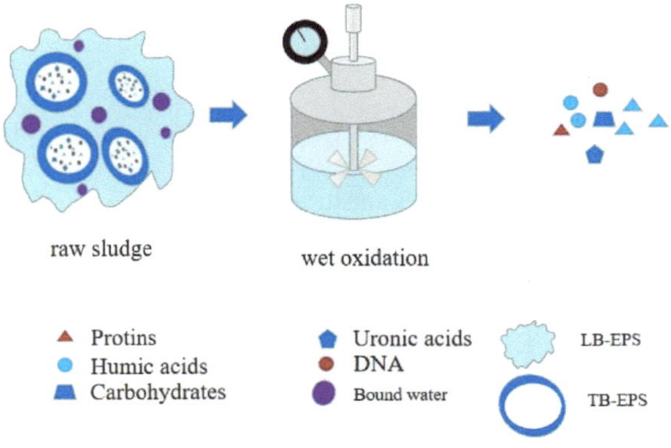

Figure 5. WAO pretreatment enhances sludge dewatering mechanism.

In wet air oxidation, insoluble organics substances are first converted into simple soluble organics by dissolution, and then further oxidized to carbon dioxide, water, and readily biodegradable intermediates [60,61]. The study found that the temperature in WAO had a significant influence on the dewatering performance of the sludge, and the dewatering performance was significantly improved at high temperature. In addition, the SCOD first increased and then decreased with increasing temperature, that is, the dewatering performance of the sludge was closely related to the soluble organic matter it contained. At a pressure of 24 MPa and under a constant amount of oxygen, the degradation rates of TSS and VSS increased from 74% and 85% to 85% and 92%, respectively, with a temperature increase from 220 °C to 240 °C, indicating an increase the temperature, thereby reducing the concentration of total/volatile suspended solids in the sludge [62]. High temperatures can promote the degradation of suspended solid SS [63]. Liang also confirmed that sludge dewatering performance is related to VSS% [32]. In addition to the reaction temperature, the oxygen content also has an influence on the sludge wet oxidation. The TSS/VSS value decreased with increasing oxygen content, indicating that changing experimental conditions affects the concentration of suspended matter. Therefore, to study the interaction between different influencing factors, the RSM method was used to analyze the effects of reaction temperature, oxygen dosage and reactor speed on the TSS, VSS and TCOD concentrations of the sludge at 5 and 60 min [64]. The three factors had a significant impact on the concentrations of TSS, VSS, and TCOD, and when the reaction reached 60 min, the rotation rate factor was $p > 0$. Throughout the course of the reaction, the reaction temperature is the main influencing factor. However, increasing the reaction temperature and oxygen dose promoted both the removal rate of TSS, VSS, and SS and the sludge yield.

A large number of studies have shown that WAO can be used directly in municipal sludge treatment, affecting sludge settling/dewatering performance, TSS and VSS content, TOC distribution, etc. Temperature and oxygen concentration have independent or synergistic effects on oxidation by affecting reaction kinetics or organic matter dissolution and gas-liquid mass transfer rates [65]. Organic carbon can be practically used as a denitrifying carbon source. Urrea studied the effects of reaction time, sludge characteristics and catalyst on TOC in the sludge supernatant, and found that the TOC concentration increased with the reaction time, and the TOC concentration was higher in the presence of the catalyst [61]. Since the supernatant contains a variety of organic carbons, especially acetic acid, Strong uses the supernatant as a denitrifying carbon source to obtain a good denitrification effect. However, WAO has the disadvantages of high energy consumption and high equipment investment. The WAO after adding the catalyst can speed up sludge conditioning and reduce the operating cost, and has a wide application background, but the catalyst of CWAO is heavy metal. CWAO faces challenges, such as catalyst separation and recovery and poisoning from by-products [66]. In the future, CWAO and WAO are planned be developed towards low cost, high stability, and easy recovery of toxic by-products.

7. Conclusions

This paper provides an overview of the application of various advanced oxidation technologies to improve sludge dewatering efficiency and demonstrates the applicability and efficiency of various advanced oxidation technologies to remove sludge contaminants. After treatment with Fenton's reagent and activated persulfate, sludge dewatering performance and sludge-water separation efficiency improved significantly. Fenton, Fenton-like, and Fenton coupling technologies can not only enhance sludge dewatering but also effectively remove micro-metals in sludge to achieve sludge detoxification. Compared to ozone and H_2O_2, activated persulfate oxidation is more efficient in sludge dewatering. It can achieve rapid sludge dewatering in 1 min and persulfate is easy to store. Moreover, the generated $\cdot SO_4^-$ has a wider range of applications. It can play a role in the pH range and allow the remaining sludge to be treated at different pH levels. In contrast, $\cdot OH$ is more beneficial in reducing SRF and sludge water content. It is worth noting that the oxidizing power of $\cdot SO_4^-$ is surpassed only to $\cdot OH$. At the same time, the external energy input significantly promoted the oxidation of free radicals and reduced the amount of the corresponding reagents that need to be added. The sludge treated with ozone had an improved sedimentation performance, which had an enhanced effect on further anaerobic fermentation. Both wet air oxidation and supercritical water oxidation were carried out at high temperature, high pressure and oxidant, and the organic matter in the sludge was directly oxidized into carbon dioxide, water, and intermediate products. The addition of permanganate and ferrate alone did not significantly improve the corresponding dehydration indices. Relevant catalytic means should be employed in these systems to exploit the potential level of free radical oxidation. There is a need to develop more conditioning and enhanced dewatering technologies for sludge dewatering, involving simple process, high efficiency, low consumption, environmentally friendly and low carbon. Therefore, optimizing the combination of advanced oxidation technology to achieve complementary advantages can maximize the advantages of advanced oxidation technology in sludge reduction applications.

Author Contributions: Conceptualization, J.X., T.R., J.J., Z.H., J.M., A.L. and Y.S.; methodology, J.X. and Y.S.; software, A.L. and Y.S.; validation, A.L. and Y.S.; formal analysis, A.L. and Y.S.; investigation, A.L. and Y.S.; resources, J.X., T.R. and Y.S.; data curation, A.L. and Y.S.; writing—original draft preparation, A.L. and Y.S.; writing—review and editing, A.L. and Y.S.; visualization, J.X., T.R., J.J., Z.H., J.M. and Y.S.; supervision, J.X., T.R., J.J. and Z.H.; project administration, J.X., T.R., J.J. and Z.H.; funding acquisition, J.X., T.R., J.J., Z.H. and J.M. All authors have read and agreed to the published version of the manuscript.

Funding: This work was supported by Natural Science Foundation of Jiangsu Province in China (No. BK20201362), and the 2018 Six Talent Peaks Project of Jiangsu Province (JNHB-038), University-enterprise joint development project (202112346).

Institutional Review Board Statement: Not applicable.

Informed Consent Statement: Not applicable.

Conflicts of Interest: The authors declare no conflict of interest. The company had no role in the design of the study; in the collection, analyses, or interpretation of data; in the writing of the manuscript, and in the decision to publish the results.

References

1. Lin, W.; Liu, X.; Ding, A.; Ngo, H.H.; Zhang, R.; Nan, J.; Ma, J.; Li, G. Advanced oxidation processes (AOPs)-based sludge conditioning for enhanced sludge dewatering and micropollutants removal: A critical review. *J. Water Process Eng.* **2022**, *45*, 102468. [CrossRef]
2. Toutian, V.; Barjenbruch, M.; Loderer, C.; Remy, C. Impact of process parameters of thermal alkaline pretreatment on biogas yield and dewaterability of waste activated sludge. *Water Res.* **2021**, *202*, 117465. [CrossRef]

3. Izydorczyk, G.; Mikula, K.; Skrzypczak, D.; Trzaska, K.; Moustakas, K.; Witek-Krowiak, A.; Chojnacka, K. Agricultural and non-agricultural directions of bio-based sewage sludge valorization by chemical conditioning. *Environ. Sci. Pollut. Res.* **2021**, *28*, 47725–47740. [CrossRef]
4. Jiang, X.; Li, Y.; Tang, X.; Jiang, J.; He, Q.; Xiong, Z.; Zheng, H. Biopolymer-based flocculants: A review of recent technologies. *Environ. Sci. Pollut. Res.* **2021**, *28*, 46934–46963. [CrossRef]
5. Hatinoğlu, M.D.; Sanin, F.D. Sewage sludge as a source of microplastics in the environment: A review of occurrence and fate during sludge treatment. *J. Environ. Manag.* **2021**, *295*, 113028. [CrossRef]
6. Liu, H.; Basar, I.A.; Nzihou, A.; Eskicioglu, C. Hydrochar derived from municipal sludge through hydrothermal processing: A critical review on its formation, characterization, and valorization. *Water Res.* **2021**, *199*, 117186. [CrossRef]
7. Nabaterega, R.; Kumar, V.; Khoei, S.; Eskicioglu, C. A review on two-stage anaerobic digestion options for optimizing municipal wastewater sludge treatment process. *J. Environ. Chem. Eng.* **2021**, *9*, 105502. [CrossRef]
8. Lu, J.; Xu, S. Post-treatment of food waste digestate towards land application: A review. *J. Clean. Prod.* **2021**, *303*, 127033. [CrossRef]
9. Cao, B.; Zhang, T.; Zhang, W.; Wang, D. Enhanced technology based for sewage sludge deep dewatering: A critical review. *Water Res.* **2021**, *189*, 116650. [CrossRef]
10. Langone, M.; Basso, D. Process waters from hydrothermal carbonization of sludge: Characteristics and possible valorization pathways. *Int. J. Environ. Res. Public Health* **2020**, *17*, 6618. [CrossRef]
11. Wu, B.; Dai, X.; Chai, X. Critical review on dewatering of sewage sludge: Influential mechanism, conditioning technologies and implications to sludge re-utilizations. *Water Res.* **2020**, *180*, 115912. [CrossRef] [PubMed]
12. Liu, D.; Edraki, M.; Fawell, P.; Berry, L. Improved water recovery: A review of clay-rich tailings and saline water interactions. *Powder Technol.* **2020**, *364*, 604–621. [CrossRef]
13. Brisolara, K.F.; Bourgeois, J. Biosolids and sludge management. *Water Environ. Res.* **2019**, *91*, 1168–1176. [CrossRef]
14. Ibrahim, A.B.A.; Akilli, H. Supercritical water gasification of wastewater sludge for hydrogen production. *Int. J. Hydrogen Energy* **2019**, *44*, 10328–10349. [CrossRef]
15. Brisolara, K.F.; Qi, Y.; Kendrick, Q.; Davis, Y. Biosolids and sludge management. *Water Environ. Res.* **2018**, *90*, 978–1006. [CrossRef] [PubMed]
16. Wen, Q.; Ma, M.; Hou, H.; Yu, W.; Gui, G.; Wu, Q.; Xiao, K.; Zhu, Y.; Tao, S.; Liang, S.; et al. Recirculation of reject water in deep-dewatering process to influent of wastewater treatment plant and dewaterability of sludge conditioned with Fe^{2+}/H_2O_2, $Fe^{2+}/Ca(ClO)_2$, and $Fe^{2+}/Na_2S_2O_8$: From bench to pilot-scale study. *Environ. Res.* **2022**, *203*, 111825. [CrossRef] [PubMed]
17. Tunçal, T.; Mujumdar, A.S. Modern techniques for sludge dewaterability improvement. *Dry. Technol.* **2022**. [CrossRef]
18. Xia, J.; Rao, T.; Ji, J.; He, B.; Liu, A.; Sun, Y. Enhanced dewatering of activated sludge by Skeleton-Assisted flocculation process. *Int. J. Environ. Res. Public Health* **2022**, *19*, 6540. [CrossRef]
19. Deng, P.; Liu, C.; Wang, M.; Lan, G.; Zhong, Y.; Wu, Y.; Fu, C.; Shi, H.; Zhu, R.; Zhou, L. Effect of dewatering conditioners on phosphorus removal efficiency of sludge biochar. *Environ. Technol.* **2022**. [CrossRef] [PubMed]
20. Feng, J.; Zhang, T.; Sun, J.; Zhu, J.; Yan, W.; Tian, S.; Xiong, Y. Improvement of sewage sludge dewatering by piezoelectric effect driven directly with pressure from pressure filtration: Towards understanding piezo-dewatering mechanism. *Water Res.* **2022**, *209*, 117922. [CrossRef]
21. Wang, L.; Shi, Y.; Chai, J.; Huang, L.; Wang, Y.; Wang, S.; Pi, K.; Gerson, A.R.; Liu, D. Transfer of microplastics in sludge upon Fe(II)-persulfate conditioning and mechanical dewatering. *Sci. Total Environ.* **2022**, *838*, 156316. [CrossRef]
22. Li, Y.; Liu, L.; Li, X.; Xie, J.; Guan, M.; Wang, E.; Lu, D.; Dong, T.; Zhang, X. Influence of alternating electric field on deep dewatering of municipal sludge and changes of extracellular polymeric substance during dewatering. *Sci. Total Environ.* **2022**, *842*, 156839. [CrossRef] [PubMed]
23. Guan, R.; Yuan, X.; Wu, Z.; Jiang, L.; Li, Y.; Zeng, G. Principle and application of hydrogen peroxide based advanced oxidation processes in activated sludge treatment: A review. *Chem. Eng. J.* **2018**, *339*, 519–530. [CrossRef]
24. Liang, J.; Zhou, Y. Iron-based advanced oxidation processes for enhancing sludge dewaterability: State of the art, challenges, and sludge reuse. *Water Res.* **2022**, *218*, 118499. [CrossRef]
25. Barati Rashvanlou, R.; Pasalari, H.; Moserzadeh, A.A.; Farzadkia, M. A combined ultrasonic and chemical conditioning process for upgrading the sludge dewaterability. *Int. J. Environ. Anal. Chem.* **2022**, *102*, 1613–1626. [CrossRef]
26. Liu, H.; Yang, J.; Shi, Y.; Li, Y.; He, S.; Yang, C.; Yao, H. Conditioning of sewage sludge by Fenton's reagent combined with skeleton builders. *Chemosphere* **2012**, *88*, 235–239. [CrossRef]
27. Yu, W.; Yang, J.; Shi, Y.; Song, J.; Shi, Y.; Xiao, J.; Li, C.; Xu, X.; He, S.; Liang, S.; et al. Roles of iron species and pH optimization on sewage sludge conditioning with Fenton's reagent and lime. *Water Res.* **2016**, *95*, 124–133. [CrossRef] [PubMed]
28. Tao, S.; Yang, J.; Hou, H.; Liang, S.; Xiao, K.; Qiu, J.; Hu, J.; Liu, B.; Yu, W.; Deng, H. Enhanced sludge dewatering via homogeneous and heterogeneous Fenton reactions initiated by Fe-rich biochar derived from sludge. *Chem. Eng. J.* **2019**, *372*, 977–996. [CrossRef]
29. Zeng, X.; Twardowska, I.; Wei, S.; Sun, L.; Wang, J.; Zhu, J.; Cai, J. Removal of trace metals and improvement of dredged sediment dewaterability by bioleaching combined with Fenton-like reaction. *J. Hazard. Mater.* **2015**, *288*, 51–59. [CrossRef]
30. Li, Y.; Zhang, A. Removal of steroid estrogens from waste activated sludge using Fenton oxidation: Influencing factors and degradation intermediates. *Chemosphere* **2014**, *105*, 24–30. [CrossRef]

31. He, D.; Sun, M.; Bao, B.; Chen, J.; Luo, H.; Li, J. Rethinking the timing of flocculant addition during activated sludge dewatering. *J. Water Process Eng.* **2022**, *47*, 102744. [CrossRef]
32. Liang, J.; Huang, S.; Dai, Y.; Li, L.; Sun, S. Dewaterability of five sewage sludges in Guangzhou conditioned with Fenton's reagent/lime and pilot-scale experiments using ultrahigh pressure filtration system. *Water Res.* **2015**, *84*, 243–254. [CrossRef] [PubMed]
33. Zhang, H.; Yang, J.; Yu, W.; Luo, S.; Peng, L.; Shen, X.; Shi, Y.; Zhang, S.; Song, J.; Ye, N.; et al. Mechanism of red mud combined with Fenton's reagent in sewage sludge conditioning. *Water Res.* **2014**, *59*, 239–247. [CrossRef]
34. Song, K.; Zhou, X.; Liu, Y.; Gong, Y.; Zhou, B.; Wang, D.; Wang, Q. Role of oxidants in enhancing dewaterability of anaerobically digested sludge through Fe (II) activated oxidation processes: Hydrogen peroxide versus persulfate. *Sci. Rep.* **2016**, *6*, 24800. [CrossRef]
35. Hu, L.; Wang, P.; Shen, T.; Wang, Q.; Wang, X.; Xu, P.; Zheng, Q.; Zhang, G. The application of microwaves in sulfate radical-based advanced oxidation processes for environmental remediation: A review. *Sci. Total Environ.* **2020**, *722*, 137831. [CrossRef] [PubMed]
36. Karadoğan, Ü.; Çevikbilen, G.; Korkut, S.; Pasaoglu, M.E.; Teymur, B. Dewatering of Golden Horn sludge with geotextile tube and determination of optimum operating conditions: A novel approach. *Mar. Georesources Geotechnol.* **2022**, *40*, 782–794. [CrossRef]
37. Sha, L.; Wu, Z.; Ling, Z.; Liu, X.; Yu, X.; Zhang, S. Investigation on the improvement of activated sludge dewaterability using different iron forms (ZVI vs. Fe(II))/peroxydisulfate combined vertical electro-dewatering processes. *Chemosphere* **2022**, *292*, 133416. [CrossRef]
38. Zhen, G.; Lu, X.; Su, L.; Kobayashi, T.; Kumar, G.; Zhou, T.; Xu, K.; Li, Y.; Zhu, X.; Zhao, Y. Unraveling the catalyzing behaviors of different iron species (Fe^{2+} vs. FeO) in activating persulfate-based oxidation process with implications to waste activated sludge dewaterability. *Water Res.* **2018**, *134*, 101–114. [CrossRef]
39. Zhou, X.; Wang, Q.; Jiang, G.; Liu, P.; Yuan, Z. A novel conditioning process for enhancing dewaterability of waste activated sludge by combination of zero-valent iron and persulfate. *Bioresour. Technol.* **2015**, *185*, 416–420. [CrossRef]
40. Song, K.; Zhou, X.; Liu, Y.; Xie, G.; Wang, D.; Zhang, T.; Liu, C.; Liu, P.; Zhou, B.; Wang, Q. Improving dewaterability of anaerobically digested sludge by combination of persulfate and zero valent iron. *Chem. Eng. J.* **2016**, *295*, 436–442. [CrossRef]
41. Liu, M.; Yuan, C.; Ru, S.; Li, J.; Lei, Z.; Zhang, Z.; Shimizu, K.; Yuan, T.; Li, F. Combined organic reagents for co-conditioning of sewage sludge: High performance in deep dewatering and significant contribution to the floc property. *J. Water Process Eng.* **2022**, *48*, 102855. [CrossRef]
42. Fan, X.; Wang, Y.; Zhang, D.; Guo, Y.; Gao, S.; Li, E.; Zheng, H. Effects of acid, acid-ZVI/PMS, Fe(II)/PMS and ZVI/PMS conditioning on the wastewater activated sludge (WAS) dewaterability and extracellular polymeric substances (EPS). *J. Environ. Sci.* **2020**, *91*, 73–84. [CrossRef]
43. Zhou, X.; Jin, W.; Wang, L.; Che, L.; Chen, C.; Li, S.; Wang, X.; Tu, R.; Han, S.; Feng, X.; et al. Alum sludge conditioning with ferrous iron/peroxymonosulfate oxidation: Characterization and mechanism. *Korean J. Chem. Eng.* **2020**, *37*, 663–669. [CrossRef]
44. Chacana, J.; Labelle, M.; Laporte, A.; Gadbois, A.; Barbeau, B.; Comeau, Y. Ozonation of Primary Sludge and Digested Sludge to Increase Methane Production in a Chemically Enhanced Primary Treatment Facility. *Ozone Sci. Eng.* **2017**, *39*, 148–158. [CrossRef]
45. Lin, N.; Zhu, W.; You, X.; Wang, X.; Zhong, J.; Sun, J.; Sun, T.; Cao, J.; Li, L. Verification of the mechanism and effect of secondary advanced dewatering promoted by selective oxidative decomposition: On pilot scale. *J. Environ. Chem. Eng.* **2022**, *10*, 107217. [CrossRef]
46. Zhang, J.; Zhang, J.; Tian, Y.; Li, N.; Kong, L.; Sun, L.; Yu, M.; Zuo, W. Changes of physicochemical properties of sewage sludge during ozonation treatment: Correlation to sludge dewaterability. *Chem. Eng. J.* **2016**, *301*, 238–248. [CrossRef]
47. Li, L.; Peng, C.; Deng, L.; Zhang, F.; Wu, D.; Ma, F.; Liu, Y. Understanding the synergistic mechanism of PAM-FeCl3 for improved sludge dewaterability. *J. Environ. Manag.* **2022**, *301*, 113926. [CrossRef]
48. Srinivasan, A.; Young, C.; Liao, P.H.; Lo, K.V. Radiofrequency-oxidation treatment of sewage sludge. *Chemosphere* **2015**, *141*, 212–218. [CrossRef]
49. Yu, H.; Gu, L.; Zhang, D.; Wen, H.; Wang, M.; Zhu, N. Enhancement of sludge dewaterability by three-dimensional electrolysis with sludge-based particle electrodes. *Sep. Purif. Technol.* **2022**, *287*, 120599. [CrossRef]
50. Ge, D.; Bian, C.; Yuan, H.; Zhu, N. An in-depth study on the deep-dewatering mechanism of waste activated sludge by ozonation pre-oxidation and chitosan re-flocculation conditioning. *Sci. Total Environ.* **2020**, *714*, 136627. [CrossRef]
51. Zeng, R.; Li, Y.; Sha, L.; Liu, X.; Zhang, S. Electro-dewatering of steel industrial sludge: Performance and metal speciation. *J. Water Process Eng.* **2022**, *46*, 102600. [CrossRef]
52. Bień, B.; Bień, J.D. Analysis of reject water formed in the mechanical dewatering process of digested sludge conditioned by physical and chemical methods. *Energies* **2022**, *15*, 1678. [CrossRef]
53. Liu, Y.F.; Dong, L.M.; Cui, G.N.; Hu, X.Y.; Yu, Z.Q.; Liang, S. Low-field nuclear magnetic resonance detection of changes in the water distribution in citric acid biosludge during dewatering. *Int. J. Environ. Sci. Technol.* **2022**, *19*, 4153–4166. [CrossRef]
54. Liang, J.; Zhang, S.; Ye, M.; Huang, J.; Yang, X.; Li, S.; Huang, S.; Sun, S. Improving sewage sludge dewaterability with rapid and cost-effective in-situ generation of Fe2+ combined with oxidants. *Chem. Eng. J.* **2020**, *380*, 122499. [CrossRef]
55. Yuan, L.; Liu, H.; Lu, Y.; Lu, Y.; Wang, D. Enhancing the dewaterability of waste activated sludge by the combined ascorbic acid and zero-valent iron/persulfate system. *Chemosphere* **2022**, *303*, 135104. [CrossRef]

56. Rai, P.K.; Lee, J.; Kailasa, S.K.; Kwon, E.E.; Tsang, Y.F.; Ok, Y.S.; Kim, K.H. A critical review of ferrate(VI)-based remediation of soil and groundwater. *Environ. Res.* **2018**, *160*, 420–448. [CrossRef]
57. Han, Q.; Dong, W.; Wang, H.; Liu, T.; Tian, Y.; Song, X. Degradation of tetrabromobisphenol A by ferrate(VI) oxidation: Performance, inorganic and organic products, pathway and toxicity control. *Chemosphere* **2018**, *198*, 92–102. [CrossRef]
58. Wu, S.; Liu, H.; Lin, Y.; Yang, C.; Lou, W.; Sun, J.; Du, C.; Zhang, D.; Nie, L.; Yin, K.; et al. Insights into mechanisms of UV/ferrate oxidation for degradation of phenolic pollutants: Role of superoxide radicals. *Chemosphere* **2020**, *244*, 125490. [CrossRef] [PubMed]
59. Ai, J.; Zhang, W.; Chen, F.; Liao, G.; Li, D.; Xia, H.; Wang, D.; Ma, T. Catalytic pyrolysis coupling to enhanced dewatering of waste activated sludge using KMnO4-Fe(II) conditioning for preparing multi-functional material to treat groundwater containing combined pollutants. *Water Res.* **2019**, *158*, 424–437. [CrossRef] [PubMed]
60. Zhang, J.; Qi, Y.; Zhang, X.; Zhang, G.; Yang, H.; Nattabi, F. Experimental investigation of sludge dewatering for single- and double-drainage conditions with a vacuum negative pressure load at the bottom. *PLoS ONE* **2021**, *16*, e253806. [CrossRef]
61. Urrea, J.L.; Collado, S.; Oulego, P.; Díaz, M. Formation and Degradation of Soluble Biopolymers during Wet Oxidation of Sludge. *ACS Sustain. Chem. Eng.* **2017**, *5*, 3011–3018. [CrossRef]
62. Zhang, X.; Tong, G.; Zhou, Y.; Li, G.; Zhang, H. Enhancing paper sludge dewatering by waste polyester fiber and FeCl3 for preparation of Fe-rich biochar. *BioResources* **2021**, *16*, 2326–2345. [CrossRef]
63. Xiao, Y.; Lu, Y.; Zheng, G.; Zhou, L. Impact of initial sludge pH on enhancing the dewaterability of waste activated sludge by zero-valent iron-activated peroxydisulphate. *Environ. Technol.* **2021**, *42*, 2573–2586. [CrossRef] [PubMed]
64. Yang, G.; Zhang, G.; Wang, H. Current state of sludge production, management, treatment and disposal in China. *Water Res.* **2015**, *78*, 60–73. [CrossRef] [PubMed]
65. Bień, B.; Bień, J.D. Conditioning of sewage sludge with physical, chemical and dual methods to improve sewage sludge dewatering. *Energies* **2021**, *14*, 5079. [CrossRef]
66. Liu, C.; Zhou, X.; Zhou, L.; Wei, Y.; Liu, J. Enhancement of sludge electro-dewatering by anthracite powder modification. *Environ. Res.* **2021**, *201*, 111510. [CrossRef]

International Journal of
Environmental Research and Public Health

MDPI

Article

Enhanced Dewatering of Activated Sludge by Skeleton-Assisted Flocculation Process

Jiahua Xia [1], Ting Rao [1], Juan Ji [1], Bijuan He [1], Ankang Liu [2] and Yongjun Sun [3,*]

1 Nanjing Jiangbei New Area Public Utilities Holding Group Co., Ltd., Nanjing 210044, China; xiayu_610@sina.com (J.X.); raot@mail.ustc.edu.cn (T.R.); jj13814186775@163.com (J.J.); he13770509319@163.com (B.H.)
2 Nanjing Water Purification Environmental Research Institute Co., Ltd., Nanjing 211100, China; lakyx2@126.com
3 College of Urban Construction, Nanjing Tech University, Nanjing 211800, China
* Correspondence: sunyongjun@njtech.edu.cn

Abstract: Sludge dewatering is the fundamental process of sludge treatment. Environmentally friendly and efficient sludge conditioning methods are the premises of sludge to achieve dehydration reduction and resource utilization. In response to sewage plant sludge dehydration, fly ash (FA), polymerized aluminum chloride (PAC), and polymer sulfate (PFS) were studied separately to determine their sludge dehydration performance, and the effects of these three conditioner composite regulations on sludge dehydration properties were studied. Compared to the sludge treated only with conditioner, the average particle size of floc increased and the organic matter content in the filtrate decreased. The sludge dewatering efficiency after the conditioning effect is better than that after conditioning a single conditioner. After PFS conditioning with fly ash, the water content and specific resistance (SRF) of the sludge cake can be reduced to 76.39% and 6.63×10^{10} m/kg, respectively. The moisture content and specific resistance (SRF) of the sludge cake can be reduced to 76.10% and 6.91×10^{10} m/kg, respectively. The pH of the sludge and filtrate changed slightly after PAC conditioning with fly ash coupling. These results indicate that fly-ash coupled with PAC and fly-ash coupled with PFS are expected to become a novel and effective environmental protection combined conditioning method for sludge dewatering.

Keywords: sludge conditioning; sludge dewatering; conditioner; fly ash; flocculation

Citation: Xia, J.; Rao, T.; Ji, J.; He, B.; Liu, A.; Sun, Y. Enhanced Dewatering of Activated Sludge by Skeleton-Assisted Flocculation Process. *Int. J. Environ. Res. Public Health* **2022**, *19*, 6540. https://doi.org/10.3390/ijerph19116540

Academic Editors: Yung-Tse Hung, Hamidi Abdul Aziz, Issam A. Al-Khatib and Paul B. Tchounwou

Received: 26 April 2022
Accepted: 24 May 2022
Published: 27 May 2022

1. Introduction

Great things were achieved in the sewage treatment industry with the completion and operation of a large number of municipal sewage treatment plants [1]. The proliferation of wastewater treatment has led to the production of a large amount of activated sludge, which has gradually increased the load on sewage treatment plants, and wastewater treatment has become a key challenge for global water management [2]. However, the wastewater treatment industry has long been "heavy water and light sludge", resulting in a serious lag of the water treatment industry in terms of sludge treatment level, technology and capital investment [3]. At present, the sludge generated by a large number of wastewater treatments cannot be effectively treated and disposed of, and the persistent heavy metals and persistent organic pollutants in the sludge will pose a great threat and challenge to the natural environment, such as atmosphere, water, and soil [4]. Among other things, the high moisture content of sludge becomes the bottleneck in sludge treatment and disposal, and the high moisture content and low bulk density of sludge pose problems in storage, transportation, and economic costs [5]. Therefore, the sludge must be dewatered immediately. Mechanical dewatering is widely used in wastewater treatment plants, but the efficiency of mechanical dewatering of sludge is limited due to the hydrophilicity of extracellular polymers [6]. Therefore, before mechanical dewatering, the sludge needs to be pre-treated to reduce the moisture content of the sludge and improve the sludge dehydration efficiency.

The commonly used sludge conditioning methods include physical conditioning, chemical conditioning, and biological conditioning. At present, physical conditioning is a sludge conditioning method that destroys the microbial cells in the sludge by physical methods, changes the structure of the sludge, and reduces the binding effect of the sludge and water, thereby releasing the internal water [7]. Although physical conditioning can reduce the moisture content of the sludge, each has its own advantages and disadvantages due to its different mode of action. Among them, thermal conditioning has obvious effects on improving the dewatering rate of most of the sludges, but the treatment effect on low-organic sludges is small [8]. Although the microwave treatment time is short, the heating temperature is low, the thermal efficiency is high, and the equipment is simple, the microwave treatment has requirements on the amount of sludge entering the sludge. Freeze conditioning can be carried out spontaneously, without chemical additives and inexpensively, but is limited in its scope. Ultrasonic conditioning has high efficiency and environmental protection and application prospects, but its working principle and complex effects need to be further discussed [9].

Overall, the high energy consumption and the high equipment costs of the physical conditioning restrict its large-scale use. Microbial conditioning promotes mutual coagulation and precipitation of colloidal suspensions in water and improves sludge dewatering performance. Microbial conditioning has powerful benefits such as: high efficiency, non-toxicity, no secondary pollution, biodegradation, compact sludge flakes and a wide range of applications [10]. However, the current research on such flocculants is not comprehensive, the research level is low, the mechanism of flocculants is not fully understood, and problems such as large dosage, high production cost, and poor suitability arise.

Chemical conditioning is widely used as the optimal choice among broad comparisons. In general, chemical conditioning is mainly divided into inorganic and organic polymer conditioning [11]. At present, the widely used inorganic chemical conditioning agents mainly include polyferric sulfate (PFS), polyaluminum chloride (PAC), polyaluminum ferric chloride (PAFC), and polyaluminum silicate (PASiC). Organic conditioners have different electrical properties. Cationic and amphoteric are more commonly used as flakes are often negatively charged [12]. The use of inorganic conditioning agents for chemical conditioning has the advantages of easy availability of raw materials, easy manufacture and low cost, and it is very effective in removing fine suspended particles in sludge [13]. These advantages greatly increase the amount of sludge, reduce the fertilization efficiency and calorific value of the sludge, and also bring heavy metals into the sludge, posing a threat to the environment. Organic polymer conditioners have many types, each with its own advantages [14]. However, the organic flocculants have the disadvantage that they are difficult to degrade and the residual monomers are toxic. There are certain disadvantages due to physical, chemical, and microbial conditioning alone, and combined conditioning technology has become a research focus in recent years [15]. Combined conditioning can reduce the compressibility, the amount of chemicals added, and the cost of conditioning, and improve the dewatering performance of the sludge [16]. The research and development of efficient, low-consumption, safe and stable flocculants is the direction and goal of flocculant research in view of the currently increasing water pollution and ever stricter environmental regulations.

This study was used for sludge pretreatment with PAC, PFS, and fly ash. Sludge ratio resistance SRF was used as an indicator of sludge dewatering properties with sludge. Filtrate pH, sludge pH, filtrate COD, sludge SEM, filtrate UV spectrum, and floc particle size were the most intuitive evaluation indices of dewatering performance. By examining the influence of different chemical conditioning agents on different indicators, the main influencing factors of the three aforementioned conditioners are optimized, the sludge conditioning mechanism of different conditioners is discussed, and the sewage treatment plant is measured in different ways. The effect of sludge dewatering performance is enhanced to provide a reference for sludge pretreatment technology of sewage treatment plants.

2. Materials and Methods

2.1. Materials

Polymeric ferric sulfate, ammonium molybdate, silver sulfate, ferric nitrate, cerium nitrate, cobalt nitrate, potassium sodium tartrate, and ascorbic acid were all of analytical grade and were purchased from Sinopharm Chemical Reagent Co., Ltd. (Shanghai, China). Mercury sulphate and potassium antimony tartrate were all of analytical grade and were purchased from Shanghai McLean Biochemical Co., Ltd. (Shanghai, China). The fly ash was taken from a coal chemical enterprise in Jiangsu. Cationic polyacrylamide (CPAM) was purchased from Jiangyin Huadong Water Treatment Co., Ltd. (Wuxi, China). The actual sludge water sample was taken from a sewage treatment plant in Jiangsu, and the original sludge quality is shown in Table 1.

Table 1. Original sludge quality index.

Sludge Index	Moisture Content of Raw Slduge (%)	Organic Matter Content (%)	SRF (m/kg)	pH Value
Numerical value	98.90	46.30	2.97×10^{12}	7.51

2.2. Sludge Conditioning and Dehydration Experiments

Sludge conditioning experiments were carried out with fly ash, PAC, and PFS. For different coagulants, six groups of samples of 500 mL each were taken each time, and different concentrations of coagulants were added into them. DS is the abbreviation for Dry Solid Content. According to the pre-experiment, different fly ash dosages (e.g., 5, 10,15, 20, 25, 30 g/g DS) were selected. Different dosages of PAC (e.g., 50, 100, 150, 200, 250, 300 mg/g DS) and PFS (e.g., 50, 100, 150, 200, 250, 300 mg/g DS) were used to improve the sludge dewatering performance. After addition of the agent, we spun at 200 rpm for 30 s, then spun at 60 rpm for 3 min and let stand for 30 min before dehydration. After selecting the optimal fly ash dosage, the above steps were repeated for fly ash combined with PAC and fly ash combined with PFS.

Sludge dewatering adopts vacuum filtration. First, the wet filter paper is spread out in the cloth funnel. After removing the air bubbles, the dosed sludge was injected into the cloth hopper and the vacuum filtration pressure is maintained at 0.05 MPa. It was filtered for 2 min and the duration of the suction and the filtrate volume suctioned off during the corresponding time were recorded. After dosing and stirring, the pH of the sludge was measured. After the sludge was dewatered, the filtrate was taken out, and the pH of the filtrate and the COD of the filtrate, and the UV spectrum of the filtrate were measured. The SRF is calculated from the measurements taken during the dewatering of the sludge. After drying the sludge cake, moisture content, SEM, fractal dimension, and micrograph of the filter cake were measured.

2.3. Sludge Index Analysis Method

COD was measured with a Hach analyzer (COD analyzer, DR1010, Hach Company, Loveland, CO, USA); the particle size analysis of the sludge cake was determined with a laser particle size analyzer (laser particle size analyzer, Winner2018, Jinan Micro-Nano Particle Instrument Co., Ltd., Jinan, China); and the UV of the filtrate was measured with a UV/visible spectrophotometer (UV/visible spectrophotometer, UV-5500PC, Shanghai Yuanyan Instrument Co., Ltd., Shanghai, China) in the entire UV band (GB 6920-86).

3. Results

3.1. Effect of PFS Dosage on Sludge Dewatering Performance

As shown in Figure 1, the SRF of the sludge first decreased and then increased with increasing PFS dosage. When the PFS dosage is 200 mg/g DS, the lowest sludge specific resistance SRF is 2.10×10^{11} m/kg. The change trend of the moisture content of the filter cake is the same as that of the sludge specific resistance SRF. At the PFS dosage of 200 mg/g DS, the minimum moisture content of the filter cake was 86.52%. With increasing PFS dosage, the COD of the filtrate first decreased and then increased. When the PFS dosage is 200 mg/g DS, the minimum COD content of the filtrate is 110 mg/L. The pH of the filtrate decreased rapidly with the increasing in PFS dosage and then stabilized. When the PFS dosage was 300 mg/g DS, the pH of the filtrate was at least 3.50. The pH of the sludge decreased gradually with the increasing in PFS dosage. When the dosage of the PFS was 300 mg/g DS, the lowest sludge pH was 3.62.

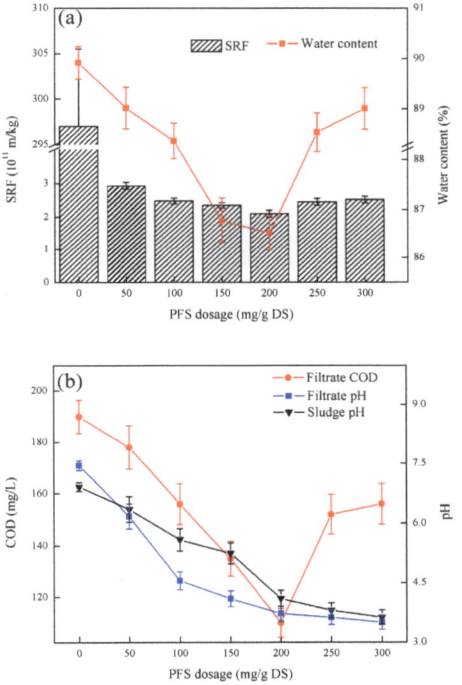

Figure 1. Effect of PFS dosage on sludge dewatering performance: (**a**) PFS dosage on SRF and filter cake moisture content, (**b**) PFS dosage on filtrate COD, filtrate pH and sludge pH.

Figure 1 shows that PFS can improve the filtration performance of the sludge. However, when the PFS dosage reached the optimum, the sludge specific resistance SRF did not improve significantly. When the PFS dosage exceeds 200 mg/g DS, the sludge resistivity SRF obviously increases, showing the inhibitory effect of sludge dewatering performance [17]. The reason is that when the PFS concentration is too high, the long-chain molecules may adhere to each other and cannot be stretched effectively, thereby reducing the total amount of adsorbed sludge colloidal particles and the flocculation efficiency. The tendency of the moisture content of the sludge cake to change was similar to the sludge specific resistivity SRF, which further indicated that excess PFS had an inhibitory effect on the sludge dewatering performance [18]. The COD of the filtrate first decreased and then increased because Fe^{3+} dissolved in water and neutralized the negative charge on the sludge surface, hence causing the sludge flocs to aggregate into large particles to settle and form hydroxide precipitates. The organic matter in the sludge remains in the sludge cake. When the PFS dosage is too high, the long chains of PFS will combine with each other, and the organic matter cannot be efficiently flocculated into large particles and becomes trapped in the sludge cake. The pH of the sludge and filtrate decreased because PFS was essentially an intermediate product of polynuclear hydroxyl complexes, and the mechanism of action was close to that of organic polymers. The adsorption of PFS and particulate matter is actually a surface complex coordination, and the surface hydroxyl groups will supplement unsaturated sites [19]. After adsorbing the particles on the surface of PFS, they will still absorb hydroxyl groups from the solution and continue the hydrolysis and precipitation process until they become saturated and become a hydroxide precipitation gel. Thus, the pH of the sludge and filtrate are acidic. After considering the three indicators of SRF, moisture content of the sludge cake, and COD in the filtrate, the optimal PFS dosage is 200 mg/g DS.

3.2. Effect of PAC Dosage on Sludge Dewatering Performance

As shown in Figure 2, with increasing the PAC dosage, the SRF of the sludge first decreased and then increased. When the PAC dosage was 200 mg/g DS, the lowest SRF of sludge specific resistance was 2.40×10^{11} m/kg. The change trend of the moisture content of the filter cake is the same as that of the sludge specific resistance SRF. When the PAC dosage was 200 mg/g DS, the minimum moisture content of the filter cake was 86.00%. With the increase in PAC dosage, the COD of the filtrate first decreased and then increased. When the PAC dosage was 200 mg/g DS, the minimum COD content of the filtrate was 67 mg/L. The pH of the filtrate decreased slowly with the increase in PAC dosage. When the PAC dosage was 300 mg/g DS, the pH of the filtrate was at least 6.05. The pH of the sludge decreased gradually with the increase in PAC dosage. When the PFS dosage was 300 mg/g DS, the lowest sludge pH was 6.66.

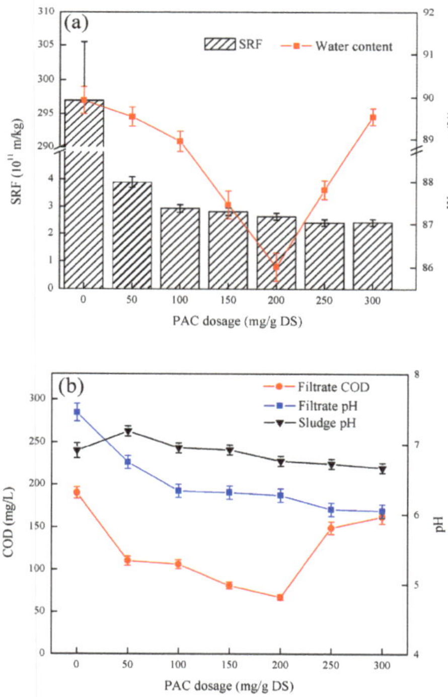

Figure 2. Effect of PAC dosage on sludge dewatering performance: (**a**) the effect of PAC dosage on SRF and filter cake moisture content, (**b**) the effect of PAC dosage on filtrate COD, filtrate pH and sludge pH.

As shown in Figure 2, the addition of PAC can greatly reduce the specific resistivity SRF of the sludge, so that the sludge has good dewatering performance. When the sludge specific resistance SRF was reduced to 2.40×10^{11} m/kg, and the PAC dosage further increased, the sludge dewatering performance did not improve significantly, and the sludge specific resistivity had a gradually increasing tendency. Similar to the addition of excess PFS, excess PAC also exhibited an inhibitory effect on sludge dewatering performance because when the dosage is too high, the surface of the colloid may be covered by the inorganic PAC macromolecule [20]. When the two colloidal particles are close to each other, they cannot aggregate due to the mutual repulsion between the PAC macromolecules, thereby resulting in a "colloidal protective" effect. The flocculation effect is reduced or even restabilized. Therefore, part of the organic matter in the sludge cannot be retained in the sludge cake, hence increasing the COD of the filtrate when the amount of PAC is too high. The change trend of the moisture content of the filter cake is similar to that of the SRF, which further indicates that an adequate amount of PAC can promote the dewatering performance of the sludge [21]. The hydrolysis intermediate of PAC has a higher positive charge, and due to the stronger electrical neutralization and pollutants, PAC has less effect on the filtrate pH and sludge pH than PFS. Compared with PFS, the specific resistance of sludge conditioned by PAC was slightly higher than that of 2.10×10^{11} m/kg after conditioning by PFS because cell lysis helps to release intracellular water under acidic conditions, thereby further improving the sludge dewatering performance. After considering the three indicators of SRF, sludge cake moisture content and filtrate COD, the optimal PAC dosage was 200 mg/g DS.

3.3. *Effect of Fly Ash Dosage on Sludge Dewatering Performance*

As shown in Figure 3, as the fly ash dosage increased, the SRF of the sludge first decreased and then increased. When the fly ash dosage is 10 g/g DS, the minimum SRF is 1.03×10^{11} m/kg. The moisture content of the filter cake gradually decreased with increasing fly ash dosing. When the fly ash dosage is 30 g/g DS, the minimum moisture content of the filter cake is 66.95%. The COD of the filtrate increased gradually with the addition of fly ash. When the fly ash dosage was 5 g/g DS, the COD content of the filtrate was at least 9 mg/L. The pH of the filtrate gradually increased with the fly ash dosing. When the fly ash dosage was 30 g/g DS, the pH of the filtrate was at its highest at 7.91. With increases in fly ash dosage, the pH of the sludge slowly increased. When the fly ash dosage was 30 g/g DS, the sludge pH was up to 7.60.

Figure 3 shows that fly ash can improve the dewatering performance of sludge. The raw sludge flocs are loose and few internal filtration channels are observed, and separating the water from the sludge is difficult, thereby making the filtration speed very slow. Forming a porous structure within the fly ash particles is easy. Due to its small particle size and strong surface adsorption capacity, a charge neutralization effect occurs after mixing with the sludge. When the surface tension is larger than the adsorption force of the sludge on the adsorbed water, the sludge adsorbed water can be separated. In addition, after mixing the fly ash with the sludge, it can absorb the sludge flocs and form a skeleton structure with its particles as the center, which reduces the compressibility of the sludge, builds a water filter channel for the sludge cake, and reduces the sludge viscosity, which significantly reduces the sludge dewatering time [22]. If the amount of fly ash is too large, the fly ash will accumulate itself, clogging the water filter channel and increasing the sludge resistivity. The trend in moisture content of fly ash and PAC and PFS conditioned sludge cakes is different, showing that the moisture content of conditioned sludge cakes continues to decrease. The reason is that the fly ash dosage is significantly increased compared to PAC and PFS, which significantly increases the total solid content in the sludge [23]. When the amount of fly ash reaches a certain level, the decrease in the moisture content of the sludge is caused only by the increase in the solid content and has nothing to do with the properties of the fly ash itself. Therefore, under high dosage, the moisture content of the filter cake shows a rapidly declining trend. The COD of the filtrate continues to increase because fly ash is a tiny ash particle discharged during the combustion of fuel, which contains unburned carbon, calcium silicate, and anorthite [24]. When the fly ash is added to the sludge, hydrolysis takes place, increasing the COD of the filtrate and making the solution weakly alkaline. Therefore, the pH of the sludge and filtrate slowly increases. With full consideration of economic benefits and dehydration performance, the fly ash was tested at 10g/g TS by coupling PFS and PAC.

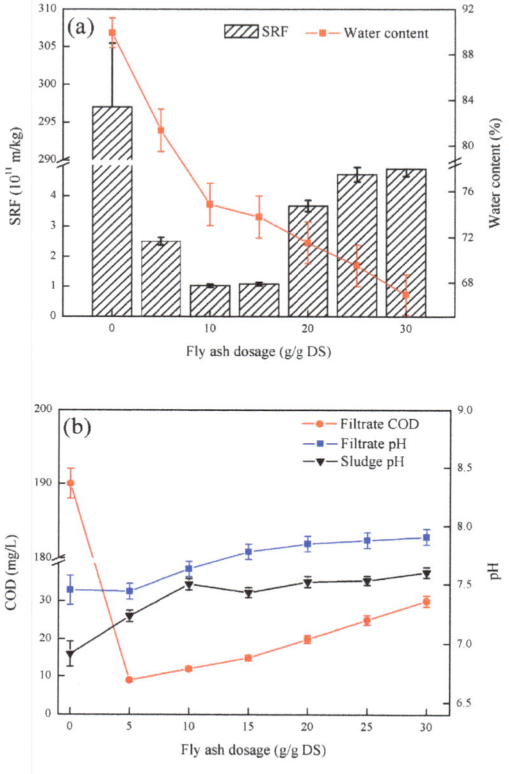

Figure 3. Effect of fly ash dosage on sludge dewatering performance: (**a**) the effect of fly ash dosage on SRF and filter cake moisture content, (**b**) fly ash dosage on filtrate COD and filtrate pH and sludge pH.

3.4. Effect of Fly Ash Coupled with PFS on Sludge Dewatering Performance

As shown in Figure 4, the sludge SRF first decreased and then increased with increasing PFS dosage. When the PFS dosage was 200 mg/g DS, the lowest specific resistance of sludge was 6.63×10^{10} m/kg. The moisture content of the filter cake first decreased and then increased again with increasing in PFS dosage. When the PFS dosage was 200 mg/g DS, the minimum moisture content of the filter cake was 76.39%. The COD of the filtrate first decreased and then increased with the PFS dosage. When the PFS dosage was 200 mg/g DS, the minimum COD content of the filtrate was 37 mg/L. The pH of the filtrate gradually decreased with increasing PFS dosage. When the PFS dosage was 300 mg/g DS, the pH of the filtrate was at least 5.51. With the increase in PFS dosage, the pH of the sludge gradually decreased and tended to be flat.

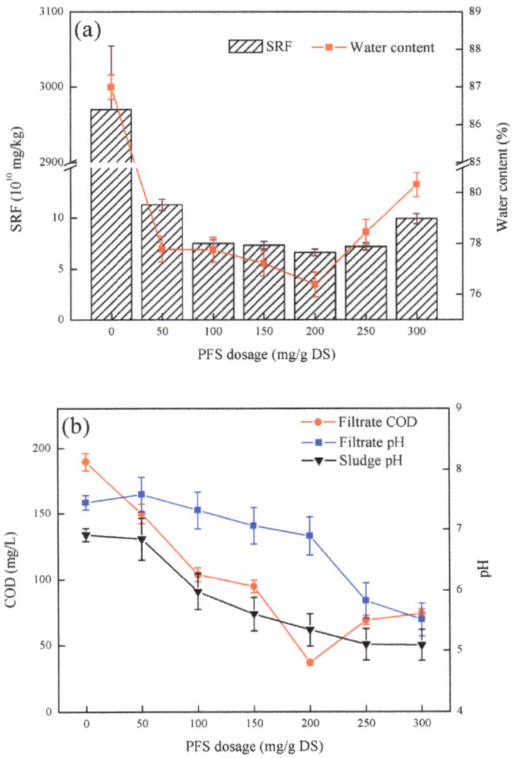

Figure 4. Effect of fly ash coupled with PFS on sludge dewatering performance: (**a**) the effect of PFS dosage on SRF and filter cake moisture content, (**b**) the effect of PFS dosage on filtrate COD, filtrate pH, and sludge pH Influence.

Figure 4 shows that fly ash coupled with PFS can significantly improve sludge filtration performance. Under the same conditions, fly ash coupled with PFS can improve the sludge dewatering effect. Fly ash provides an unobstructed water filtration channel and allows the PFS to play a better role by changing the microstructure of the sludge and its own adsorption performance [25]. When the PFS dosage is too high, the long-chain structure of the PFS itself sticks together, so that the electrostatic charge of the particles is positive, which leads to the stability of the particles, the specific resistance of the sludge, the moisture content of the filter cake, and the COD of the filtrate [26]. The changing trends of the filter cake pH and sludge pH were consistent with the addition of PFS alone, but the pH of sludge and pH of filtrate after PFS conditioning in combination with fly ash were greater than those after PFS conditioning alone. This phenomenon shows that PFS adsorbs hydroxyl groups from the solution to make the sludge acidic, and the calcium silicate and anorthite in the fly ash can neutralize a certain acidity after dissolving in water, which will affect the sludge pH and pH of the sludge can increase filtrate [27]. After considering the three indicators of SRF, sludge cake moisture content, and filtrate COD, the optimal PFS dosage became 200 mg/g DS.

3.5. Effect of FA-Coupled PAC on Sludge Dewatering Performance

As shown in Figure 5, with the increase in PAC dosage, the sludge SRF first decreased and then increased. When the PAC dosage was 150 mg/g DS, the lowest SRF value was 6.91×10^{10} m/kg. With increasing PAC dosage, the moisture content of the filter cake first decreased and then increased again. When the PAC dosage was 150 mg/g DS, the minimum moisture content of the filter cake was 76.10%. The COD of the filtrate first decreased and then increased with the increase in PFS dosage. When the PAC dosage was 200 mg/g DS, the minimum COD of the filtrate was 26 mg/L. The pH of the filtrate decreased gradually with the increase in the PAC dosage and tended to be gentle. When the PAC dosage was 300 mg/g DS, the pH of the filtrate was at least 7.31. The pH of the sludge decreased gradually with the increase in PAC dosage. When the PAC dosage was 300 mg/g DS, the pH of the sludge was the lowest at 7.04.

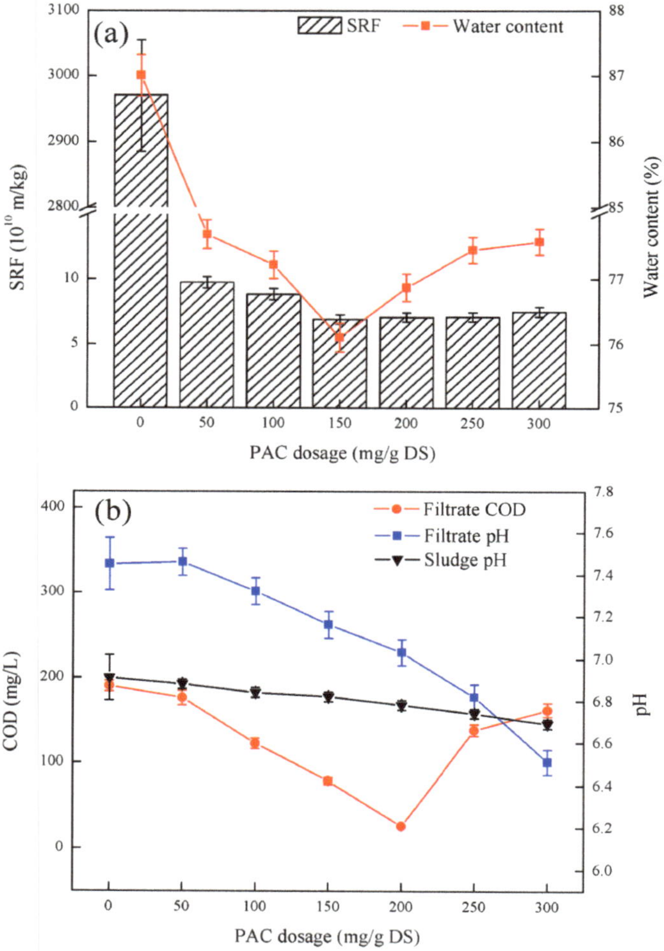

Figure 5. Effect of fly ash coupled with PAC on sludge dewatering performance: (**a**) the effect of PAC dosage on SRF and filter cake moisture content, (**b**) the effect of PAC dosage on filtrate COD, filtrate pH, and sludge pH Influence.

Figure 5 shows that the fly ash coupled with PAC can improve the filtration performance of the sludge. In the process of sludge conditioning by fly ash coupled with PAC, the hydrolyzate interacts with the colloid in the sludge through charge neutralization and adsorption bridging, which reduces the surface charge of the sludge particles, aggregates the sludge particles into clusters, and increases the concentration of sludge particles. Larger sludge particle size increases sludge density. By forming a "skeleton", the fly ash reconstructs the sludge structure, plays a supporting role in the subsequent dehydration process, and provides a water filter channel [28]. When the PAC dosage is too high, the long-chain structure of the PAC itself sticks to each other, which makes the electrostatic charge of the particles positive, which leads to the stability of the particles, blocks the fine pores formed by fly ash, and increases the specific resistance of the sludge [29]. The trend of change in the moisture content of the sludge cake was similar to that of the sludge resistivity SRF, further indicating that excess PFS had an inhibitory effect on sludge dewatering performance. With a suitable PAC dosage, the organic matter can be retained effectively and the COD of the filtrate gradually decreases [30]. If the PAC dosage is too high, the organic matter in the sludge cannot be retained effectively. In addition, the fly ash contains unburned carbon. When the fly ash is added to the sludge, hydrolysis takes place, increasing the COD of the filtrate. The change trend of filter cake pH and sludge pH is consistent with the trend of adding PAC alone and coupled with PFS by fly ash, but the pH of the sludge and filtrate after the conditioning of fly ash coupled with PAC is greater than that of adding PFS alone and coupled with fly ash. The sludge pH and filtrate pH after PFS conditioning indicate that the hydrolysis intermediate of PAC has a higher positive charge than PFS, which is mainly due to the strong electric neutralization and contaminant interaction [31]. Calcium silicate and anorthite in fly ash can neutralize certain acidity after being dissolved in water and can also increase the pH of sludge and filtrate. After considering the three indicators of SRF, sludge cake moisture content, and filtrate COD, the optimal PAC dosage was 200 mg/g DS.

3.6. Floc Structure after Sludge Conditioning

As shown in Figure 6, after adding different conditioners, the particle size distribution follows the same rule: with the increase in particle size, it all shows a trend of first increasing and then decreasing. The D50 of the sludge conditioned by PFS, PAC, FA, fly ash coupled with PFS, and fly ash coupled with PAC were 2.82, 1.68, 0.39, 2.92, and 1.74 μm, respectively. The median particle size of the sludge conditioned by fly ash coupled with PFS was significantly smaller than other conditioners.

Figure 6 shows that the flocs of the raw sludge are loose. The addition of the compound conditioner can neutralize the negative charge on the surface of the flocs, reduce the repulsive force between colloidal ions, and make the flocs prone to collision and aggregation, thus forming flocculation particles with large particle size and dense structure. By comparing the size of the sludge flocs under the optimal dosage of the five conditioners, the size relationship of the median particle size of the sludge particles after conditioning by the five conditioners is: fly ash coupled with PFS > fly ash Coupling PAC > PFA > PFS > PAC. The size of the flocs conditioned by the PFS conditioner is larger than that of the PAC conditioner. During the coagulation process, PFS tends to hydrolyze to form insoluble hydroxide precipitates because the hydrolysis rate of PFS is higher than that of PAC [32]. The order of the median particle size of flocs does not completely correspond to the sludge conditioning performance, indicating that the particle size formed during the sludge conditioning process is not the only factor that affects the sludge dewatering performance [33]. The long-chain structure of PFS and PAC helps to form larger floc size.

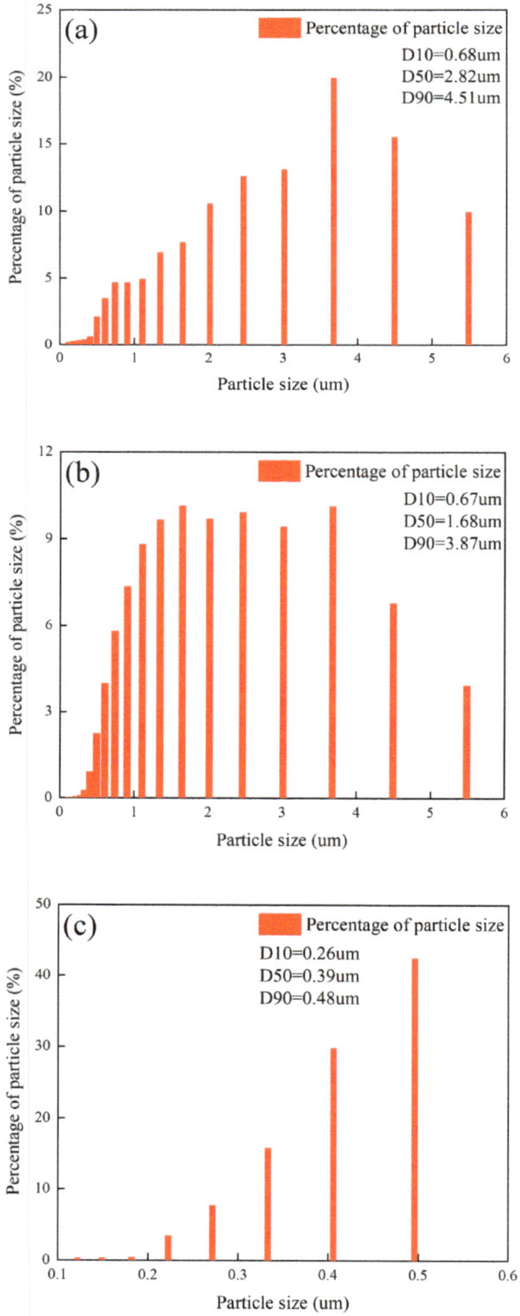

Figure 6. *Cont.*

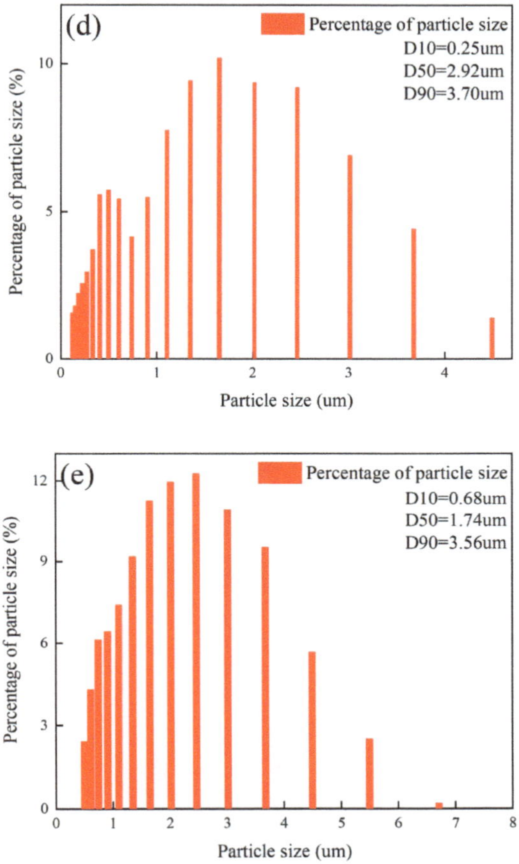

Figure 6. Floc size distribution of sludge conditioned with various flocculants under the optimal dosage conditions: (**a**) PFS, (**b**) PAC, (**c**) fly ash, (**d**) fly ash + PFS, (**e**) Fly Ash + PAC.

3.7. Effects of Different Conditioners on UV of Filtrate

As shown in Figure 7, the raw sludge and the conditioned sludge filtrate show clear absorption peaks between 190 nm and 230 nm. The peak of the original sludge is 192.9 nm, and the absorbance is 4.0223. The peak of the sludge filtrate after PFS conditioning was 197.9 nm, and the absorbance was 3.3556. The peak of the sludge filtrate after PAC conditioning was 193.6 nm, and the absorbance was 2.5271. The peak of the sludge filtrate conditioned by fly ash is 194.1, and the absorbance is 3.342. The peak of the sludge filtrate conditioned by fly ash coupled with PFS is 192.7 nm, and the absorbance is 2.3378. The peak of the sludge filtrate conditioned by fly ash coupled with PAC was at 193 nm, and the absorbance was 2.1510. The absorbance of the conditioned sludge was less than 4.0223. Among them, the absorbance of sludge filtrate conditioned by fly ash coupled with PAC was lowest.

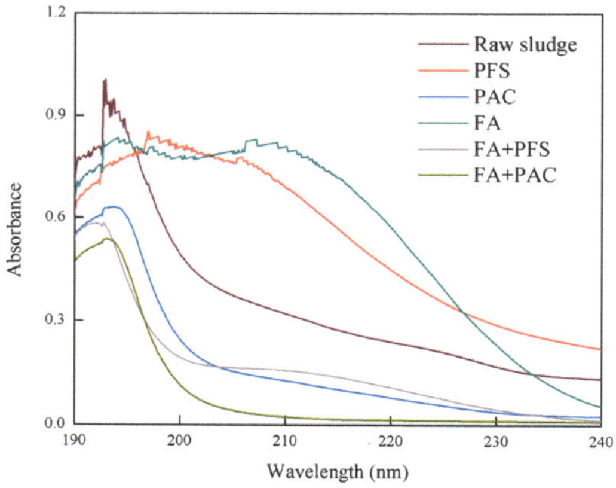

Figure 7. Effects of different conditioners on UV of filtrate.

Figure 7 shows that the raw sludge has clear UV absorption peaks at 190–230 nm, of which 190–230 nm is the characteristic band of organic acids and aromatic compounds. Therefore, many refractory aromatic compounds were observed in the raw sludge. After treatment with the conditioner, the absorption peaks at 190–230 nm were significantly reduced, indicating that most of the aromatic compounds had been removed. The absorption peaks of the PAC and FA conditioned sludge filtrate at 200,230 nm exceed those of the original sludge, which may be due to the destruction of unsaturation during aromatics degradation [34]. The results show that fly ash coupled with PFS and fly ash coupled with PAC can effectively reduce organic acids and aromatic compounds in sludge filtrate.

3.8. Effect of Conditioner on Fractal Dimension of Sludge Floc

Figure 8 and Table 2 show the sludge cake with rough surface and tight internal structure after dehydration by PFS alone. The sludge cake prepared by fly ash alone has a loose structure and a spongy structure with spherical particles of different sizes on its surface. The sludge cake treated with PFS and PAC has a large number of pore structures and cracks and spherical particles on the surface. The sludge conditioned only by PAC and PFS has a dense structure, large floc volume, and less pore structure, which is not conducive to reducing the moisture content of the sludge cake. The surface of fly ash has many spherical particles with smooth surface. Microspheres play a crucial role in the adsorption of sludge flocs. The skeletal structure of fly ash maintains the permeability of the filter cake during compression and dehydration, which helps resist the deformation of the cake body, but the amount of fly ash is large. From the analysis of conditioning ability, to use only a single conditioner for sludge conditioning is not an ideal method of removing organic matter and sludge structure [35]. The pore structure of sludge conditioned by PAC-coupled fly ash and PFS-coupled fly ash was significantly improved and the number of pores increased, providing water filtration channels and reducing the compressibility of the sludge organic matter. PAC and PFS increase floc size through charge neutralization and net sweeping, which aids floc settling and drying [36]. Therefore, the fractal dimension of sludge conditioned by PAC-coupled fly ash and PFS-coupled fly ash is smaller. The results show that the sludge cake surface micromorphology results correspond to the sludge dewatering performance.

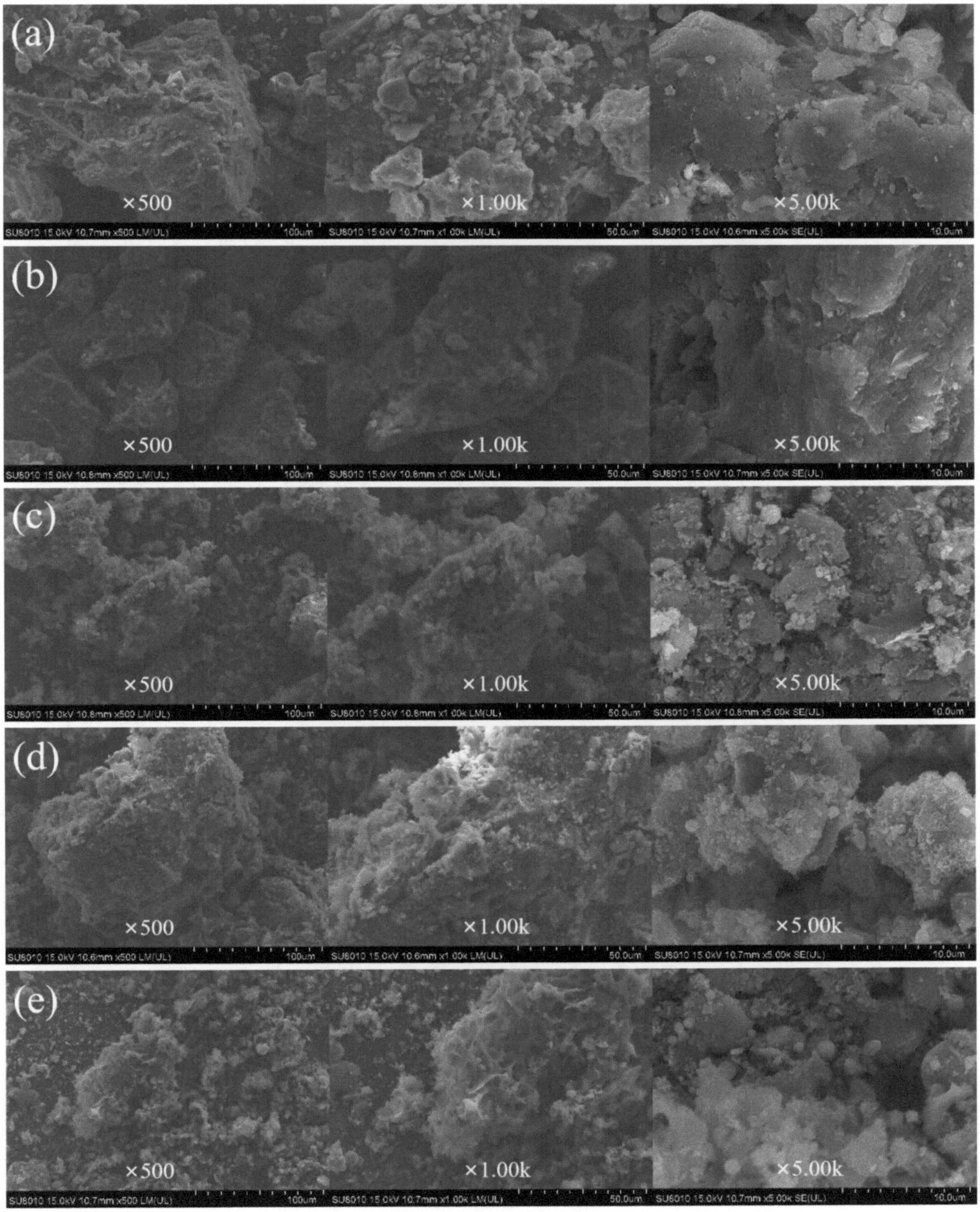

Figure 8. SEM figure of sludge conditioning with optimal dosage: (**a**) PFS, (**b**) PAC, (**c**) fly ash, (**d**) fly ash coupled with PFS, (**e**) fly ash coupled with PAC.

Table 2. Fractal dimension of sludge cake under optimal dosage.

Conditioner	PFS	PAC	Fly Ash	Fly Ash +PFS	Fly Ash + PAC
Fractal dimension	2.03	1.99	1.96	1.91	1.86

4. Conclusions

Fly ash can effectively reduce the moisture content of the sludge, and the fly ash dosage of 10 g/g DS can reduce the sludge specific resistance SRF to 1.03×10^{11} m/kg. After continuously increasing the fly ash dosage, it no longer had a significant impact on reducing the sludge specific resistance. The dosage of PAC and PFS, which can effectively reduce the sludge resistivity, is small when used as a single agent for sludge conditioning, but the moisture content of the filter cake is high. Considering the economic factors of sludge treatment and disposal, fly ash coupled with PAC and fly ash coupled with PFS were selected as composite conditioners for deep dewatering of sludge. The fly ash of 10 g/g DS coupled with PAC of 200 mg/g DS can reduce the sludge-specific resistance SRF, the moisture content of the filter cake, and the COD of the filtrate to 6.91×10^{10} m/kg, 76.10%, and 26 mg/L, respectively. The fly ash of 10 g/g DS coupled with the PFC of 200 mg/g DS can reduce the sludge-specific resistance SRF, the filter-cake moisture content, and the filtrate COD to 6.63×10^{10} m/kg, 76.39%, and 37 mg/L, respectively. Compared to a single conditioner, the addition of a compound conditioner can greatly improve the dewatering effect of the sludge and greatly reduce the amount of conditioner added, effectively increase the particle size of the floc, and reduce the organic matter in the filtrate. The effect on the pH of the sludge is small, which is beneficial for the subsequent treatment and disposal of the sludge and filtrate. Compared with the separate flocculation technology and oxidation technology, the skeleton-assisted flocculation method has the advantages of smaller chemical dosage and low cost. An improved dewatering system for activated sludge by scaffolding and flocculants was designed to provide a guide to practical engineering.

Author Contributions: Conceptualization, J.X., T.R., J.J., B.H., A.L. and Y.S.; methodology, J.X., Y.S.; software, A.L., Y.S.; validation, A.L., Y.S.; formal analysis, A.L., Y.S.; investigation, A.L., Y.S.; resources, J.X., T.R. and Y.S.; data curation, A.L., Y.S.; writing—original draft preparation, A.L., Y.S.; writing—review and editing, A.L., Y.S.; visualization, J.X., T.R., J.J., B.H. and Y.S.; supervision, J.X., T.R., J.J. and B.H.; project administration, J.X., T.R., J.J. and B.H.; funding acquisition, J.X., T.R., J.J. and B.H. All authors have read and agreed to the published version of the manuscript.

Funding: This work was supported by Natural Science Foundation of Jiangsu Province in China (No. BK20201362), and the 2018 Six Talent Peaks Project of Jiangsu Province (JNHB-038), University-enterprise joint development project (202112346).

Institutional Review Board Statement: Not applicable.

Informed Consent Statement: Not applicable.

Data Availability Statement: Data is contained within the article.

Conflicts of Interest: The authors declare no conflict of interest. The company had no role in the design of the study; in the collection, analyses, or interpretation of data; in the writing of the manuscript, and in the decision to publish the results.

References

1. Yang, G.; Zhang, G.; Wang, H. Current state of sludge production, management, treatment and disposal in China. *Water Res.* **2015**, *78*, 60–73. [CrossRef] [PubMed]
2. Christensen, M.L.; Keiding, K.; Nielsen, P.H.; Jørgensen, M.K. Dewatering in biological wastewater treatment: A review. *Water Res.* **2015**, *82*, 14–24. [CrossRef] [PubMed]
3. Lin, H.; Zhang, M.; Wang, F.; Meng, F.; Liao, B.; Hong, H.; Chen, J.; Gao, W. A critical review of extracellular polymeric substances (EPSs) in membrane bioreactors: Characteristics, roles in membrane fouling and control strategies. *J. Membr. Sci.* **2014**, *460*, 110–125. [CrossRef]

4. Zhao, P.; Shen, Y.; Ge, S.; Yoshikawa, K. Energy recycling from sewage sludge by producing solid biofuel with hydrothermal carbonization. *Energ. Convers. Manag.* **2014**, *78*, 815–821. [CrossRef]
5. Mowla, D.; Tran, H.N.; Allen, D.G. A review of the properties of biosludge and its relevance to enhanced dewatering processes. *Biomass Bioenergy* **2013**, *58*, 365–378. [CrossRef]
6. Chen, R.; Sheng, Q.; Chen, S.; Dai, X.; Dong, B. The three-stage effect of hydrothermal treatment on sludge physical-chemical properties: Evolution of polymeric substances and their interaction with physicochemical properties. *Water Res.* **2022**, *211*, 118043. [CrossRef]
7. Zheng, H.; Sun, Y.; Zhu, C.; Guo, J.; Zhao, C.; Liao, Y.; Guan, Q. UV-initiated polymerization of hydrophobically associating cationic flocculants: Synthesis, characterization, and dewatering properties. *Chem. Eng. J.* **2013**, *234*, 318–326. [CrossRef]
8. Geng, N.; Wang, Y.; Zhang, D.; Fan, X.; Li, E.; Han, Z.; Zhao, X. An electro-peroxone oxidation-Fe(III) coagulation sequential conditioning process for the enhanced waste activated sludge dewatering: Bound water release and organics multivariate change. *Sci. Total Environ.* **2022**, *833*, 155272. [CrossRef]
9. Zheng, H.; Sun, Y.; Guo, J.; Li, F.; Fan, W.; Liao, Y.; Guan, Q. Characterization and evaluation of dewatering properties of PADB, a highly efficient cationic flocculant. *Ind. Eng. Chem. Res.* **2014**, *53*, 2572–2582. [CrossRef]
10. Li, Y.; Yuan, X.; Wu, Z.; Wang, H.; Xiao, Z.; Wu, Y.; Chen, X.; Zeng, G. Enhancing the sludge dewaterability by electrolysis/electrocoagulation combined with zero-valent iron activated persulfate process. *Chem. Eng. J.* **2016**, *303*, 636–645. [CrossRef]
11. Zhou, L.; Ying, G.; Liu, S.; Zhao, J.; Yang, B.; Chen, Z.; Lai, H. Occurrence and fate of eleven classes of antibiotics in two typical wastewater treatment plants in South China. *Sci. Total Environ.* **2013**, *452–453*, 365–376. [CrossRef] [PubMed]
12. Kim, M.S.; Lee, K.; Kim, H.; Lee, H.; Lee, C.; Lee, C. Disintegration of waste activated sludge by Thermally-Activated persulfates for enhanced dewaterability. *Environ. Sci. Technol.* **2016**, *50*, 7106–7115. [CrossRef] [PubMed]
13. Jia, A.; Wan, Y.; Xiao, Y.; Hu, J. Occurrence and fate of quinolone and fluoroquinolone antibiotics in a municipal sewage treatment plant. *Water Res.* **2012**, *46*, 387–394. [CrossRef] [PubMed]
14. Hyrycz, M.; Ochowiak, M.; Krupińska, A.; Włodarczak, S.; Matuszak, M. A review of flocculants as an efficient method for increasing the efficiency of municipal sludge dewatering: Mechanisms, performances, influencing factors and perspectives. *Sci. Total Environ.* **2022**, *820*, 153328. [CrossRef]
15. Rao, B.; Su, J.; Xu, J.; Xu, S.; Pang, H.; Zhang, Y.; Xu, P.; Wu, B.; Lian, J.; Deng, L. Coupling mechanism and parameter optimization of sewage sludge dewatering jointly assisted by electric field and mechanical pressure. *Sci. Total Environ.* **2022**, *817*, 152939. [CrossRef]
16. Cai, M.; Qian, Z.; Xiong, X.; Dong, C.; Song, Z.; Shi, Y.; Wei, Z.; Jin, M. Cationic polyacrylamide (CPAM) enhanced pressurized vertical electro-osmotic dewatering of activated sludge. *Sci. Total Environ.* **2022**, *818*, 151787. [CrossRef]
17. Sun, Y.; Zheng, H.; Zhai, J.; Teng, H.; Zhao, C.; Zhao, C.; Liao, Y. Effects of surfactants on the improvement of sludge dewaterability using cationic flocculants. *PLoS ONE* **2014**, *9*, e111036. [CrossRef]
18. Sun, Y.; Fan, W.; Zheng, H.; Zhang, Y.; Li, F.; Chen, W. Evaluation of dewatering performance and fractal characteristics of alum sludge. *PLoS ONE* **2015**, *10*, e130683. [CrossRef]
19. Guan, Q.Q.; Zheng, H.L.; Zhai, J.; Zhao, C.; Zheng, X.K.; Tang, X.M.; Chen, W.; Sun, Y.J. Effect of template on structure and properties of cationic polyacrylamide: Characterization and mechanism. *Ind. Eng. Chem. Res.* **2014**, *53*, 5624–5635. [CrossRef]
20. Liao, Y.; Zheng, H.L.; Qian, L.; Sun, Y.J.; Dai, L.; Xue, W.W. UV-Initiated polymerization of hydrophobically associating cationic polyacrylamide modified by a Surface-Active monomer: A comparative study of synthesis, characterization, and sludge dewatering performance. *Ind. Eng. Chem. Res.* **2014**, *53*, 11193–11203. [CrossRef]
21. Zhu, J.; Zheng, H.; Jiang, Z.; Zhang, Z.; Liu, L.; Sun, Y.; Tshukudu, T. Synthesis and characterization of a dewatering reagent: Cationic polyacrylamide (P(AM-DMC-DAC)) for activated sludge dewatering treatment. *Desalin. Water Treat.* **2013**, *51*, 2791–2801. [CrossRef]
22. Zhen, G.; Lu, X.; Wang, B.; Zhao, Y.; Chai, X.; Niu, D.; Zhao, A.; Li, Y.; Song, Y.; Cao, X. Synergetic pretreatment of waste activated sludge by Fe(II)–activated persulfate oxidation under mild temperature for enhanced dewaterability. *Bioresour. Technol.* **2012**, *124*, 29–36. [CrossRef] [PubMed]
23. Guan, Q.; Zheng, H.; Zhai, J.; Liu, B.; Sun, Y.; Wang, Y.; Xu, Z.; Zhao, C. Preparation, characterization, and flocculation performance of P(acrylamide-co-diallyldimethylammonium chloride) by UV-initiated template polymerization. *J. Appl. Polym. Sci.* **2015**, *132*, 41747. [CrossRef]
24. Sun, Y.; Ren, M.; Zhu, C.; Xu, Y.; Zheng, H.; Xiao, X.; Wu, H.; Xia, T.; You, Z. UV-Initiated graft copolymerization of cationic Chitosan-Based flocculants for treatment of zinc Phosphate-Contaminated wastewater. *Ind. Eng. Chem. Res.* **2016**, *55*, 10025–10035. [CrossRef]
25. Yan, C.; Zhan, M.; Xv, K.; Zhang, S.; Liang, T.; Yu, R. Sludge dewaterability enhancement under low temperature condition with cold-tolerant Bdellovibrio sp. CLL13. *Sci. Total Environ.* **2022**, *820*, 153269. [CrossRef] [PubMed]
26. Sun, Y.; Zhu, C.; Zheng, H.; Sun, W.; Xu, Y.; Xiao, X.; You, Z.; Liu, C. Characterization and coagulation behavior of polymeric aluminum ferric silicate for high-concentration oily wastewater treatment. *Chem. Eng. Res. Des.* **2017**, *119*, 23–32. [CrossRef]
27. Sun, W.; Tang, M.; Sun, Y.; Xu, Y.; Zheng, H. Effective sludge dewatering technique using the combination of Fenton's reagent and CPAM. *Can. J. Chem. Eng.* **2018**, *96*, 1256–1263. [CrossRef]
28. Xiao, H.; Liu, H.; Jin, M.; Deng, H.; Wang, J.; Yao, H. Process control for improving dewatering performance of sewage sludge based on carbonaceous skeleton-assisted thermal hydrolysis. *Chemosphere* **2022**, *296*, 134006. [CrossRef]

29. Chen, L.; Zhu, H.; Sun, Y.; Chiang, P.; Sun, W.; Xu, Y.; Zheng, H.; Shah, K.J. Characterization and sludge dewatering performance evaluation of the photo-initiated cationic flocculant PDD. *J. Taiwan Inst. Chem. E.* **2018**, *93*, 253–262. [CrossRef]
30. Li, Z.; Li, Y.; Wang, D.; Yuan, L.; Liu, X.; Pan, C.; Zhang, X. Insights into cetyl trimethyl ammonium bromide improving dewaterability of anaerobically fermented sludge. *Chem. Eng. J.* **2022**, *435*, 134968. [CrossRef]
31. Lu, X.; Xu, Y.; Sun, W.; Sun, Y.; Zheng, H. UV-initiated synthesis of a novel chitosan-based flocculant with high flocculation efficiency for algal removal. *Sci. Total Environ.* **2017**, *609*, 410–418. [CrossRef] [PubMed]
32. Sun, Y.; Chen, A.; Pan, S.; Sun, W.; Zhu, C.; Shah, K.J.; Zheng, H. Novel chitosan-based flocculants for chromium and nickle removal in wastewater via integrated chelation and flocculation. *J. Environ. Manag.* **2019**, *248*, 109241. [CrossRef] [PubMed]
33. Ding, A.; Lin, W.; Chen, R.; Ngo, H.H.; Zhang, R.; He, X.; Nan, J.; Li, G.; Ma, J. Improvement of sludge dewaterability by energy uncoupling combined with chemical re-flocculation: Reconstruction of floc, distribution of extracellular polymeric substances, and structure change of proteins. *Sci. Total Environ.* **2022**, *816*, 151646. [CrossRef] [PubMed]
34. Zeng, R.; Li, Y.; Sha, L.; Liu, X.; Zhang, S. Electro-dewatering of steel industrial sludge: Performance and metal speciation. *J. Water Process Eng.* **2022**, *46*, 102600. [CrossRef]
35. Deng, H.; Liu, H.; Jin, M.; Xiao, H.; Yao, H. Phosphorus transformation during the carbonaceous skeleton assisted thermal hydrolysis of sludge. *Sci. Total Environ.* **2022**, *827*, 154252. [CrossRef] [PubMed]
36. Raynaud, M.; Vaxelaire, J.; Olivier, J.; Dieudé-Fauvel, E.; Baudez, J. Compression dewatering of municipal activated sludge: Effects of salt and pH. *Water Res.* **2012**, *46*, 4448–4456. [CrossRef] [PubMed]

International Journal of
Environmental Research and Public Health

MDPI

Article

Micro- and Macroelements Content of Plants Used for Landfill Leachate Treatment Based on *Phragmites australis* and *Ceratophyllum demersum*

Aleksandra Wdowczyk * and **Agata Szymańska-Pulikowska**

Institute of Environmental Engineering, Wroclaw University of Environmental and Life Sciences, pl. Grunwaldzki 24, 50-363 Wroclaw, Poland; agata.szymanska-pulikowska@upwr.edu.pl
* Correspondence: aleksandra.wdowczyk@upwr.edu.pl

Abstract: One of the key problems associated with the functioning of landfills is the generation of leachate. In order to reduce their negative impact on the environment, various treatment technologies are applied. Among them, solutions based on the use of phytotechnology deserve special attention. The aim of this study was to evaluate the impact of landfill leachate on the content of micro- and macroelements in plant material. The research was carried out in four municipal waste landfills located in Poland. Emergent macrophytes (*P. australis*) and submergent macrophytes (*C. demersum*) were used in this research. The migration and distribution of pollutants reaching the roots and shoots of *P. australis* from water solutions were also studied. The concentrations of heavy metals in the studied plants were low in all analysed cases. Higher metal contents could often be observed in roots rather than in shoots, but these differences were insignificant. The chemical composition of the studied plant samples was primarily related to the source of origin of the treated leachate (landfill), as clearly demonstrated by cluster analysis. In the conducted studies, no important differences were noted in the accumulation of the studied components between submergent plants (*C. demersum*) and emergent macrophytes (*P. australis*).

Keywords: municipal solid waste (MSW); landfills leachate treatment; phytoremediation; bioconcentration factor (BCF); translocation factor (TF)

Citation: Wdowczyk, A.; Szymańska-Pulikowska, A. Micro- and Macroelements Content of Plants Used for Landfill Leachate Treatment Based on *Phragmites australis* and *Ceratophyllum demersum*. *Int. J. Environ. Res. Public Health* **2022**, *19*, 6035. https://doi.org/10.3390/ijerph19106035

Academic Editors: Yung-Tse Hung, Hamidi Abdul Aziz and Issam A. Al-Khatib

Received: 4 May 2022
Accepted: 13 May 2022
Published: 16 May 2022

1. Introduction

The landfilling of municipal waste is a common waste management practice [1,2]. One of the key problems associated with the functioning of landfills is the generation of leachate [3]. Leachate may be a potential environmental hazard due to the high content of inorganic compounds and dissolved organic compounds, suspended solids, heavy metals and hazardous substances [4–8]. The most frequent pollutants in leachate are ammonium nitrogen, chlorides, sulphates and heavy metals. In natural ecosystems, heavy metals are characterised by high mobility, relatively high chemical stability and carcinogenicity [9]. Less common are humic acids, dioxins, furans, pesticides, phthalates or pharmaceutical compounds [10–13]. Long-term contact with the substances contained in leachate may result in their uptake by aquatic organisms and bioaccumulation in subsequent links of the food chain [14,15]. Migration in the environment may also lead to serious threats to human health [16].

In order to reduce the negative impact of landfill leachate on the environment, various treatment technologies are applied. Due to the complexity and variety of leachates, as well as the disadvantages of the technologies used, despite many years of research, there is still no agreement as to the optimal method of their treatment [17].

Among the methods used to treat leachate from landfills, biological, chemical and physical ones can be distinguished. Physical and chemical methods (e.g., coagulation/flocculation, membrane treatment or chemical precipitation) are recommended mainly in the treatment of mature leachate with low biodegradability [18]. However, biological methods are recommended for young leachates due to their high biodegradability. Biological processes are preferred for the treatment of landfill leachate, but it should be taken into account that they have their limitations and do not always provide the expected effectiveness in the treatment of mature leachate [19].

Among them, solutions based on the use of phytotechnology deserve special attention [20,21]. Phytoremediation is used to eliminate the pollutants present in leachate, including organic compounds (petroleum hydrocarbons, pesticides) and trace elements (metals, metalloids), in water and soil [22]. It is a set of techniques or processes in which plants are used to extract, store, degrade or immobilise pollutants from a site (soil, water or sediment) [23].

Depending on the mechanisms used by plants, phytoremediation may include: phytodegradation (the decomposition of pollutants), phytoextraction (the accumulation of pollutants in plant tissues) and phytostabilisation (the transformation of toxins into non-toxic or less toxic forms that leads to the reduced mobility of pollutants through accumulation in roots or immobilisation in the rhizosphere) [22,24,25].

Phytoremediation is considered to be an effective, low-cost and environmentally friendly pollutant removal technology. Some plants have the ability to take up and bind biogenic compounds and metals in less harmful forms in tissues that are metabolically inactive [26]. Differences in the efficiency of the extraction of nutrients and metals from wastewater can be observed among different plant species [27,28]. Constructed wetlands are a well-known technology for wastewater treatment. They are recognized to be effective in the removal of dissolved organics and suspended solids [29].

To date, constructed wetlands have usually been planted with Phragmites spp. (Poaceae), Scirpus spp. (Cyperaceae), Iris spp. (Iridaceae), Typha spp. (Typhaceae), Eleocharis spp. (Cyperaceae), Juncus spp. (Juncaceae) and Salix [30] worldwide. Unfortunately, most plant species act selectively and do not show efficiency in removing pollutants from multi-component mixtures [31]. Therefore, the selection of suitable plant species is a very important issue, as they should survive the potentially toxic effects of the pollutants present in leachate and their variability over time [31]. In addition, these plants should have the ability to produce large amounts of biomass and accumulate high concentrations of pollutants [32].

Recently, the most frequently used aquatic plant in phytoremediation has been *Phragmites australis* [33], which is one of the most widely distributed emergent plant species in the world. It is commonly found in Europe but also in North America and various regions of South America and Australia [34,35]. In addition to being a species with a very wide geographical range, its advantages include its low cost and minimal requirements [36]. *P. australis* is not a hyperaccumulator, but it is characterised by a fast growth rate, high biomass production and a deep root system [37]. All these characteristics make it one of the favourite test species and therefore, since the 1970s, it has been widely used in the phytoremediation of different types of wastewaters, soils and sediments [38]. The efficiency of *P. australis* in the bioaccumulation of heavy metals and other pollutants has been confirmed in numerous studies [37,39,40].

Among aquatic plants, submergent macrophytes are also considered very effective in the phytoremediation of pollutants [41]. They are among the major primary producers in aquatic ecosystems and, due to their large contact area, they show a higher potential for removing pollutants from liquids [42,43]. *Ceratophyllum demersum* is one of the representatives of submergent macrophytes, common throughout the world. It has no roots but can anchor itself by means of rhizoids—modified root-like leaves [44]. The conducted studies show that *C. demersum* is able to produce an internal concentration of metals and other components several times higher than in the surrounding environment. The high biomass growth, high removal efficiency and wide range of tolerance to metals make these plants an excellent choice for the phytoremediation process [45,46].

To date, a large amount of research has been carried out on leachate treatment using phytoremediation. However, to the best of our knowledge, few studies conducted so far have included analyses of the content of micro- and macroelements in the context of the effectiveness of their collection from leachate from landfills. Moreover, after analysing the literature, the authors observed that few studies compared the effectiveness of phytoremediation with the use of emerged and submerged plants on leachate from various landfills, taking into account the effect of the applied dose of the leachate.

Therefore, this paper will present the range of content and movement of macro- and microelements in the plant material of *Phragmites australis* and *Ceratophyllum demersum*. The plants were exposed to different concentrations of leachate from municipal waste landfills at different stages of operation (active and closed). The first part of the study included the evaluation of the leachate treatment efficiency by *P. australis* and *C. demersum* using physicochemical property analyses and toxicity tests [47]. The results presented here complement the picture of the components of the experimental system by presenting data on micro- and macroelement concentrations in plants.

The aim of this study was to evaluate the impact of landfill leachate on the content of micro- and macroelements in plant material. Emergent macrophytes (*P. australis*) and submergent macrophytes (*C. demersum*) were used in this research. The migration and distribution of pollutants reaching the roots and shoots of *P. australis* from water solutions were also studied.

2. Materials and Methods

2.1. The Study Location and Leachate Collection Points

This study was conducted at four municipal waste landfills, situated in Poland, in the Lower Silesian Voivodeship. Leachate samples were collected in June 2020. Two non-operational landfills (located in Wrocław and Bielawa) and two operational landfills (located in Legnica and Jawor) were selected for the study. Figure 1 shows the location of the leachate collection points.

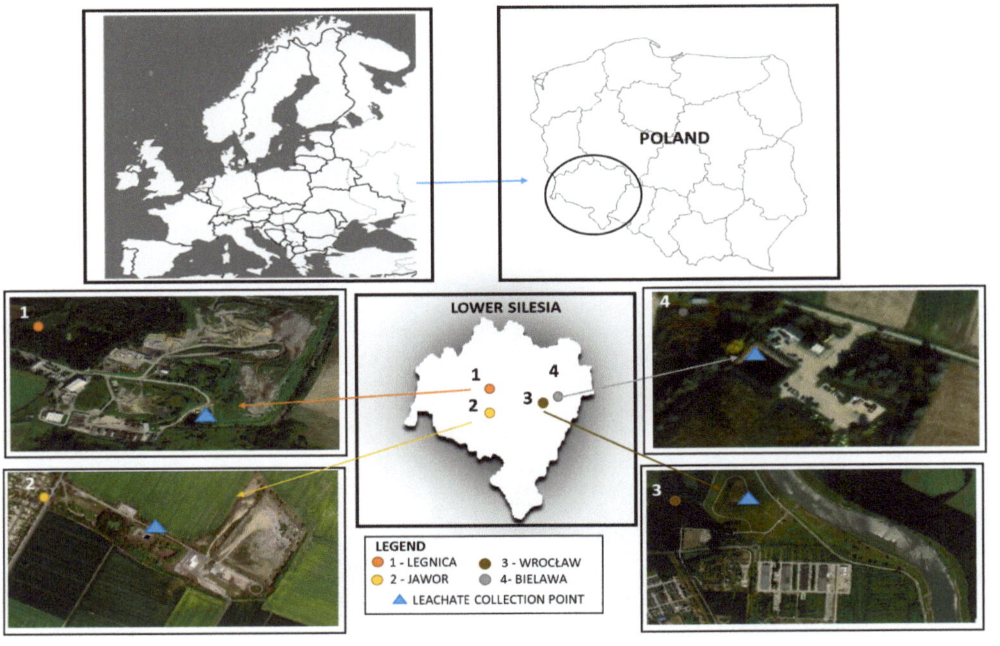

Figure 1. Location of landfills and distribution of leachate collection points.

The oldest landfill site was established in Wrocław (51°10′23.784″ N 16°55′40.74″ E) in 1966 and operated until 2000. It has an area of 11.7 ha and a capacity of about 2 million m^3. The second disused landfill is located in Bielawa (51°9′21.485″ N 17°14′18.03″ E). The facility was in operation from 2001 to 2011. The surface area of the landfill is 0.86 ha and its capacity is 37.8 thousand m^3. The largest operational landfill is located in Legnica (51°14′21.317″ N 16°11′0.251″ E), with an area of 14.12 ha and a total capacity of 2.34 million m^3. The landfill has been in operation since 1977. The second active landfill in Jawor (51°3′56.112″ N 16°12′38.927″ E) has been in operation since 1997. The area occupied by waste is 3.37 ha. The total capacity of the landfill is 231.3 thousand m^3. The exact characteristics of the landfills where the research was conducted are presented in the publication [10].

2.2. Physicochemical Composition of Leachate

Analyses of the physicochemical properties of the leachate were performed at the Environmental Research Laboratory of the Institute of Environmental Engineering at Wrocław University of Environmental and Life Sciences. The studies were performed using commonly used methods, in accordance with ISO (International Organization for Standardization) standards. Laboratory analyses not requiring mineralisation of the samples were performed within 24 h of their collection [48], followed by the analyses requiring mineralisation.

The range of research of the raw and post-treatment by *Phragmites australis* and *Ceratophyllum demersum* included: pH, electrical conductivity (EC), total Kjeldahl nitrogen (TKN), organic nitrogen (ON), ammonium nitrogen (AN), total phosphorus (TP), Chemical Oxygen Demand (COD), biochemical oxygen demand (BOD_5) and concentrations of: total dissolved solids (TDS), total suspended solids (TSS), total solids (TS), chlorides, sulphates, potassium, sodium, magnesium, calcium, iron, manganese, nickel, zinc, cadmium, copper, lead and chromium.

2.3. Conditions of Conducting an Experiment with Phragmites australis and Ceratophyllum demersum

Laboratory tests involved the exposure of *C. demersum* (rigid hornwort), *P. australis* (common reed) and seedlings to increasing concentrations of landfill leachate. The experiment was designed to check the accumulation of micro- and macronutrients and heavy metals in the plant material. The migration and distribution of pollutants from the aqueous solutions to the roots and shoots of *P. australis* were also studied.

The transferred *P. australis* and *C. demersum* plants were flooded with tap water for a period of 14 days to adapt to laboratory conditions. Then, 60 plants of similar sizes were selected from the *P. australis* seedlings that had acclimatised, and they were relocated individually to 1.5 dm^3 containers. From the *C. demersum* seedlings, 120 plants with an average length of about 20 cm were selected. In each of the 60 containers with a volume of 0.5 dm^3, two plants were placed.

The containers thus prepared (with the *P. australis* and *C. demersum* seedlings) were completed with the landfills leachate from the four facilities. The series for each landfill consisted of solutions with increasing concentrations: from 0% (tap water), through 6.25%, 12.5%, 25% and 50% to 100% [49]. The leachate exposure lasted for another 14 days [32,50,51]. No additional aeration was applied during the experiment. Each variant was performed in 3 repetitions. The study included an evaluation of the efficiency of the leachate treatment by *P. australis* and *C. demersum* using physicochemical analyses and toxicity tests and is detailed in the publication [52].

2.4. Analysis of Selected Components in the Plant Material

After completing the first part of the experiment, the plants were removed from the prepared leachate solutions. *P. australis* was separated into the aboveground part (the stem and leaves) and the underground part (the roots).

The collected plants were dried at room temperature until they reached a constant weight [53]. The dried mass was ground in a laboratory grinder and subjected to chemical analyses. The samples were analysed for: total nitrogen (TN), total phosphorus (TP), sodium, potassium, calcium, magnesium, iron, manganese, zinc, lead, nickel, cadmium, copper and chromium. The results of the analyses are given in mg/g d.m.

The content of the macro- and microelements and heavy metals in the plant material was determined after wet digestion in a mixture of concentrated perchloric acid, sulphuric acid and nitric acid (4:1:10 ratio) [54]. All analyses were performed in three repetitions. The control samples (plants placed in tap water) were subjected to the same procedure.

2.5. Bioconcentration Factor (BCF) and Translocation Factor (TF)

The bioconcentration factor (BCF) is used to characterise the behaviour of a chemical substance in the environment [55]. In particular, the BCF is used to verify the ability of a plant to accumulate elements from the substrate or the external solution [56]. A higher BCF value indicates a better phytoaccumulation capacity. The BCF was calculated as follows [46,57]:

$$BCF = \frac{C_P}{C_W} \tag{1}$$

where:

C_P—the pollutant concentration in the plant (mg/kg),
C_W—the concentration in the external solution (mg/dm^3).

In calculating the BCF values, the results of the composition analyses of the analysed plants were converted from mg/g to mg/kg.

Apart from checking the ability of the plant to accumulate elements from the substrate or the external solution, it is important to verify the ability to transfer pollutants from the underground parts of the plant to the aboveground parts. It can be characterised by the translocation factor (TF), which is the quotient of the concentration of the pollutants accumulated in the aboveground parts of the plant to the concentration present in the underground parts:

$$TF = C_a/C_u \tag{2}$$

where:

C_a—the pollutant concentration in the aboveground tissues (mg/kg, mg/g),
C_u—the pollutant concentration in the underground tissues (mg/kg, mg/g).

TF values above 1 indicate the translocation ability [1].

2.6. Data Treatment and Statistical Analysis

The obtained results from the plant material, i.e., *Phragmites australis* and *Ceratophyllum demersum*, were subjected to statistical analysis, using Statistica 13.1 software (StatSoft Polska, StatSoft, Inc., Tulsa, OK, USA). Because different concentrations of the leachate were used in the experiment, the chemical composition of the plants was presented in the form of basic descriptive statistics (minimum and maximum).

The assessment of similarity between the plant samples was performed using cluster analysis. It is one of the methods of multivariate analysis, useful in the case of large amounts of data. Hierarchical cluster analysis allows for grouping observations into clusters. Similar observations are placed inside clusters and different clusters contain observations that differ from each other. Clustering is performed on the basis of similarity or distance (dissimilarity). Clusters are aggregated according to decreasing degrees of similarity (or increasing degrees of dissimilarity) until an individual, single cluster with a tree-like structure, called a dendrogram, is formed [58].

The clustering of observations (agglomeration) was performed using Ward's method (the minimisation of the increase in sum of squares MISSQ). This method is based on minimising the heterogeneity (variance) in clusters and finding the highest possible similarity between observations. It has been shown in many studies to be accurate and useful for creating a primary cluster structure [59,60]. The distances between objects were determined based on the Euclidean distance.

3. Results and Discussion

3.1. Content of Selected Macro- and Microelements in Landfill Leachate and Plants

3.1.1. Physicochemical Composition of Leachate from Landfills

Table 1 presents the results of the physicochemical analyses of the raw leachate collected from the active and non-operational landfills.

Table 1. Physicochemical characteristics of raw landfill leachate.

Pollution Indicators	Unit	Non-Operational Landfills		Active Landfills	
		Bielawa	Wrocław	Legnica	Jawor
pH	-	8.42	8.16	8.84	8.11
EC	μS/cm	2318	1800	7791	3919
TKN	mg N/dm^3	51.12	1.09	269.85	310.16
TP	mg P/dm^3	3.11	0.22	3.936	12.04
Sodium	mg Na/dm^3	151.8	91.2	177.8	162.9
Potassium	mg K/dm^3	256.2	61.2	507.6	410
Calcium	mg Ca/dm^3	150.3	309.4	68.1	171.5
Magnesium	mg Mg/dm^3	79.3	52.5	87.9	91.2
Iron	mg Fe/dm^3	0.5	0.15	2.16	3.81
Manganese	mg Mn/dm^3	0.35	0.02	0.47	1.61
Copper	mg Cu/dm^3	0.0457	0.0486	0.1448	0.059
Zinc	mg Zn/dm^3	0.1977	0.2241	0.4987	0.3049
Chromium	mg Cr/dm^3	0.00025	0.00025	0.1927	0.00025
Lead	mg Pb/dm^3	0.0163	0.0054	0.0866	0.004
Nickel	mg Ni/dm^3	0.00025	0.00025	0.1716	0.0209
Cadmium	mg Cd/dm^3	0.0033	0.0047	0.0033	0.0039

The pH value of the leachate from all landfills ranged from 8.11 (Jawor) to 8.84 (Legnica). An alkaline reaction of the leachate is characteristic of older and mature facilities (i.e., operating for more than 10 years) [61,62], which is the case for the studied leachate.

pH changes could provide a more or less favourable environment for plant growth, and this will vary with the species of the plants [63].

In Europe, the pH of the solid substrate on which *P. australis* grows is generally seven but plants also develop well in a pH range of 5.5 to 7.5 [64]. It has been shown that *P. australis* can tolerate a wide pH range even between 2.5 and 9.8 [65]. *C. demersum* thrives in neutral and alkaline environments, but a pH that is too high can lead to stress in plants. Gao et al. showed that plants experience stress at a pH above nine and with an elevated nitrogen content [66].

The highest EC values were recorded at the active landfills, ranging from 3919 μS/cm (Jawor) to 7794 μS/cm (Legnica). In non-operational landfills, the values were lower and ranged from 1800 μS/cm (Wrocław) to 2318 μS/cm (Bielawa). The EC is related to the amount of ions available to the plants in the root zone. The optimum EC value depends on the species and on the environmental conditions. A high EC value hinders nutrient uptake by increasing the osmotic pressure of the nutrient solution, while an EC concentration that is too low can negatively affect plant health and yield [67,68]. As shown by Huang et al., the pH and EC of the external substrate have a significant effect on nutrient accumulation in plants [69].

Nitrogen is one of the most important nutrients for plants [69] but its too high levels can disrupt the dynamic balance of the antioxidant mechanism, causing the accumulation of reactive oxygen inside plant cells [66,70]. High nitrogen contents were recorded in the leachate from the active landfills (from 269.85 mg/dm^3 to 310.16 mg/dm^3). In non-operational landfills, these values were much lower: from 1.09 mg/dm^3 (Wrocław) to 51.12 mg/dm^3 (Bielawa).

Phosphorus is essential for plant physiological functions. It plays an important role in key processes such as photosynthesis, respiration, storage and the transfer of energy. It also improves the yield quality and is essential for seed formation [71]. The phosphorus content in the leachate from the active landfills were higher than those from non-operational ones and ranged from 0.22 mg/dm^3 (Wrocław) to 12.04 mg/dm^3 (Jawor).

The sodium and potassium content in the samples from the active landfills were also higher than those in the leachate from non-operational landfills. The sodium concentration in the active landfills reached 177.8 mg/dm^3 (Legnica), whereas in closed ones the values were lower, up to 151.8 mg/dm^3 (Bielawa). Potassium concentrations were similar; the highest values were observed in the active landfills (up to 507.6 mg/dm^3 in Legnica), whereas in the closed landfills, the maximum value was recorded in the Bielawa landfill (256.2 mg/dm^3). These ions play an important role in plant physiology [72].

The Fe and Mn content in the closed landfills were low and, in all cases, amounted to <1 mg/dm^3. In the active landfills, the values were slightly higher. The Mn content ranged up to 1.61 mg/dm^3 (Jawor), while the Fe content ranged from 2.16 mg/dm^3 (Legnica) to 3.81 mg/dm^3 (Jawor). Fe and Mn in appropriate concentrations have positive effects on plants. Manganese is an important component of various metabolic enzymes [73], and Fe is needed by enzymes to catalyse reactions in the cytochrome [74].

The content of the heavy metals in the leachate from the studied landfills remained at a very low level (i.e., <1 mg/dm^3), which may indicate that the landfilled waste was mainly municipal waste (not containing these components) or that they may have already leached from the landfilled waste. The analyses performed confirm that heavy metal content is not currently the most serious problem in the management of landfill leachate [5,72,75]. Heavy metals in high concentrations can cause toxic effects in plants, accumulate in aquatic organisms and move up the food chain [76,77]. However, the presence of metals in appropriate concentrations can also have positive effects on plant growth and development [78]. Cu plays an important role in the composition of enzymes for protein synthesis and photosynthesis. Although Cr is not an essential element for plants, there are studies indicating that, at low concentrations, it can have a positive effect on growth stimulation in plants [79].

3.1.2. Chemical Composition of Plants

The results of the chemical composition analysis of *C. demersum* after exposure to the leachate from the four municipal waste landfills and in the control sample are shown in Figure 2. The following parameters were analysed: TN, TP, sodium, potassium, calcium, magnesium, iron, manganese, zinc, lead, nickel, cadmium, copper and chromium.

C. demersum is a nitrophile that tolerates high nitrogen concentrations and has a very good nitrogen removal effect; therefore, it is often used for nitrogen removal from wastewater [80]. Only in the case of the plants after exposure to the leachate from the Wrocław landfill were lower TN contents found than those found in the control sample. In the remaining cases, the TN content of the control sample was within the content range corresponding to the plants after contact with the leachate. The TN values ranged from 23.7 mg/g (Wrocław) to 53.1 mg/g (Legnica).

The Na content in the tissues of *C. demersum* remained at a similar level, both in the control sample and after exposure to the leachate. The obtained results did not show any major fluctuations; only in the case of exposure to the leachate from Legnica, the maximum Na content in the plant was significantly higher (amounted to 39.4 mg/g), with the minimum content similar to the other landfills and the control sample. Additionally, the magnesium and TP content in the plant samples after exposure to the leachate from the landfills included in the study did not show marked differences. The contents found were also similar to the results of the control sample.

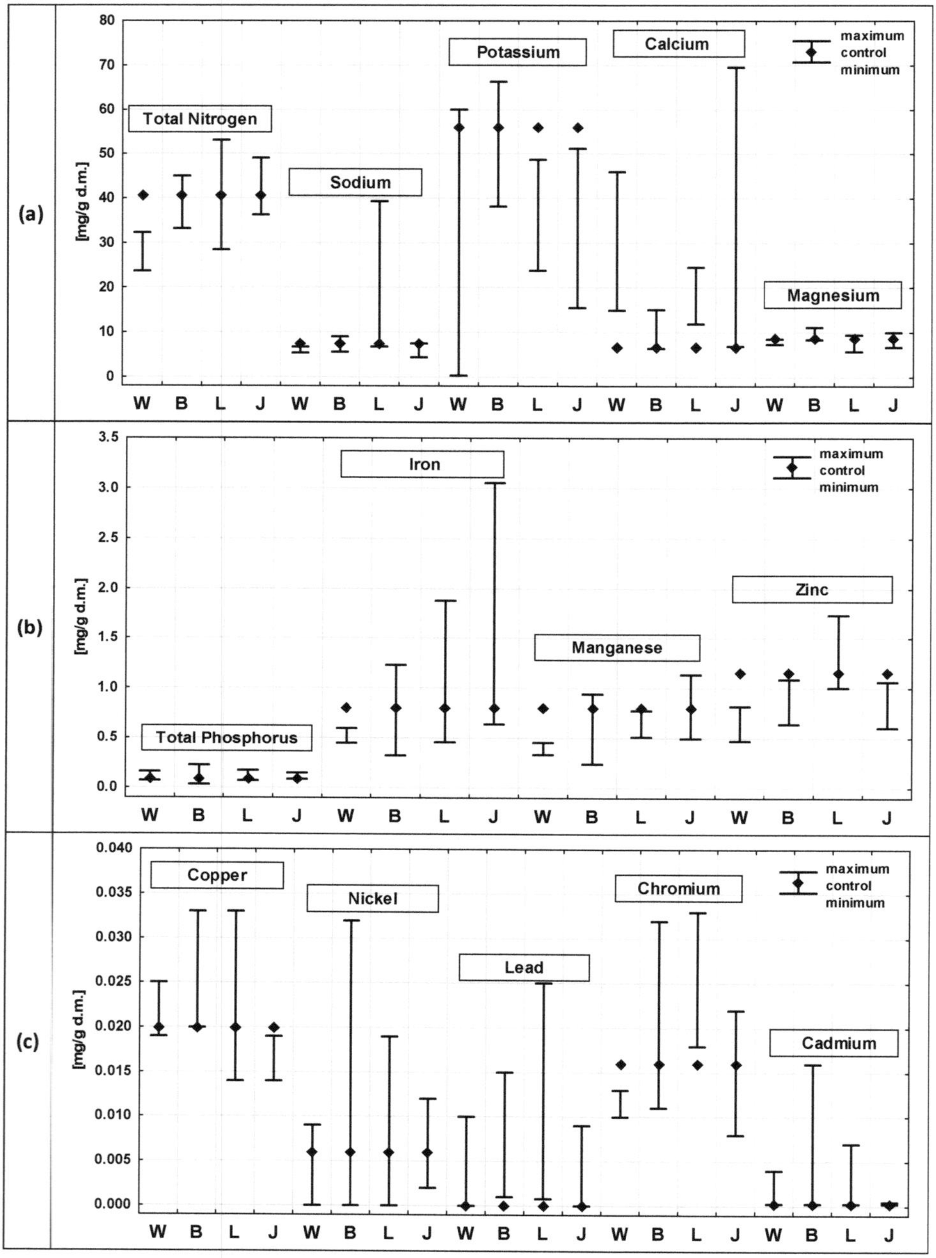

Figure 2. Comparison of chemical composition of plants after exposure to leachate from active (L—Legnica, J—Jawor) and closed (W—Wrocław, B—Bielawa) landfills with the composition of the control sample—results of analyses for *Ceratophyllum demersum*: (**a**) contents of TN, sodium, potassium, calcium and magnesium; (**b**) contents of TP, iron, manganese and zinc; (**c**) contents of copper, nickel, lead, chromium and cadmium.

The potassium content in the *C. demersum* tissues showed rather high variability. The maximum values were higher than those found in the control sample in the case of the Wrocław (60.1 mg/g) and Bielawa (66.4 mg/g) landfills. However, in the case of the plants exposed to the leachate from the landfills in Legnica and Jawor, all the values were lower than those found in the control sample.

The calcium content in the *C. demersum* tissues in most of the analysed cases exceeded those in the control sample. Only the lowest values for the landfills in Bielawa and Jawor were at the level of the control sample. Similar relations are visible in case of the lead and cadmium content. Most of the samples showed an increase in the content of those metals after exposure to the leachate; only minimal values were similar to those of the control sample.

The iron content in the studied plant samples showed quite high variability. The samples after exposure to the leachate from Jawor were characterised with the highest values. Additionally, the majority of the samples from Legnica and Bielawa contained more iron than the control sample. Only the samples after exposure to the leachate from the Wrocław landfill were characterised by a low iron content, below the value for the control sample. Similar relationships were evident for the contents of manganese and chromium (whose highest contents were found in the plants after exposure to the leachate from the landfill in Legnica).

The zinc contents of the tested plants were different. They were higher than the control sample only after the exposure to the Legnica leachate. In the case of copper, however, the contents were lower than those in the control sample only after the exposure of the plants to the leachate from the Jawor landfill. On the other hand, the nickel contents in the control sample were within the range of the values obtained for the samples after contact with the landfill leachate. The highest nickel contents were found in the landfill in Bielawa.

The contents of the majority of the analysed heavy metals (Cu, Ni, Pb, Cr and Cd) did not exceed the level of 1 mg/g. The highest contents of Cu, Ni, Pb, Cr and Cd were recorded after exposure to the leachate from the Legnica and Bielawa landfills and were higher than the values found in the control samples. The data presented in the literature also show that *C. demersum* is able to accumulate significant amounts of various metals, including: Mn, Cu, Cr, Fe, Cd and Pb [57].

The bioavailability of heavy metals is influenced by various factors, especially pH and redox potential [81]. During the conducted studies, it was found that metal concentrations in water and soil tended to be negatively correlated with pH, which affects the amount of metals taken up by plants [82] and may explain the rather low heavy metal content in the plants studied.

In the case of the TP, TN, Na, K, Ca, Mg and Cd contents, in each variant in the external solution, higher contents were observed than in the *C. demersum* plants. The reverse was true for Zn, where higher concentrations were observed in *C. demersum* than in the external solution. Additionally, others, during studies on *C. demersum,* have observed that they are able to produce internal concentrations of metals and other components several times higher than the surrounding environment [45,46]. For Fe, Mn, Cu, Ni, Pb and Cr, the situation was variable once higher values were recorded in the external solution and once they were in the plants.

Figures 3 and 4 show the results of the chemical composition analyses of *P. australis* after exposure to the leachate from the four municipal waste landfills and in the control samples, divided into roots (Figure 3) and shoots (Figure 4). The analysis of the chemical composition of the plants included content determinations of: TN, TP, sodium, potassium, calcium, magnesium, iron, manganese, zinc, lead and nickel, cadmium, copper and chromium.

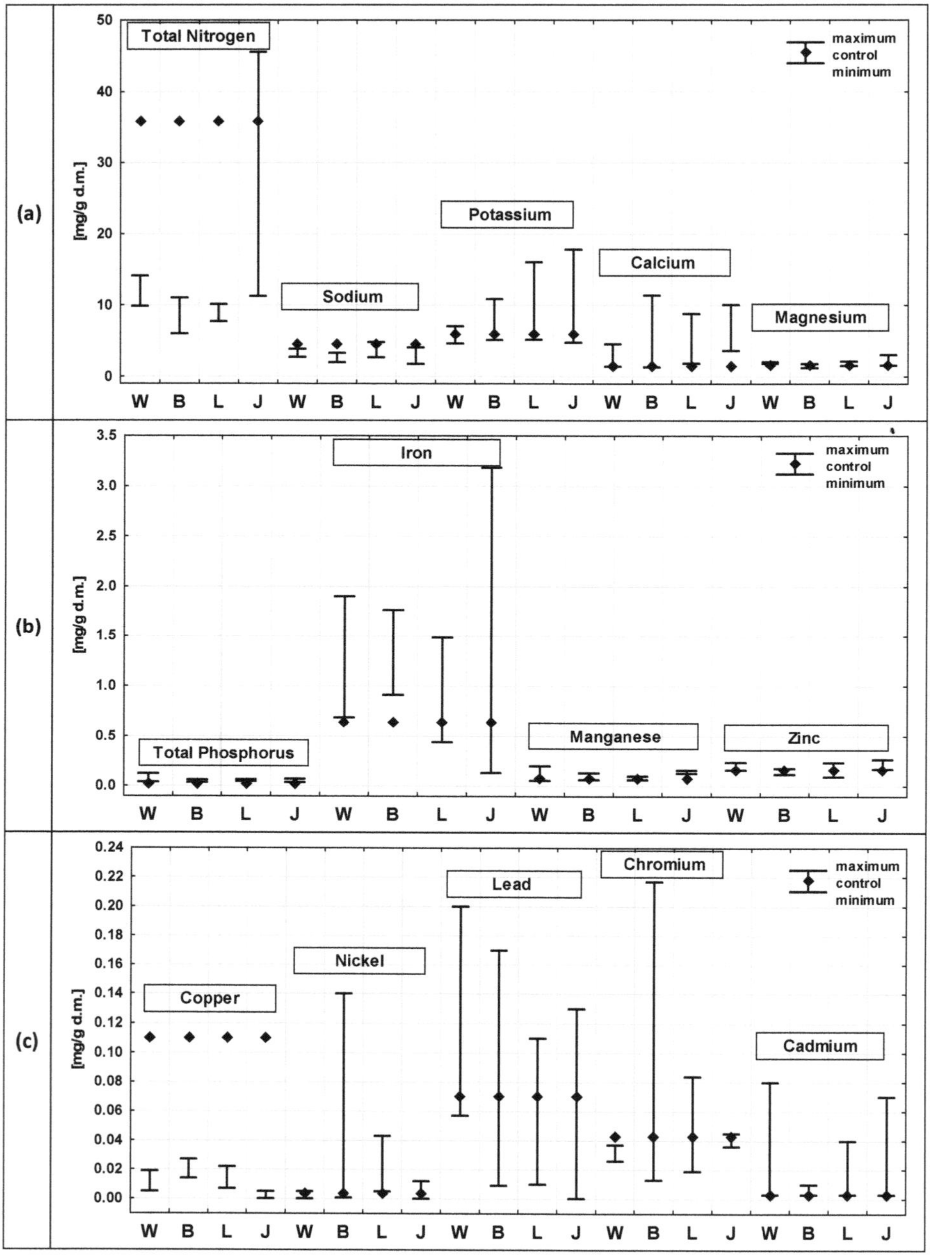

Figure 3. Comparison of chemical composition of plants after exposure to leachate from active (L—Legnica, J—Jawor) and closed (W—Wrocław, B—Bielawa) landfills with the composition of the control sample—results of analyses for *Phragmites australis* roots: (**a**) contents of TN, sodium, potassium, calcium and magnesium; (**b**) contents of TP, iron, manganese and zinc; (**c**) contents of copper, nickel, lead, chromium and cadmium.

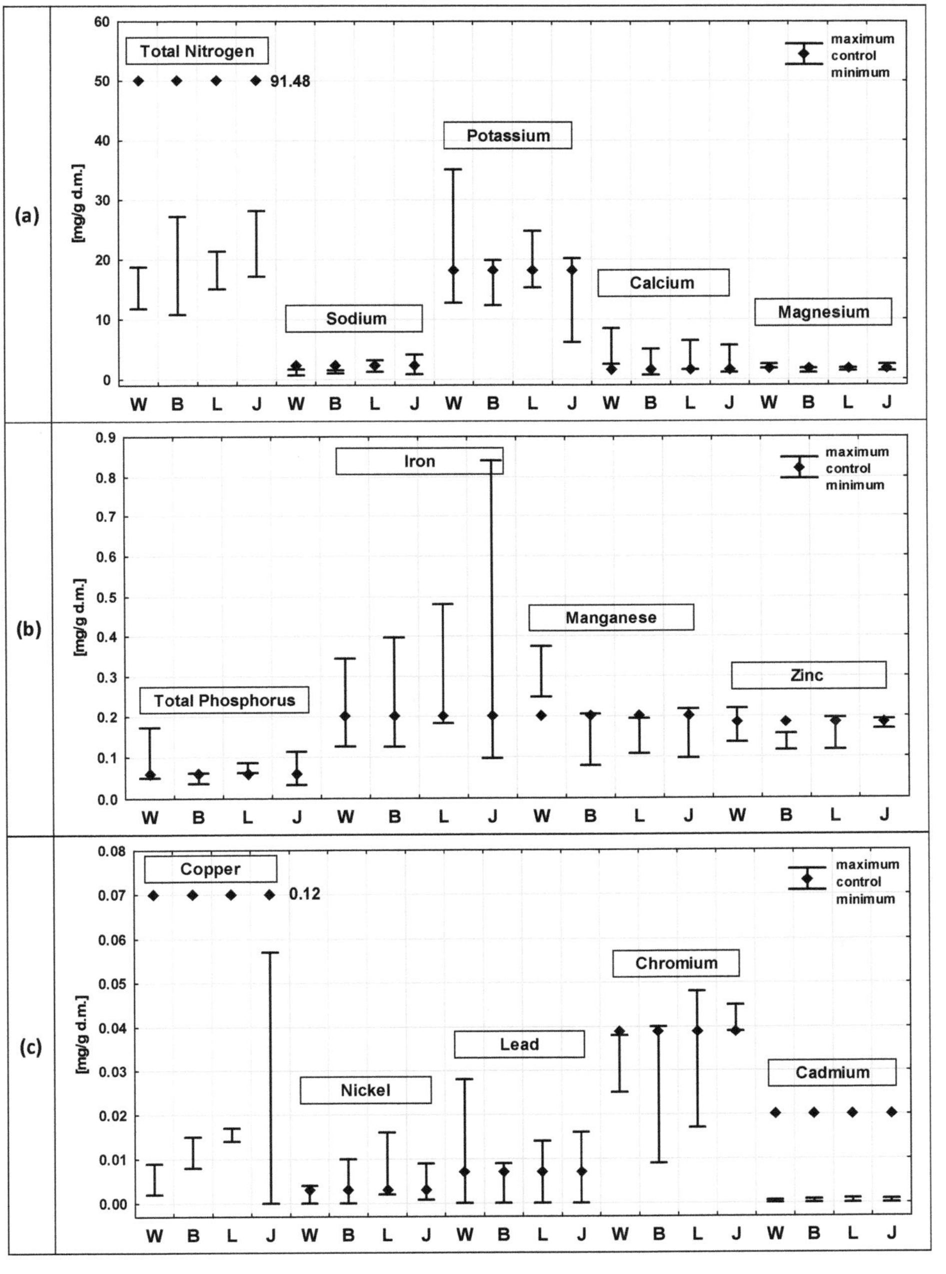

Figure 4. Comparison of chemical composition of plants after exposure to leachate from active (L—Legnica, J—Jawor) and closed (W—Wrocław, B—Bielawa) landfills with the composition of the control sample—results of analyses for *Phragmites australis* shoots: (**a**) contents of TN, sodium, potassium, calcium and magnesium; (**b**) contents of TP, iron, manganese and zinc; (**c**) contents of copper, nickel, lead, chromium and cadmium.

The TN content was, in almost all cases, lower in the roots than in the shoots, except for the exposure to the Jawor leachate, where the maximum TN content in the root (45.6 mg/g) exceeded that observed in the shoots (28.2 mg/g). The highest TN content was recorded in the reed shoots in the control sample (91.5 mg/g) and it was more than 2.5 times higher than the value recorded in the roots.

In the case of Na, no such clear differentiation was found, and in all cases the contents did not exceed 4.13 mg/g. In the shoots of the control samples, three times higher Na contents were recorded than in the roots. Similarly, for the Ca content, the tendency was variable. The highest value was recorded in the roots after exposure to the leachate from the Bielawa landfill (11.4 mg/g) and from the Jawor landfill (10.1 mg/g). In the case of the shoots, the value was lower and amounted to 8.4 mg/g (in the leachate from the Wrocław landfill).

Higher K concentrations were observed in the *P. australis* shoots than in the roots. The highest K value was recorded in the *P. australis* shoots after exposure to the leachate from Wroclaw. The Mg content in the roots and shoots were similar, ranging from 1.08 mg/g (Bielawa, shoots) to 3.11 mg/g (roots, Jawor). A slightly higher value was recorded in the shoots in the control sample of *P. australis* than in the roots. No significant differences between the TP content in the roots and shoots were observed either; they ranged from 0.02 mg/g (roots, control sample) to 0.17 mg/g (shoots, Wrocław). In most cases, the TP content was higher in the shoots than in roots (except for the leachate from the Jawor landfill).

The Fe content in the *P. australis* roots were higher than in the shoots in each of the analysed cases. The values ranged from 0.1 mg/g (Bielawa, shoots) to 3.2 mg/g (Jawor, roots). According to the data presented in the literature, the roots and rhizomes of *P. australis* can accumulate higher amounts of heavy metals than the remaining organs, which is determined by the fact that they have large, airy intercellular spaces in the parenchyma [83].

The Mn contents were, in all cases, higher in the shoots than in roots, except for the Jawor landfill, where a slightly higher value was recorded in the roots. The Mn contents were low and did not pose any potential toxicity to the plants—they ranged from 0.063 mg/g to 0.37 mg/g. As reported in the literature, the toxic concentration of Mn for different plants ranges from 0.4 mg/g to 1 mg/g [84].

The Zn contents in the roots and shoots of *P. australis* were similar in all cases, ranging from 0.09 mg/g to 0.27 mg/g. Similar contents were reported by Vymazal et al. [85], where the Zn content in the *P. australis* roots ranged between 0.0135 mg/g and 0.202 mg/g dry mass.

Slightly higher values were observed in the roots than in the shoots, except for the control sample and the minimum value, recorded after exposure to the Legnica leachate. As reported in the literature, the toxic concentration of Zn for different plants ranges from 0.1 mg/g to 0.4 mg/g [84]. The Zn content of the leachate remained at levels that did not pose potential toxicity to *P. australis*.

The contents of the heavy metals (Cu, Ni, Pb, Cr and Cd) were low in all cases and did not exceed 0.06 mg/g. In most cases, higher metal contents were observed in the roots than in the shoots, but these differences were insignificant. In studies conducted on *P. australis*, it was proven that underground organs show a higher capacity to accumulate heavy metals compared to shoots [81]. The increased accumulation of heavy metals in underground organs may result, for example, from a defence mechanism that protects the species from the harmful effects of toxic concentrations on photosynthetic processes and prevents the translocation of toxic concentrations from roots to aboveground organs [86].

In the conducted studies, only in some cases a higher accumulation was observed in the roots, while in the remaining cases, similar metal contents were recorded in these parts of the plants.

As reported in the literature, the toxic concentration of Pb for different plants ranges from 0.01 mg/g to 0.1 mg/g [84]. *P. australis*, under natural conditions, grows on substrates where the heavy metal content does not exceed: Cd (<0.04 mg/g), Pb (<15.9 mg/g), Cu

(<0.275 mg/g), Cr (<0.218 mg/g) and Ni (<0.082 mg/g) [65]. Common reed has the ability to take up large amounts of micronutrients due to its extensive tissue system and defence mechanisms [64]. The ability of plants to accumulate nutrients and toxic elements, including heavy metals, has been confirmed by numerous researchers [87,88]. It has been shown that higher aquatic plants are able to accumulate metal ions to levels well beyond their physiological needs, but excess can cause deleterious effects in plants [88].

The conducted analysis of the chemical composition of *P. australis* roots and shoots showed that the levels of Na, Mg, Zn and TP differed slightly in all the analysed variants. Most of the analysed parameters were higher in the external solution than in the *P. australis* roots. Higher concentrations in the external solution were observed in both *P. australis* and *C. demersum*, which can be justified by the fact that only a limited number of elements move easily, thus being absorbed by the plants [40,89]. However, it also happened that higher values were recorded in *P. australis* roots than in the external solution, e.g., for: Cr, Pb and Ni (after exposure to the leachate from Bielawa, Wrocław and Jawor) as well as Zn and Mn (after exposure to the leachate from Wrocław) and Fe (after exposure to the leachate from Bielawa). Additionally, in the case of the *P. australis* shoots, higher contents were recorded than in the external solution in several cases. Higher concentrations in shoots were recorded for the same parameters as in the roots, i.e., Fe, Mn, Zn, Cr, Pb and Ni. In addition, a several times higher concentration of TN was observed in the *P. australis* shoots after exposure to the leachate from the Wrocław landfill than in the external solution.

3.2. Analysis of Similarities between the Composition of *C. demersum* and *P. australis*

The assessment of the similarity between the chemical composition of the plant samples after the end of the exposure to different concentrations of leachate from the landfills included in the study was carried out using cluster analysis. The dendrogram diagram for the analysed samples of rigid hornwort (Figure 5a) shows two main clusters (A and B). Cluster A includes only samples of plants treated with the leachate from the Wrocław and Jawor landfills. The greatest similarity is between the samples 6.25 and 12.5 W and 12.5, 50 and 100 J, which form the smallest clusters. Cluster B is divided into B′ and B″. Cluster B′ is dominated by the samples of the plants treated with the leachate from the Legnica landfill. The smallest clusters are formed by the samples 25 L, 25 J and 12.5 L and 50, 100 L and 50 W. Cluster B″ includes all the samples of plants treated with the leachate from the landfill in Bielawa and also the control sample (0). The highest similarity is found between the samples 25 B, 6.25 J and 0, 6.25 B and 25 W and between the samples 12.5B, 6.25 L, 50 and 100 B and 100 W. The analysis of the clusters forming the dendrogram for the examined rigid hornwort samples shows a fairly clear influence of the source of origin of the treated leachate (landfill) on the formation of individual clusters. The similarity between the composition of the plants treated with the most diluted leachate from the oldest, non-operational landfill in Wrocław and the plants subjected to the highest concentrations of the leachate from the Jawor landfill, active during the study, is also evident. The low degree of the contamination of the leachate from the Jawor landfill is confirmed by the similarity between the plants treated with the sample with the lowest concentration and the control sample, shown in the figure. The samples of the plants treated with the leachate from the other landfills (in Bielawa and Legnica) were found in separate clusters and the distances between those clusters prove high differences in the composition of those samples.

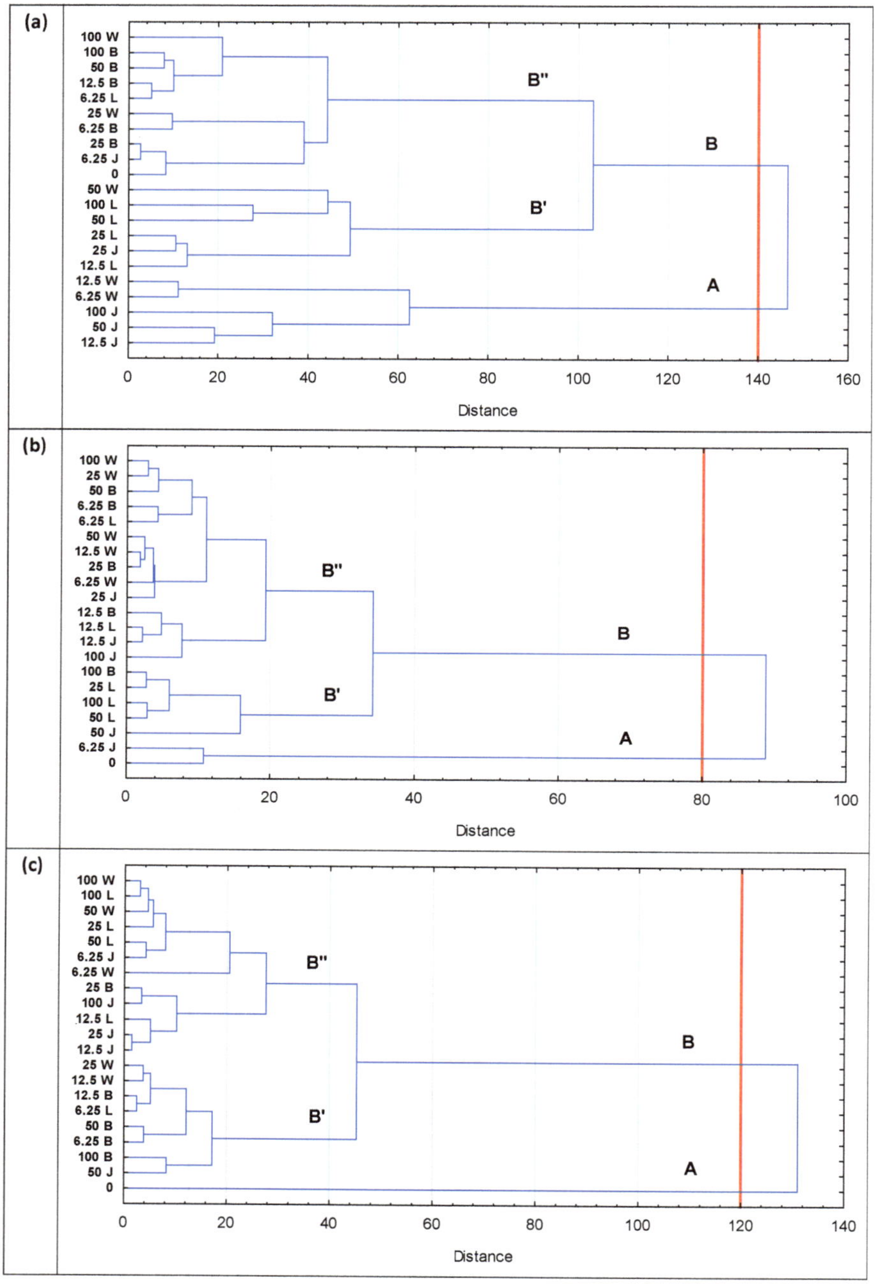

Figure 5. Dendrogram showing the similarities between the analysed samples: (**a**) *Ceratophyllum demersum*, (**b**) *Phragmites australis*—roots, and (**c**) *Phragmites australis*—shoots. Agglomeration was carried out using Ward's method.

Figure 5b shows a dendrogram illustrating the results of the cluster analysis for the reed root samples used to treat the different concentrations of leachate from the four landfills included in the study. On the basis of similarities, the samples were divided into two main clusters (A and B). Cluster A includes the control sample (0) and the roots of the reed used to treat the leachate from the Jawor landfill at the lowest concentration (6.25 J). All the other samples differed from the two mentioned above significantly enough to be included in the second cluster (B). Cluster B is divided into B′ and B″. Cluster B′ mainly includes samples of the reed roots used for the treatment of the leachate from the Legnica landfill (25, 50 and 100 L) as well as the samples of 100 B and 50 J. The biggest part of cluster B′ are the samples of the reed roots used for the treatment of the leachate from the Wrocław landfill (all) and the samples for the Bielawa landfill (except for 100 B). The presented dendrogram illustrates clear differences between the control sample and the reed root samples used for landfill leachate treatment. A difference is also evident between the majority of the samples for the Legnica landfill (25, 50 and 100 L, cluster B′) and the remaining samples, mainly corresponding to the Wrocław and Bielawa landfills (cluster B″).

Figure 5 c shows the results of the cluster analysis for the reed shoot samples used to treat the different concentrations of leachate from the four landfills included in the study. The largest differences were between the control sample (0, cluster A) and the other samples (cluster B). In cluster B, there are also two smaller differing samples (B′ and B″). Cluster B′ mainly includes samples of the reed shoots used for treating the leachate from the Bielawa landfill (except for 25 B) and corresponds to the lower concentrations of the leachate from the landfills in Wrocław (12.5 and 25 W), Legnica (6.25 L), and also sample 50 J. Cluster B″ consists mainly of the samples of the reed shoots used for treating the leachate from the landfills in Legnica (100, 50, 25 and 12.5 L), Jawor (100, 25, 12.5 J) and Wrocław (6.25, 50 and 100 W). The presented dendrogram illustrates in particular the clear difference between the samples of the reed shoots used for the landfill leachate treatment and the control sample (0). This difference is even greater than in the analysis performed for the reed roots, which is related to the placement of the cut-off level (defining the number of significant clusters) at a greater distance. The remaining clusters show a greater mixing of samples, which indicates the occurrence of similarities in chemical composition.

3.3. Bioconcentration Factor (BCF)

The bioconcentration factor (BCF) determines the ability of a plant to accumulate elements from the substrate [90]. Table 2 shows the calculated bioconcentration factor (BCF) values for two plants, i.e., *Phragmites australis* and *Ceratophyllum demersum*, after exposure to the leachate from the four municipal waste landfills and in the control samples.

The (BCF) values in the presented studies ranged from 0 to 10^1. A BCF above one is a key feature of hyperaccumulators [91,92]. The BCF can reach different orders of magnitude. Maderra-Parra et al. [31], in studies on the phytoremediation of landfill leachate, obtained BCFs ranging from 10^0 to 10^2. Pandey et al. [82] obtained similar BCF values. It is reported in the literature that a good bioaccumulator should exhibit a BCF above 1000 [93,94].

The BCF for *P. australis* and *C. demersum*, after exposure to the leachate from the landfills in Legnica and Jawor, was <1 in each variant. However, in the case of the other two landfills, i.e., Bielawa and Wrocław, the BCF values above one were obtained only for Cr for *P. australis* and TN for *C. demersum*.

In the conducted studies, for *C. demersum*, the BCF for Cr was <1 in all cases. However, Abdallah [46], in a study on the phytoremediation of metals from aqueous solutions by *C. demersum*, proved that it is a good accumulator of Cr, obtaining BCF values > 1000 in all cases. The BCF for Cu in all cases for *P.australis* was below one. However, after the exposure of *C. demersum*, the BCF for Cu only in the control sample was above one (i.e., BCF—4.59). The highest BCF value for *Ceratophyllum demersum* was recorded for Mn in the control sample and it amounted to 32.25. A higher BCF value was also obtained for Fe after exposure to the leachate from the Wrocław landfill (BCF—23.91).

Table 2. Bioconcentration factor (BCF) for *Phragmites australis* and *Ceratophyllum demersum* (for *P. australis*, the average for roots and stems was taken).

Pollution Indicators	*Phragmites australis*					*Ceratophyllum demersum*				
	Active Landfills		Non-Operational Landfills		Control Sample	Active Landfills		Non-Operational Landfills		Control Sample
	Legnica	Jawor	Bielawa	Wrocław		Legnica	Jawor	Bielawa	Wrocław	
TP	0.00	0.00	0.00	0.00	0.04	0.00	0.00	0.00	0.00	0.09
TKN	0.00	0.00	0.00	0.01	**12.74**	0.00	0.00	0.00	0.01	**8.13**
Sodium	0.00	0.00	0.00	0.00	0.00	0.00	0.00	0.00	0.00	0.00
Potassium	0.00	0.00	0.00	0.00	0.01	0.00	0.00	0.00	0.00	0.03
Calcium	0.00	0.00	0.00	0.00	0.00	0.00	0.00	0.00	0.00	0.00
Magnesium	0.00	0.00	0.00	0.00	0.00	0.00	0.00	0.00	0.00	0.00
Iron	0.00	0.00	0.01	0.01	0.21	0.00	0.00	0.01	**23.91**	0.40
Manganese	0.00	0.00	0.00	0.00	**5.59**	0.00	0.00	0.01	0.01	**32.25**
Copper	0.00	0.09	0.60	0.30	**4.59**	0.00	0.01	0.00	0.02	0.80
Zinc	0.00	0.00	0.00	0.00	0.00	0.00	0.00	0.00	0.00	0.00
Chromium	0.00	0.00	**4.62**	**1.02**	**1.65**	0.00	0.00	0.45	0.41	0.64
Cadmium	0.00	0.00	0.00	0.00	0.39	0.00	0.00	0.00	0.00	0.00
Nickel	0.00	0.00	0.00	0.00	0.00	0.00	0.00	0.00	0.00	0.00
Lead	0.00	0.00	0.00	0.00	0.00	0.00	0.00	0.00	0.00	0.00

BCF values above 1 are shown in bold in the table.

In the case of the *Phragmites australis* exposure, the highest BCF value was obtained for TN in the control sample (BCF—12.74). Similar values to those obtained in this study were obtained by Sochacki et al. during a study conducted on *P. australis*, where the BCF ranged up to 10^1 and they concluded that this species does not play an important role as a metal accumulator [95]. On the other hand, Daud et al. [32], investigating the potential of *Lemna minor* in the phytoremediation of landfill leachate, obtained much lower BCF values; in each case, these were less than one. The discrepancies in the BCF values obtained indicate that different plant species develop different mechanisms for tolerating or accumulating heavy metals, depending on specific environmental conditions [96]. In turn, the ability to accumulate heavy metals in plants depends on their ability to uptake and transport intracellularly, which includes, inter alia, mobilization in the rhizosphere and the transport of metals across the plasma membrane of root cells [97].

It is reported in the literature that submergent plant species are able to accumulate a higher amount of heavy metals compared to emergent species due to the fact that they can remove heavy metals from the water with their entire surface [21,98]. However, no significant differences were observed between *C. demersum* and *P. australis* in the study conducted. Additionally, Vymazal et al. [85] did not observe any significant differences between the plants, and the element concentrations found in both plants (*Phragmites and Phalaris*) were at very similar levels.

3.4. Translocation Factor (TF)

The ability to translocate pollutants in different parts of the plant can be assessed using the translocation factor (TF). Figure 6 shows the calculated translocation factor (TF) values for *P. australis* after exposure to the leachate from the four municipal waste landfills and in the control sample.

The uptake and translocation of individual components by plants from water or soil, in addition to their availability, are also influenced by ongoing reactions (antagonistic and/or synergistic) between individual elements [81]. A shoot:root translocation ratio (S/R) greater than one indicates the efficient transport of components from roots to shoots and is a key feature of all hyperaccumulators [99]. In contrast, a TF below one indicates that pollutants are accumulated in the lower part of the plant [100].

The translocation factor varies depending on the type of pollutant [101], as confirmed by the analyses performed. Its magnitude is also influenced by the plant species. It has been observed that some plant species store components in roots (i.e., root accumulators) and some in shoots (i.e., shoot accumulators). The translocation factor may also be influenced by changes in the epidermis and mesoderm of plants, e.g., caused by an increased content of metals, which hinders the migration of nutrients from the roots to the shoots [97].

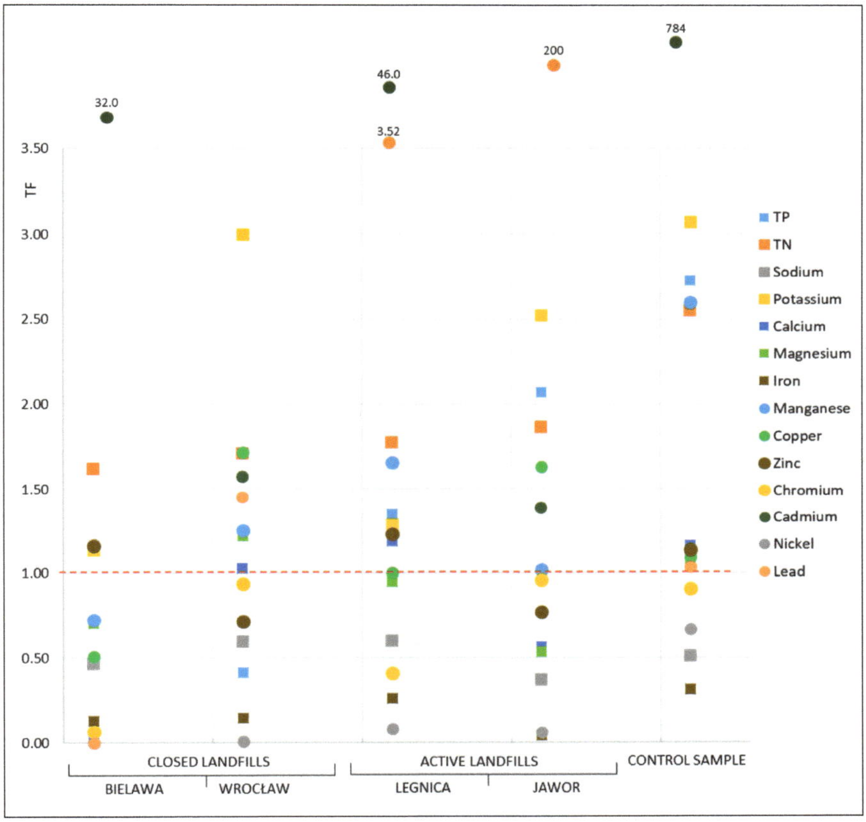

Figure 6. Translocation factor (TF) of pollutants of *Phragmites australis* in leachate from four municipal waste landfills.

In the case of Na, Fe, Cr and Ni, no transposition into the aboveground parts of *P. australis* was observed, either after exposure to the leachate from both active and closed facilities or in the control sample. The low TF for Na, Fe, Cr and Ni in the aboveground parts of *P. australis* can be explained by the absorption and/or adsorption of ions from the aqueous solution through the roots, as the shoots under normal conditions do not play such a role [92]. The low TF values for nickel can be further justified by the fact that nickel is preferentially adsorbed by mineral components [102]. Additionally, Bonanno et al. [81] observed that Cr shows low mobility from roots to shoots in *P. australis* tissues. It is also reported in the literature that Fe shows high accumulation in roots but low translocation from roots to the remaining organs [40], which is consistent with the results obtained in the present study.

Other authors, during studies conducted on hydrophytic species, also found no or a minimal transposition of components to the aboveground parts of plants, observing the highest accumulation of metals in the roots [24,25]. Stoltz and Greger [103], in studies conducted, among others, on *P. australis,* observed that most of the micronutrients they studied (i.e., Cd, Cu, Zn and As) accumulated only in the plant roots and did not move to the shoots, with the exception of Pb, which accumulated both in the roots and shoots. Additionally, Bonanno and Giudice [81], during their study on the bioaccumulation of metals (i.e., Cd, Cr, Cu, Hg, Mn, Ni, Pb and Zn) by *P. australis,* showed low mobility from the roots to the aboveground organs, proving that underground organs are the main zones of metal accumulation. It has been observed in many studies that the concentration of elements in the roots is much higher than in the leaves and shoots, which confirms that the roots are the main site for metal uptake [104].

In the presented study, only in the case of cadmium transposition to the aboveground parts of *P. australis* was recorded in each of the analysed variants. The highest TF for cadmium was recorded in the control sample, where it was 784.

In the case of Mn, a low TF was recorded only after exposure to the leachate from the closed landfill in Bielawa, which means accumulation in the roots. In all other cases, the movement of Mn from the roots to the shoots was observed.

The Zn accumulated both in the roots (in the case of exposure to the leachate from the landfills in Wrocław and Jawor) and moved to the aboveground parts (in the case of exposure to the leachate from the landfill in Bielawa, Legnica and the control sample).

4. Conclusions

The analysis of micro- and macroelements in two selected plants, i.e., emerging macrophytes (*P. australis*) and submerged macrophytes (*C. demersum*), treated with leachate from four municipal landfills, showed that:

1. The concentrations of the heavy metals (Cu, Ni, Pb, Cr and Cd) in the studied plants were low in all the analysed cases. Higher metal contents could often be observed in the roots rather than in the shoots, but these differences were insignificant.
2. The chemical composition of the studied plant samples was primarily related to the source of origin of the treated leachate (landfill), as clearly demonstrated by cluster analysis for the *C. demersum* samples. However, in the case of the roots and shoots of *P. australis*, significant differences were observed between the control and leachate-treated samples. In many cases, similarities were found between plant samples after exposure to different leachate concentrations, which was due to minor differences in their composition.
3. The conducted studies showed that *P. australis* was not effective in translocating certain elements (i.e., Na, Fe, Cr and Ni). The TF in all cases was below 1.0, indicating that translocation with *P. australis* may not be an effective option for removing high concentrations of pollutants contained in leachate. Cadmium showed the highest translocation to the aboveground parts of *P. australis* (in each of the analysed variants).
4. The bioconcentration factor BCF, for the majority of the analysed parameters in *P. australis* and *C. demersum*, remained at a low level. It reached values > 1 only in a few cases, which demonstrates that both plants do not exhibit the features of a good hyperaccumulator of the pollutants contained in leachate.
5. In the conducted studies, no significant differences were observed in the accumulation of the studied components between submergent plants (*C. demersum*) and emergent macrophytes (*P. australis*).

Author Contributions: Conceptualization, A.W. and A.S.-P.; methodology, A.W. and A.S.-P.; software, A.S.-P.; validation, A.S.-P.; formal analysis, A.W. and A.S.-P.; investigation, A.W.; resources, A.W. and A.S.-P.; data curation, A.W.; writing—original draft preparation, A.W.; writing—review and editing, A.S.-P.; visualization, A.W. and A.S.-P.; supervision, A.S.-P.; funding acquisition, A.W. All authors have read and agreed to the published version of the manuscript.

Funding: This research was funded by B010/0004/20 Maintenance and development of the research potential of chairs and institutes—subsidy 2020 -IKSIG—Institute of Environmental Engineering. Wrocław University of Environmental and Life Sciences (Poland) as the Ph.D. research programme "Innovative Doctorate", no. N070/0001/21. The APC is financed by Wroclaw University of Environmental and Life Sciences.

Institutional Review Board Statement: Not applicable.

Informed Consent Statement: Not applicable.

Data Availability Statement: Not applicable.

Conflicts of Interest: The authors declare no conflict of interest.

References

1. Madera-Parra, C.A.; Peña, M.R.; Peña, E.J.; Lens, P.N.L. Cr(VI) and COD removal from landfill leachate by polyculture constructed wetland at a pilot scale. *Environ. Sci. Pollut. Res.* **2015**, *22*, 12804–12815. [CrossRef] [PubMed]
2. Wdowczyk, A.; Szymańska-Pulikowska, A.; Domańska, M. Analysis of the Bacterial Biocenosis of Activated Sludge Treated with Leachate from Municipal Landfills. *Int. J. Environ. Res. Public Health* **2022**, *19*, 1801. [CrossRef] [PubMed]
3. Ziyang, L.; Youcai, Z.; Tao, Y.; Yu, S.; Huili, C.; Nanwen, Z.; Renhua, H. Natural attenuation and characterization of contaminants composition in landfill leachate under different disposing ages. *Sci. Total Environ.* **2009**, *407*, 3385–3391. [CrossRef] [PubMed]
4. Lee, A.H.; Nikraz, H.; Hung, Y.T. Influence of Waste Age on Landfill Leachate Quality. *Int. J. Environ. Sci. Dev.* **2010**, *1*, 347–350. [CrossRef]
5. Naveen, B.P.; Mahapatra, D.M.; Sitharam, T.G.; Sivapullaiah, P.V.; Ramachandra, T.V. Physico-chemical and biological characterization of urban municipal landfill leachate. *Environ. Pollut.* **2017**, *220*, 1–12. [CrossRef]
6. Aziz, H.A.; Umar, M.; Yusoff, M.S. Variability of parameters involved in leachate pollution index and determination of LPI from four landfills in Malaysia. *Int. J. Chem. Eng.* **2010**, *2010*, 747953. [CrossRef]
7. Umar, M.; Aziz, H.A.; Yusoff, M.S. Trends in the use of Fenton, electro-Fenton and photo-Fenton for the treatment of landfill leachate. *Waste Manag.* **2010**, *30*, 2113–2121. [CrossRef]
8. Wdowczyk, A.; Szymańska-Pulikowska, A.; Gałka, B. Removal of selected pollutants from landfill leachate in constructed wetlands with different filling. *Bioresour. Technol.* **2022**, *353*, 127136. [CrossRef]
9. Charazińska, S.; Lochyński, P.; Burszta-Adamiak, E. Removal of heavy metal ions form acidic electrolyte for stainless steel electropolishing via adsorption using Polish peats. *J. Water Process Eng.* **2021**, *42*, 102169. [CrossRef]
10. Wdowczyk, A.; Szymańska-Pulikowska, A. Analysis of the possibility of conducting a comprehensive assessment of landfill leachate contamination using physicochemical indicators and toxicity test. *Ecotoxicol. Environ. Saf.* **2021**, *221*, 112434. [CrossRef]
11. Lu, M.-C.; Chen, Y.Y.; Chiou, M.-R.; Chen, M.Y.; Fan, H.-J. Occurrence and treatment efficiency of pharmaceuticals in landfill leachates. *Waste Manag.* **2016**, *55*, 257–264. [CrossRef] [PubMed]
12. Moody, C.M.; Townsend, T.G. A comparison of landfill leachates based on waste composition. *Waste Manag.* **2017**, *63*, 267–274. [CrossRef] [PubMed]
13. Wowkonowicz, P.; Kijeńska, M. Phthalate release in leachate from municipal landfills of central Poland. *PLoS ONE* **2017**, *12*, 0174986. [CrossRef] [PubMed]
14. Gupta, A.; Paulraj, R. Leachate composition and toxicity assessment: An integrated approach correlating physicochemical parameters and toxicity of leachates from MSW landfill in Delhi. *Environ. Technol.* **2017**, *38*, 1599–1605. [CrossRef] [PubMed]
15. Long, Y.Y.; Shen, D.S.; Wang, H.T.; Lu, W.J. Migration behavior of Cu and Zn in landfill with different operation modes. *J. Hazard. Mater.* **2010**, *179*, 883–890. [CrossRef] [PubMed]
16. Gworek, B.; Dmuchowski, W.; Koda, E.; Marecka, M.; Baczewska, A.H.; Bragoszewska, P.; Sieczka, A.; Osiński, P. Impact of the municipal solid waste Lubna landfill on environmental pollution by heavy metals. *Water* **2016**, *8*, 470. [CrossRef]
17. Tejera, J.; Hermosilla, D.; Gascó, A.; Miranda, R.; Alonso, V.; Negro, C.; Blanco, Á. Treatment of mature landfill leachate by electrocoagulation followed by Fenton or UVA-LED photo-Fenton processes. *J. Taiwan Inst. Chem. Eng.* **2021**, *119*, 33–44. [CrossRef]
18. Tejera, J.; Gascó, A.; Hermosilla, D.; Alonso-Gomez, V.; Negro, C.; Blanco, Á. Uva-led technology's treatment efficiency and cost in a competitive trial applied to the photo-fenton treatment of landfill leachate. *Processes* **2021**, *9*, 1026. [CrossRef]
19. Tejera, J.; Hermosilla, D.; Gascó, A.; Negro, C.; Blanco, Á. Combining coagulation and electrocoagulation with uva-led photo-fenton to improve the efficiency and reduce the cost of mature landfill leachate treatment. *Molecules* **2021**, *26*, 6425. [CrossRef]
20. Bello, A.O.; Tawabini, B.S.; Khalil, A.B.; Boland, C.R.; Saleh, T.A. Phytoremediation of cadmium-, lead- and nickel-contaminated water by Phragmites australis in hydroponic systems. *Ecol. Eng.* **2018**, *120*, 126–133. [CrossRef]
21. Gałczyńska, M.; Mańkowska, N.; Milke, J.; Buśko, M. Possibilities and limitations of using Lemna minor, Hydrocharis morsus-ranae and Ceratophyllum demersum in removing metals with contaminated water. *J. Water Land Dev.* **2019**, *40*, 161–173. [CrossRef]
22. Guidi Nissim, W.; Palm, E.; Pandolfi, C.; Mancuso, S.; Azzarello, E. Willow and poplar for the phyto-treatment of landfill leachate in Mediterranean climate. *J. Environ. Manage.* **2021**, *277*, 111454. [CrossRef] [PubMed]
23. Bauddh, K.; Singh, B.; Korstad, J. *Phytoremediation Potential of Bioenergy Plants*; Springer: Berlin/Heidelberg, Germany, 2017; ISBN 9789811030840.
24. Kennen, K.; Kirkwood, N. *Phyto: Principles and Resources for Site Remediation and Landscape Design*; Routledge: London, UK, 2015.
25. Kumar Yadav, K.; Gupta, N.; Kumar, A.; Reece, L.M.; Singh, N.; Rezania, S.; Ahmad Khan, S. Mechanistic understanding and holistic approach of phytoremediation: A review on application and future prospects. *Ecol. Eng.* **2018**, *120*, 274–298. [CrossRef]
26. Küpper, H.; Parameswaran, A.; Leitenmaier, B.; Trtílek, M.; Šetlík, I. Cadmium-induced inhibition of photosynthesis and long-term acclimation to cadmium stress in the hyperaccumulator Thlaspi caerulescens. *New Phytol.* **2007**, *175*, 655–674. [CrossRef] [PubMed]

27. Milke, J.; Gałczyńska, M.; Wróbel, J. The importance of biological and ecological properties of Phragmites Australis (Cav.) Trin. Ex steud., in phytoremendiation of aquatic ecosystems-The review. *Water* **2020**, *12*, 1770. [CrossRef]
28. Kumar, V.; Singh, J.; Saini, A.; Kumar, P. Phytoremediation of copper, iron and mercury from aqueous solution by water lettuce (*Pistia stratiotes* L.). *Environ. Sustain.* **2019**, *2*, 55–65. [CrossRef]
29. Gajewska, M.; Skrzypiec, K.; Jóźwiakowski, K.; Mucha, Z.; Wójcik, W.; Karczmarczyk, A.; Bugajski, P. Kinetics of pollutants removal in vertical and horizontal flow constructed wetlands in temperate climate. *Sci. Total Environ.* **2020**, *718*, 137371. [CrossRef]
30. Gizińska-Górna, M.; Jóźwiakowski, K.; Marzec, M. Reliability and Efficiency of Pollutant Removal in Four-Stage Constructed Wetland of SSVF-SSHF-SSHF-SSVF Type. *Water* **2020**, *12*, 3153. [CrossRef]
31. Madera-Parra, C.A.; Peña-Salamanca, E.J.; Peña, M.R.; Rousseau, D.P.L.; Lens, P.N.L. Phytoremediation of Landfill Leachate with Colocasia esculenta, Gynerum sagittatum and Heliconia psittacorum in Constructed Wetlands. *Int. J. Phytoremediation* **2015**, *17*, 16–24. [CrossRef]
32. Daud, M.K.; Ali, S.; Abbas, Z.; Zaheer, I.E.; Riaz, M.A.; Malik, A.; Hussain, A.; Rizwan, M.; Zia-Ur-Rehman, M.; Zhu, S.J. Potential of Duckweed (*Lemna minor*) for the Phytoremediation of Landfill Leachate. *J. Chem.* **2018**, *2018*, 3951540. [CrossRef]
33. Zhang, C.B.; Wang, J.; Liu, W.L.; Zhu, S.X.; Ge, H.L.; Chang, S.X.; Chang, J.; Ge, Y. Effects of plant diversity on microbial biomass and community metabolic profiles in a full-scale constructed wetland. *Ecol. Eng.* **2010**, *36*, 62–68. [CrossRef]
34. Ghassemzadeh, F.; Yousefzadeh, H.; Arbab-Zavar, M.H. Arsenic phytoremediation by Phragmites australis: Green technology. *Int. J. Environ. Stud.* **2008**, *65*, 587–594. [CrossRef]
35. Mykleby, P.M.; Lenters, J.D.; Cutrell, G.J.; Herrman, K.S.; Istanbulluoglu, E.; Scott, D.T.; Twine, T.E.; Kucharik, C.J.; Awada, T.; Soylu, M.E.; et al. Energy and water balance response of a vegetated wetland to herbicide treatment of invasive Phragmites australis. *J. Hydrol.* **2016**, *539*, 290–303. [CrossRef]
36. Hechmi, N.; Aissa, N.B.; Abdenaceur, H.; Jedidi, N. Evaluating the phytoremediation potential of Phragmites australis grown in pentachlorophenol and cadmium co-contaminated soils. *Environ. Sci. Pollut. Res.* **2014**, *21*, 1304–1313. [CrossRef]
37. Bragato, C.; Schiavon, M.; Polese, R.; Ertani, A.; Pittarello, M.; Malagoli, M. Seasonal variations of Cu, Zn, Ni and Cr concentration in *Phragmites australis* (Cav.) *Trin ex steudel* in a constructed wetland of North Italy. *Desalination* **2009**, *246*, 35–44. [CrossRef]
38. Rezania, S.; Park, J.; Rupani, P.F.; Darajeh, N.; Xu, X. Phytoremediation potential and control of Phragmites australis as a green phytomass : An overview. *Environ. Sci. Pollut. Res.* **2019**, *26*, 7428–7441. [CrossRef]
39. Duman, F.; Cicek, M.; Sezen, G. Seasonal changes of metal accumulation and distribution in common club rush (*Schoenoplectus lacustris*) and common reed (*Phragmites australis*). *Ecotoxicology* **2007**, *16*, 457–463. [CrossRef]
40. Bonanno, G. Trace element accumulation and distribution in the organs of Phragmites australis (common reed) and biomonitoring applications. *Ecotoxicol. Environ. Saf.* **2011**, *74*, 1057–1064. [CrossRef]
41. Zhao, C.; Zhang, G.; Jiang, J. Enhanced phytoremediation of bisphenol a in polluted lake water by seedlings of ceratophyllum demersum and myriophyllum spicatum from in vitro culture. *Int. J. Environ. Res. Public Health* **2021**, *18*, 810. [CrossRef]
42. Zhang, G.; Wang, Y.; Jiang, J.; Yang, S. Bisphenol A Removal by Submerged Macrophytes and the Contribution of Epiphytic Microorganisms to the Removal Process. *Bull. Environ. Contam. Toxicol.* **2017**, *98*, 770–775. [CrossRef]
43. Dhir, B.; Sharmila, P.; Saradhi, P.P. Potential of aquatic macrophytes for removing contaminants from the environment. *Crit. Rev. Environ. Sci. Technol.* **2009**, *39*, 754–781. [CrossRef]
44. Lombardo, P.; Dennis Cooke, G. Ceratophyllum demersum—Phosphorus interactions in nutrient enriched aquaria. *Hydrobiologia* **2003**, *497*, 79–90. [CrossRef]
45. Maine, M.A.; Duarte, M.V.; Suñé, N.L. Cadmium uptake by floating macrophytes. *Water Res.* **2001**, *35*, 2629–2634. [CrossRef]
46. Abdallah, M.A.M. Phytoremediation of heavy metals from aqueous solutions by two aquatic macrophytes, *Ceratophyllum demersum* and *Lemna gibba* L. *Environ. Technol.* **2012**, *33*, 1609–1614. [CrossRef] [PubMed]
47. Szymańska-Pulikowska, A. Changes in the content of selected heavy metals in groundwater exposed to the impact of a municipal landfill site. *J. Elem.* **2012**, *17*, 689–702. [CrossRef]
48. Tomczyk, P.; Wiatkowski, M. The Effects of Hydropower Plants on the Physicochemical Parameters of the Bystrzyca River in Poland. *Energies* **2021**, *14*, 2075. [CrossRef]
49. Białowiec, A.; Koziel, J.A.; Manczarski, P. Stomatal conductance measurement for toxicity assessment in zero-effluent constructed-wetlands: Effects of landfill leachate on hydrophytes. *Int. J. Environ. Res. Public Health* **2019**, *16*, 468. [CrossRef]
50. Antić, K.; Pap, S.; Novaković, M.; Sekulić, M.T.; Adamović, D.; Radonić, J. Removal of Heavy Metal Ions from Landfill Leachate by Phytoremediation Using Eichhornia crassipes. *Ind. Wastewater Treat. Other Top. Remov.* **2018**, 427–433.
51. Tahmasbian, I.; Safari Sinegani, A.A. Improving the efficiency of phytoremediation using electrically charged plant and chelating agents. *Environ. Sci. Pollut. Res.* **2016**, *23*, 2479–2486. [CrossRef]
52. Szymańska-Pulikowska, A.; Wdowczyk, A. Changes of a landfill leachate toxicity as a result of treatment with Phragmites australis and Ceratophyllum demersum—A case study. *Front. Environ. Sci.* **2021**, *9*, 1–15. [CrossRef]
53. Huang, W.; Ratkowsky, D.A.; Hui, C.; Wang, P.; Su, J.; Shi, P. Leaf fresh weight versus dry weight: Which is better for describing the scaling relationship between leaf biomass and leaf area for broad-leaved plants? *Forests* **2019**, *10*, 256. [CrossRef]
54. Saifullah, W.; Khan, N. Profiling of various elements in Haloxylon griffithii and Convolvulus leiocalycinus using atomic absorption spectroscopy and flame photometry. *Pure Appl. Biol.* **2019**, *8*, 1535–1542. [CrossRef]
55. Toropov, A.A.; Toropova, A.P.; Lombardo, A.; Roncaglioni, A.; Benfenati, E.; Gini, G. CORAL: Building up the model for bioconcentration factor and defining it's applicability domain. *Eur. J. Med. Chem.* **2011**, *46*, 1400–1403. [CrossRef] [PubMed]

56. Wu, Q.; Wang, S.; Thangavel, P.; Li, Q.; Zheng, H.; Bai, J.; Qiu, R. Phytostabilization potential of *Jatropha curcas* L. in polymetallic acid mine tailings. *Int. J. Phytoremediation* **2011**, *13*, 788–804. [CrossRef] [PubMed]
57. Zayed, A.; Gowthaman, S.; Terry, N. Phytoaccumulation of trace elements by wetland plants: I. Duckweed. *J. Environ. Qual.* **1998**, *27*, 715–721. [CrossRef]
58. Mongi, C.E.; Langi, Y.A.R.; Montolalu, C.E.J.C.; Nainggolan, N. Comparison of hierarchical clustering methods (case study: Data on poverty influence in North Sulawesi). In Proceedings of the IOP Conference Series: Materials Science and Engineering, the 3rd Indonesian Operations Research Association—International Conference on Operations Research 2018, Manado, Indonesia, 20–21 September 2018; IOP Publishing: Bristol, UK, 2018; Volume 567.
59. Majerova, I.; Nevima, J. The measurement of human development using the ward method of cluster analysis. *J. Int. Stud.* **2017**, *10*, 239–257. [CrossRef]
60. Eszergár-Kiss, D.; Caesar, B. Definition of user groups applying Ward's method. In *Transportation Research Procedia*; ScienceDirect: Amsterdam, The Netherlands, 2017; Volume 22.
61. Wdowczyk, A.; Szymańska-Pulikowska, A. Comparison of Landfill Leachate Properties by LPI and Phytotoxicity—A Case Study. *Front. Environ. Sci.* **2021**, *9*, 1–14. [CrossRef]
62. Jorstad, L.B.; Jankowski, J.; Acworth, R.I. Analysis of the distribution of inorganic constituents in a landfill leachate-contaminated aquifer: Astrolabe Park, Sydney, Australia. *Environ. Geol.* **2004**, *46*, 263–272. [CrossRef]
63. Gillingham, M.D.; Gomes, R.L.; Ferrari, R.; West, H.M. Sorption, separation and recycling of ammonium in agricultural soils: A viable application for magnetic biochar? *Sci. Total Environ.* **2021**, *812*, 151440. [CrossRef]
64. Nawrot, N.; Wojciechowska, E.; Pazdro, K.; Szmagli, J.; Pempkowiak, J. Science of the Total Environment Uptake, accumulation, and translocation of Zn, Cu, Pb, Cd, Ni, and Cr by P. australis seedlings in an urban dredged sediment mesocosm : Impact of seedling origin and initial trace metal content. *Sci. Total Environ.* **2021**, *768*, 144983. [CrossRef]
65. Packer, J.G.; Meyerson, L.A.; Skálová, H.; Pyšek, P.; Kueffer, C. Biological Flora of the British Isles: Phragmites australis. *J. Ecol.* **2017**, *105*, 1123–1162. [CrossRef]
66. Gao, J.; Ma, N.; Zhou, J.; Wang, W.; Xiong, Z.; Mba, F.O.; Chen, N. Peroxidation damage and antioxidative capability of Ceratophyllum demersum under NH4+-N stress. *J. Freshw. Ecol.* **2012**, *27*, 539–549. [CrossRef]
67. Ding, X.; Jiang, Y.; Zhao, H.; Guo, D.; He, L.; Liu, F.; Zhou, Q.; Nandwani, D.; Hui, D.; Yu, J. Electrical conductivity of nutrient solution influenced photosynthesis, quality, and antioxidant enzyme activity of pakchoi (*Brassica campestris* L. ssp. Chinensis) in a hydroponic system. *PLoS ONE* **2018**, *13*, 0202090. [CrossRef] [PubMed]
68. Signore, A.; Serio, F.; Santamaria, P. A targeted management of the nutrient solution in a soilless tomato crop according to plant needs. *Front. Plant Sci.* **2016**, *7*, 391. [CrossRef] [PubMed]
69. Huang, L.; Liu, X.; Wang, Z.; Liang, Z.; Wang, M.; Liu, M.; Suarez, D.L. Interactive effects of pH, EC and nitrogen on yields and nutrient absorption of rice (*Oryza sativa* L.). *Agric. Water Manag.* **2017**, *194*, 48–57. [CrossRef]
70. Wang, C.; Zhang, S.H.; Wang, P.F.; Hou, J.; Li, W.; Zhang, W.J. Metabolic adaptations to ammonia-induced oxidative stress in leaves of the submerged macrophyte *Vallisneria natans* (Lour.) Hara. *Aquat. Toxicol.* **2008**, *87*, 88–98. [CrossRef]
71. Mullins, G. *Phosphorus, Agriculture & The Environment*; Virginia Tech.: Blacksburg, VA, USA, 2009.
72. Christensen, T.H.; Kjeldsen, P.; Bjerg, P.L.; Jensen, D.L.; Christensen, J.B.; Baun, A.; Albrechtsen, H.J.; Heron, G. Biogeochemistry of landfill leachate plumes. *Appl. Geochem.* **2001**, *16*, 659–718. [CrossRef]
73. Millaleo, R.; Reyes-Díaz, M.; Ivanov, A.G.; Mora, M.L.; Alberdi, M. Manganese as essential and toxic element for plants: Transport, accumulation and resistance mechanisms. *J. Soil Sci. Plant Nutr.* **2010**, *10*, 470–481. [CrossRef]
74. Thomine, S.; Lanquar, V. Iron Transport and Signaling in Plants. In *Transporters and Pumps in Plant Signaling*; Springer: Berlin/Heidelberg, Germany, 2011; pp. 99–131.
75. Wdowczyk, A.; Szymańska-Pulikowska, A. Differences in the composition of leachate from active and non-operational municipal waste landfills in poland. *Water* **2020**, *12*, 3129. [CrossRef]
76. Fauziah, S.H.; Izzati, M.N.; Agamuthu, P. Toxicity on Anabas Testudineus: A Case Study of Sanitary Landfill Leachate. *Procedia Environ. Sci.* **2013**, *18*, 14–19. [CrossRef]
77. Alkassasbeh, J.Y.M.; Heng, L.Y.; Surif, S. Toxicity testing and the effect of landfill leachate in malaysia on behavior of common carp (*Cyprinus carpio* L., 1758; Pisces, Cyprinidae). *Am. J. Environ. Sci.* **2009**, *5*, 209–217. [CrossRef]
78. Arif, N.; Yadav, V.; Singh, S.; Singh, S.; Ahmad, P.; Mishra, R.K.; Sharma, S.; Tripathi, D.K.; Dubey, N.K.; Chauhan, D.K. Influence of High and Low Levels of Plant-Beneficial Heavy Metal Ions on Plant Growth and Development. *Front. Environ. Sci.* **2016**, *4*, 69. [CrossRef]
79. Berkaoui, M.E.; Adnani, M.E.; Hakkou, R.; Ouhammou, A.; Bendaou, N.; Smouni, A.; Bouranis, L. Phytostabilization of Phosphate Mine Wastes Used as a Store-and-Release Cover to Control Acid Mine Drainage in a Semiarid Climate. *Plants* **2021**, *10*, 900. [CrossRef]
80. Foroughi, M.; Najafi, P.; Toghiani, S.; Toghiani, A.; Honarjoo, N. Nitrogen Removals by Ceratophyllum Demersum from Wastewater. *J. Residuals Sci. Technol.* **2013**, *10*, 63–68.
81. Bonanno, G.; Lo Giudice, R. Heavy metal bioaccumulation by the organs of Phragmites australis (common reed) and their potential use as contamination indicators. *Ecol. Indic.* **2010**, *10*, 639–645. [CrossRef]

82. Samecka-Cymerman, A.; Kempers, A.J. Concentrations of heavy metals and plant nutrients in water, sediments and aquatic macrophytes of anthropogenic lakes (former open cut brown coal mines) differing in stage of acidification. *Sci. Total Environ.* **2001**, *281*, 87–98. [CrossRef]
83. Sawidis, T.; Chettri, M.K.; Zachariadis, G.A.; Stratis, J.A. Heavy metals in aquatic plants and sediments from water systems in macedonia, greece. *Ecotoxicol. Environ. Saf.* **1995**, *32*, 73–80. [CrossRef]
84. Nouri, J.; Lorestani, B.; Yousefi, N.; Khorasani, N.; Hasani, A.H.; Seif, F.; Cheraghi, M. Phytoremediation potential of native plants grown in the vicinity of Ahangaran lead-Zinc mine (Hamedan, Iran). *Environ. Earth Sci.* **2011**, *62*, 639–644. [CrossRef]
85. Vymazal, J.; Švehla, J.; Kröpfelová, L.; Chrastný, V. Trace metals in Phragmites australis and Phalaris arundinacea growing in constructed and natural wetlands. *Sci. Total Environ.* **2007**, *380*, 154–162. [CrossRef]
86. Bonanno, G.; Vymazal, J.; Cirelli, G.L. Translocation, accumulation and bioindication of trace elements in wetland plants. *Sci. Total Environ.* **2018**, *631*, 252–261. [CrossRef]
87. Rofkar, J.R.; Dwyer, D.F.; Bobak, D.M. Uptake and Toxicity of Arsenic, Copper, and Silicon in Azolla caroliniana and Lemna minor. *Int. J. Phytoremediation* **2014**, *16*, 155–166. [CrossRef] [PubMed]
88. Borisova, G.; Chukina, N.; Maleva, M.; Kumar, A.; Prasad, M.N.V. Thiols as biomarkers of heavy metal tolerance in the aquatic macrophytes of Middle Urals, Russia. *Int. J. Phytoremediation* **2016**, *18*, 1037–1045. [CrossRef] [PubMed]
89. Madejón, P.; Murillo, J.M.; Marañón, T.; Lepp, N.W. Factors affecting accumulation of thallium and other trace elements in two wild Brassicaceae spontaneously growing on soils contaminated by tailings dam waste. *Chemosphere* **2007**, *67*, 20–28. [CrossRef]
90. Mishra, T.; Pandey, V.C. *Phytoremediation of Red Mud Deposits Through Natural Succession*; Elsevier Inc.: Amsterdam, The Netherlands, 2018; ISBN 9780128139134.
91. Cluis, C. Junk-greedy greens: Phytoremediation as a new option for soil decontamination. *BioTeach J.* **2004**, *2*, 1–67.
92. Capozzi, F.; Sorrentino, M.C.; Caporale, A.G.; Fiorentino, N.; Giordano, S.; Spagnuolo, V. Exploring the phytoremediation potential of Cynara cardunculus: A trial on an industrial soil highly contaminated by heavy metals. *Environ. Sci. Pollut. Res.* **2020**, *27*, 9075–9084. [CrossRef]
93. Zhu, Y.L.; Zayed, A.M.; Qian, J.; Souza, M.; Terry, N. Phytoaccumulation of Trace Elements by Wetland Plants: II. Water Hyacinth. *J. Environ. Qual.* **1999**, *28*, 339–344. [CrossRef]
94. Hasan, S.H.; Talat, M.; Rai, S. Sorption of cadmium and zinc from aqueous solutions by water hyacinth (*Eichchornia crassipes*). *Bioresour. Technol.* **2007**, *98*, 918–928. [CrossRef]
95. Sochacki, A.; Guy, B.; Faure, O.; Surmacz-Górska, J. Accumulation of Metals and Boron in Phragmites australis Planted in Constructed Wetlands Polishing Real Electroplating Wastewater. *Int. J. Phytoremediation* **2015**, *17*, 1068–1072. [CrossRef]
96. Liu, J.G.; Li, G.H.; Shao, W.C.; Xu, J.K.; Wang, D. Variations in Uptake and Translocation of Copper, Chromium and Nickel Among Nineteen Wetland Plant Species. *Pedosphere* **2010**, *20*, 96–103. [CrossRef]
97. Minkina, T.; Fedorenko, G.; Nevidomskaya, D.; Fedorenko, A.; Chaplygin, V.; Mandzhieva, S. Morphological and anatomical changes of Phragmites australis Cav. due to the uptake and accumulation of heavy metals from polluted soils. *Sci. Total Environ.* **2018**, *636*, 392–401. [CrossRef]
98. Kara, Y. Bioaccumulation of Cu, Zn and Ni from the wastewater by treated Nasturtium officinale. *Int. J. Environ. Sci. Technol.* **2005**, *2*, 63–67. [CrossRef]
99. Zhao, F.J.; Jiang, R.F.; Dunham, S.J.; McGrath, S.P. Cadmium uptake, translocation and tolerance in the hyperaccumulator Arabidopsis halleri. *New Phytol.* **2006**, *172*, 646–654. [CrossRef]
100. Singh, R.; Singh, D.P.; Kumar, N.; Bhargava, S.K.; Barman, S.C. Accumulation and translocation of heavy metals in soil and plants from fly ash contaminated area. *J. Environ. Biol.* **2010**, *31*, 421–430.
101. Pandey, S.K.; Bhattacharya, T.; Chakraborty, S. Metal phytoremediation potential of naturally growing plants on fly ash dumpsite of Patratu thermal power station, Jharkhand, India. *Int. J. Phytoremediation* **2016**, *18*, 87–93. [CrossRef]
102. de Oliveira Mesquita, F.; Pedrosa, T.D.; Batista, R.O.; de Andrade, E.M. Translocation factor of heavy metals by elephant grass grown with varying concentrations of landfill leachate. *Environ. Sci. Pollut. Res.* **2021**, *28*, 43831–43841. [CrossRef]
103. Stoltz, E.; Greger, M. Accumulation properties of As, Cd, Cu, Pb and Zn by four wetland plant species growing on submerged mine tailings. *Environ. Exp. Bot.* **2002**, *47*, 271–280. [CrossRef]
104. Galletti, A.; Verlicchi, P.; Ranieri, E. Removal and accumulation of Cu, Ni and Zn in horizontal subsurface flow constructed wetlands: Contribution of vegetation and filling medium. *Sci. Total Environ.* **2010**, *408*, 5097–5105. [CrossRef]

International Journal of
Environmental Research and Public Health

MDPI

Article

Effects of Stepwise Temperature Shifts in Anaerobic Digestion for Treating Municipal Wastewater Sludge: A Genomic Study

Gede Adi Wiguna Sudiartha [1,2], Tsuyoshi Imai [1,*] and Yung-Tse Hung [3]

1 Graduate School of Sciences and Technology for Innovation, Yamaguchi University, Yamaguchi 755-8611, Japan; adiwiguna.sudiartha@unud.ac.id
2 Environmental Engineering Study Program, Faculty of Engineering, Udayana University, Bali 80361, Indonesia
3 Department of Civil and Environmental Engineering, Cleveland State University, FH 112, 2121 Euclid Ave, Cleveland, OH 44115, USA; y.hung@csuohio.edu
* Correspondence: imai@yamaguchi-u.ac.jp

Abstract: In wastewater treatment plants (WWTP), anaerobic digester (AD) units are commonly operated under mesophilic and thermophilic conditions. In some cases, during the dry season, maintaining a stable temperature in the digester requires additional power to operate a conditioning system. Without proper conditioning systems, methanogens are vulnerable to temperature shifts. This study investigated the effects of temperature shifts on CH_4 gas production and microbial diversity during anaerobic digestion of anaerobic sewage sludge using a metagenomic approach. The research was conducted in lab-scale AD under stepwise upshifted temperature from 42 to 48 °C. The results showed that significant methanogen population reduction during the temperature shift affected the CH_4 production. With 70 days of incubation each, CH_4 production decreased from 4.55 $L{\cdot}g^{-1}$-chemical oxygen demand (COD) at 42 °C with methanogen/total population ($M{\cdot}TP^{-1}$) ratio of 0.041 to 1.52 $L{\cdot}g^{-1}$ COD ($M{\cdot}TP^{-1}$ ratio 0.027) and then to 0.94 $L{\cdot}g^{-1}$ COD ($M{\cdot}TP^{-1}$ ratio 0.026) after the temperature was shifted to 45 °C and 48 °C, respectively. *Methanosaeta* was the most prevalent methanogen during the thermal change. This finding suggests that the *Methanosaeta* genus was a thermotolerant archaea. *Anaerobaculum*, *Fervidobacterium*, and *Tepidanaerobacter* were bacterial genera and grew well in shifted-up temperatures, implying heat-resistant characteristics.

Keywords: anaerobic digestion; biogas production; genomic analysis; shifted-up temperature; sludge treatment and disposal; thermotolerant bacteria

check for updates

Citation: Sudiartha, G.A.W.; Imai, T.; Hung, Y.-T. Effects of Stepwise Temperature Shifts in Anaerobic Digestion for Treating Municipal Wastewater Sludge: A Genomic Study. *Int. J. Environ. Res. Public Health* **2022**, *19*, 5728. https://doi.org/10.3390/ijerph19095728

Academic Editor: Paul B. Tchounwou

Received: 23 April 2022
Accepted: 6 May 2022
Published: 8 May 2022

1. Introduction

Water resources and environmental protection policies worldwide have mandated thorough treatment of wastewater prior to discharge into water bodies [1]. Activated sludge treatment is a common wastewater treatment method [2]. However, the main issue regarding this type of treatment is sludge generated from primary sedimentation (PS) and activated sludge (AS). Sludge produced from wastewater treatment plants (WWTP) is produced in large volumes worldwide with up to 8910, 6510, 2960, 650, 580, 550, and 370 thousand metric tons of dry sludge produced annually by EU countries, the United States, China, Iran, Turkey, Canada, and Brazil, respectively [3]. Considering the substantial volumes of waste production, it is not surprising that WWTP sludge management and disposal have become an area of significant concern globally [4]. As a result of its high water content, low dewaterability, and rigorous regulations for sludge reuse and disposal, sludge management is a demanding and complicated issue in wastewater treatment plants [5].

In some countries, landfill is the most favorable disposal method [6]. However, due to the volume of sludge produced and the availability of the land area, attention has shifted to the development of other potential usable products. Currently, the wastewater treatment paradigm has shifted to an environmentally friendly process to reduce the volume of sludge disposed and convert it into a bioenergy source. The dry bulk of WWTP sludge contains organic components that can be utilized to generate a significant quantity of biomass energy [7]. Anaerobic digestion (AD) is one of the most reliable and promising technologies [8], with several biological wastewater treatment plants applying it as an end-treatment for sewage sludge, primary sludge, and waste-activated sludge [9–11]. In Japan, especially in Ube wastewater treatment plants, sludge generated in PS and AS is dewatered and delivered to an anaerobic digester.

AD has several benefits compared to other biological processes, such as effortless operation, potentially creating an organic by-product that may be utilized in agriculture and managed to provide an appropriate treatment process [12,13]. During anaerobic digestion, organic compounds are hydrolyzed into soluble fermentable substrates, which are subsequently fermented to acetate, carbon dioxide (CO_2), and hydrogen gas (H_2) by acetogenic and acidogenic bacteria. These products are then consumed by methanogens to generate methane (CH_4) [9]. Silva et al. [14] investigated the CH_4 and biohydrogen production from a mixture of food waste, anaerobic sewage sludge, and glycerol. The maximum yield of CH_4 and biohydrogen obtained from the mixture was 342.0 mL $CH_4{\cdot}g^{-1}$ vs. and 179.3 mL $H_2{\cdot}g^{-1}$ VS, respectively. Without any substrate addition, anaerobic sludge was also capable of generating biogas at 230 ± 29 $mL{\cdot}L^{-1}{\cdot}d^{-1}$ with CH_4 production of 153 $mL{\cdot}L^{-1}{\cdot}d^{-1}$ [15].

There are several temperature conditions where anaerobic digestion frequently occurs at psychrophilic (<30 °C), mesophilic (30–40 °C), and thermophilic (50–60 °C) [16]. Recent studies have attempted to investigate the potential of biogas production from anaerobic sludge at various temperatures. Mirmasoumi et al. [17] investigated the biogas production of anaerobic sludge under two different conditions: mesophilic condition (37 °C) and thermophilic condition (55 °C). Under mesophilic conditions, the maximum CH_4 produced was 0.246 m^3 $CH_4{\cdot}m^{-3}$ per digester per day. Meanwhile, a greater CH_4 productivity was obtained under thermophilic conditions, up to 0.64 m^3 CH_4/m^3 per digester per day. Kasinski [18] also found higher CH_4 yields under thermophilic conditions than under mesophilic conditions (0.56 L $CH_4{\cdot}g^{-1}$ vs. −0.70 L $CH_4{\cdot}g^{-1}$ VS and 0.25 L $CH_4{\cdot}g^{-1}$ vs. −0.32 L $CH_4{\cdot}g^{-1}$ vs., respectively). These findings suggest that biogas production using anaerobic digestion generally favors high temperatures. However, in a large-scale WWTP, maintaining thermophilic conditions in the reactor during anaerobic digestion will require significant energy, which will lead to higher operational expenses, especially in four-seasoned countries. In some cases, fermentation failure can occur owing to transient temperature increases caused by power outages, mechanical faults, or human errors during the fermentation process [19]. This motivates further extensive studies of thermotolerant microorganisms for the anaerobic digestion process.

Thermotolerant microorganisms are microbial consortia that are robustly adapted to harsh conditions during industrial applications [20]. Thermotolerant microorganisms are mostly mesophilic, with optimum growth temperatures of 35–45 °C, which are 5–10 °C higher than typical mesophilic strains of the same genus [21–24]. These strains cannot be classified as thermophilic microorganisms, which are characterized by an optimum growth temperature greater than 50 °C [19]. Few studies have examined the potential of thermotolerant microorganisms during anaerobic digestion. Suksong et al. [25] studied the gas production potential of thermotolerant microorganisms from the anaerobic digestion of oil palm empty fruit bunches. The maximum CH_4 yields identified for *Clostridiaceae* and *Lachnospiraceae* with prehydrolysis empty fruit bunches were 252 mL $CH_4{\cdot}g^{-1}$ vs. and 349 mL $CH_4{\cdot}g^{-1}$ VS, respectively. Su et al. [20] discovered the potential of a thermotolerant methanotrophic consortium for producing methanol from biogas. To date, there have been no reports on the potential of thermotolerant microorganisms to produce biogas from the anaerobic digestion of anaerobic sludge.

Notably, investigation of biogas production and genomic analysis from the anaerobic digestion process under shifted-up temperatures has not been performed to date. Therefore, there is an urgent need to expand our understanding of this field. Owing to the heat-resistant characteristics and possible benefits of AD in WWTP-scale applications, further research regarding thermotolerant microorganisms needs to be performed, especially to investigate their potential for biogas production. Therefore, the objectives of this study were to investigate the potential of biogas production (especially CH_4) and to identify the most biogas-producing microorganisms among the cultures after being shifted to several temperature levels.

2. Materials and Methods

2.1. Inoculum and Substrates

The inoculum (I) originated from anaerobically digested sludge obtained from a municipal sewage treatment plant located in Ube City, Yamaguchi Prefecture, Japan. The inoculum characteristics are listed in Table 1. The inoculum samples were then mixed with the substrate (S) solution before being placed in an incubator and exposed to gradually elevated temperature conditions. The substrate used for biogas production in this research was a glucose-based synthetic wastewater consisting of 1.5 $g \cdot L^{-1}$ glucose, 2 $mg \cdot L^{-1}$ $NaHCO_3$, 2 $mg \cdot L^{-1}$ K_2HPO_4, 1 $g \cdot L^{-1}$ yeast extract, 0.7 $g \cdot L^{-1}$ $(NH_4)_2HPO_4$, 0.75 $g \cdot L^{-1}$ KCl, 0.85 $g \cdot L^{-1}$ NH_4Cl, 0.42 $g \cdot L^{-1}$ $FeCl_3 \cdot 6H_2O$, 0.82 $g \cdot L^{-1}$ $MgCl_2 \cdot 6H_2O$, 0.25 $g \cdot L^{-1}$ $MgSO_4 \cdot 7H_2O$, 0.018 $g \cdot L^{-1}$ $CoCl_2 \cdot 6H_2O$, and 0.15 $g \cdot L^{-1}$ $CaCl_2 \cdot 2H_2O$. Glucose was chosen as the ideal carbon source for microbial metabolic transformations in the fermentation process, and is also a readily biodegradable substance abundantly found in municipal wastewater [26,27]. To obtain the maximum CH_4 potential, the appropriate ratio between the microorganisms and substrate must be determined [28]. The CH_4 yield, in theory, is independent of the inoculum to substrate ratio ($I \cdot S^{-1}$), and the $I \cdot S^{-1}$ ratio should influence only the kinetics of CH_4 production [29].

Table 1. Characteristics of anaerobic sludge as inoculum.

Parameters	Anaerobic Sludge	Units
pH	7.09	$pH = -\log_{10}[a(H^+)]$
Total Solid (TS)	8000	mg/L
Volatile Solid (VS)	3000	mg/L
Fixed Solid (FS)	5000	mg/L
VS/TS ratio	0.37	-

In contrast, previous studies have demonstrated that the $I \cdot S^{-1}$ ratio can impact both the CH_4 yield and production rate, as significant evidence suggested that the ratio directly influences the microorganism growth pattern [30–32]. Referring to the German Standard VDI4630, the $I \cdot S^{-1}$ ratio should be adjusted to more than two [33]. Other studies have discovered that a higher $I \cdot S^{-1}$ ratio generates more biogas on a consistent basis during the AD process, while a lower $I \cdot S^{-1}$ ratio produces less biogas due to the lower pH and accumulation of volatile fatty acids (VFAs) [34,35]. In this study, the $I \cdot S^{-1}$ ratio maintained at approximately 3.0, which indicates that 1 mL of substrate was added for every 3 mL of inoculum.

2.2. Experimental Procedures

Laboratory-scale anaerobic digester containers were later defined as vials. A total volume of 160 mL was prepared. To ensure obligate anaerobic conditions, the vial was filled with pure nitrogen gas to flush the remaining oxygen. Subsequently, the vials were capped with butyl rubber stoppers and aluminum caps. Since the laboratory-scale vials were used as the reactor in this research, there were some potential risks that may emerge, such as lower capacity to contain total biogas production, low sample provision to perform several monitoring parameters, and possibility that the sensitivities and instabilities in the laboratory scale reactors do not represent that in the full-scale digesters [36]. This research was divided into two categories: temperature shifted-up (shift-up) and controlled temperature condition. For the shift-up research, in phase 1, batch experiments were performed by adding 110 mL of sludge as inoculum to a vial mixed with 40 mL of substrate. The mixture was then incubated at 42 °C with shaking at 50 rpm for the first two weeks. This action was intended to enhance the growth of microorganisms and ensure the availability of nutrients during acclimatization.

After the first two weeks, phase 2 was initiated. Every time gas production declined sharply, up to 2 mL of substrate with a chemical oxygen demand (COD) of 2000 mg·L^{-1} was injected into the vial. This treatment altered the reactor system from a batch reactor to a fed-batch reactor system. The incubation was continued for 70 days of incubation period in the fed-batch reactor system. Every 70 days, the temperature was increased by 3 °C for the shift-up research until it reached 45 °C and 48 °C. Meanwhile, the controlled temperature research was carried out at 45 °C and 48 °C from the beginning of incubation without any temperature shifts. This study was intended to compare the biogas production and microbial communities among the shifted condition and stabilized condition. During the fermentation period, the total gas volume and composition were measured daily using gas chromatography.

A glass syringe was used to measure the volume of the biogas produced. The gas composition of the samples, such as H_2, N_2, CH_4, and CO_2, was determined using gas chromatography (GC-8APT/TCD; Shimadzu Co., Kyoto, Japan) with 60/80 activated charcoal mesh column (1.5 m × 3.0 mm internal diameter) and argon gas as the carrier gas. During operation, the temperatures of the injector, column, and detector were adjusted to 50 °C, 60 °C, and 50 °C, respectively.

2.3. Deoxyribonucleic Acid (DNA) Extraction and Sequencing

DNA was extracted according to the NucleoSpin® soil manual. Sludge samples were prepared using an MN Bead Tube Type A (MACHEREY-NAGEL GmbH & Co., Düren, Germany). KG was mixed with lysis buffer SL1 and lysed using Enhancer SX. Contaminants were precipitated using lysis buffer SL3, and the lysate was filtered using a NucleoSpin® Inhibitor Removal Column. Subsequently, the binding conditions were adjusted using Binding Buffer SB. DNA was bound by loading 550 µL sample on the NucleoSpin® Soil Column. After the binding phase, the silica membrane was washed with binding buffer SB, wash buffer SW1, and SW2. Finally, the DNA was eluted using SE elution buffer. DNA samples were then delivered to the Faculty of Medicine, Yamaguchi University, Japan, for next-generation sequencing.

Next generation sequencing (NGS) was performed to acquire a broad range of genes or gene regions from phylum to genus using the 16 s ribosomal ribonucleic acid (RNA) gene amplicons for the Illumina MiSeq System, wherein DNA or RNA are sequenced using hybrid capture or amplicon-based approaches (previously transcribed into complementary DNA). Using these approaches, the genome (all 3 billion base pairs), all coding genes (exome; 1% of the genome or 30 million base pairs—that is 20,000 genes made of 180,000 exons), all RNA produced from genes (transcriptome), and any subset of these can be sequenced [37].

2.4. Microbial Diversity Analysis

Diversity index analysis was conducted to determine possible changes in the microbial communities during the anaerobic digestion process. A diversity index is a numerical measure of how many distinct types (such as species) are present in a dataset (a community), as well as the evolutionary relationships among individuals dispersed throughout those types, such as richness, divergence, and evenness [38]. In this study, Simpson's diversity index, Shannon's diversity index, and Shannon's equitability index were utilized.

3. Results

3.1. Biogas Production under Shifted-Up Temperature

The fluctuation in daily CH_4 production and cumulative CH_4 production during the first incubation at 42 °C is shown in Figure 1. During the batch anaerobic digestion period, biogas production increased gradually with the cumulative volume of biogas generated being up to 316.5 mL on day 14, with cumulative CH_4 production being up to 120.76 mL CH_4. The CH_4 content in the biogas increased rapidly in the first 8 days, reaching 69% on the 8th day, and 64.7% on average until day 13. Cumulative CH_4 production increased gradually in the first 3 days, and then showed a substantial increase on day 7 (66.59 mL CH_4). On the final day of the batch period, the cumulative methane production reached 120.76 mL CH_4 which was equal to 1.43 $L \cdot g^{-1}$ COD feed. Theoretically, the energy recovery (expressed as CH_4 production) from digested wastewater sludge through anaerobic process was 0.38 $L \cdot g^{-1}$ COD [39]. From this study, it took approximately four days of anaerobic digestion to exceed the theoretical CH_4 production with the aforementioned inoculum and substrates, with an $I \cdot S^{-1}$ ratio of 2.75.

However, on the 14th day, methane production sharply decreased to 0 mL. Declining CH_4 production indicates a dead phase of methanogenic activity [40]. This drawdown in CH_4 production may also be linked to a decrease in pH caused by the interaction of VFAs with other fragmented precursors during oxidative processes [41]. To maintain CH_4 production, 2 mL of the substrate was injected into the vial when CH_4 production started diminishing due to the scarcity of nutrients. As illustrated in Figure 1a, substrate injection led to a spike in CH_4 production as the activity of methanogenic microorganisms increased due to the availability of glucose as a carbon source and other nutrients that expedited microbial growth. Consequently, as shown in Figure 1a*, the cumulative volume of methane produced increased significantly after substrate addition. At the end of the incubation period at 42 °C, the cumulative CH_4 production was observed to increase to 454.5 mL CH_4, with a yield of 4.55 $L \cdot g^{-1}$ COD feed.

After incubation for 70 days at 42 °C, the incubation temperature was shifted to 45 °C with the same treatment conditions and hydraulic retention times. Even though the same sample was utilized, the cumulative CH_4 calculation was restarted from 0 mL CH_4. As presented in Figure 1b, the daily CH_4 production peaked at 17.2 mL CH_4 which was obtained on day 60 after receiving the 4th feed. From the Figure 1b*, the cumulative CH_4 produced and yield after 70 days of incubation was 152.2 mL CH_4 and 1.52 $L \cdot g^{-1}$ COD feed, respectively. This was less than the volume of CH_4 produced during incubation at 42 °C by a factor of three. There was also a significant difference in CH_4 production behavior after the temperature was increased to 45 °C, e.g., a shorter period required to shift from peak days to trough days, which denotes faster methanogenesis and death phase for methanogenic bacteria. This shorter methanogenesis phase can be attributed to the higher concentration of CO_2 produced during the incubation period.

As shown in Figure 1c, CH_4 production decreased further when the temperature was increased to 48 °C. After 70 days of incubation, only 86.57 mL of CH_4 was produced, with a CH_4 yield of 0.94 $L \cdot g^{-1}$ COD feed. The CH_4 production trend after the 3rd feed at shifted-up 45 °C is a good illustration of the potential inhibition of methanogenic bacterial activity by the presence of high CO_2 levels (Figure 2). The high CO_2 levels indicated the occurrence of acetoclastic methanogenesis in the AD process, which later led to the abundant presence of VFAs, particularly acetic acid [42]. The decrease in CH_4 was also parallel to the decrease in overall biogas production consisting of H_2, N_2, CH_4 and CO_2, which is illustrated in Figure 3a. The total biogas production decreased from 1161.93 mL (11.61 $L \cdot g^{-1}$ COD feed) at 42 °C to 672 mL (6.7 $L \cdot g^{-1}$ COD feed) and then to 505 mL (5.49 $L \cdot g^{-1}$ COD feed) after the temperature was shifted to 45 °C and 48 °C, respectively. The decreasing CH_4 production volume after being shifted-up to the higher temperature at 45 °C and 48 °C was followed by the decline in CH_4 content in biogas compositions as seen in Figure 3b–d.

At the end of incubation period, the concentration of COD, total suspended solids (TSS), volatile suspended solids (VSS), and pH were measured for each temperature condition. As seen in Table 2, the concentration of COD was increasing up to 3-fold, in contrast to TSS and VSS that significantly decreased every temperature shift. This finding indicates that the number of microbial communities (represented by VSS) declined every upshifted thermal condition and subsequently causing a depletion on microbial activity, which eventually resulted in the lower organic matter consumed by microorganisms in the reactor. The pH is another parameter that affects digestion process. The pH increased substantially from 7.64 at 42 °C to 8.20 and 8.33 when the temperature was shifted-up to 45 °C and 48 °C, respectively. Subsequently, the biogas production decreased along with the rising pH. This finding supports a research study from Kouzi et al. [43] who discovered that the optimum pH range for sewage sludge AD was 7.0, while the biogas production was considerably lower in the reactors with higher pH of 8.0, 9.0, and 10.0.

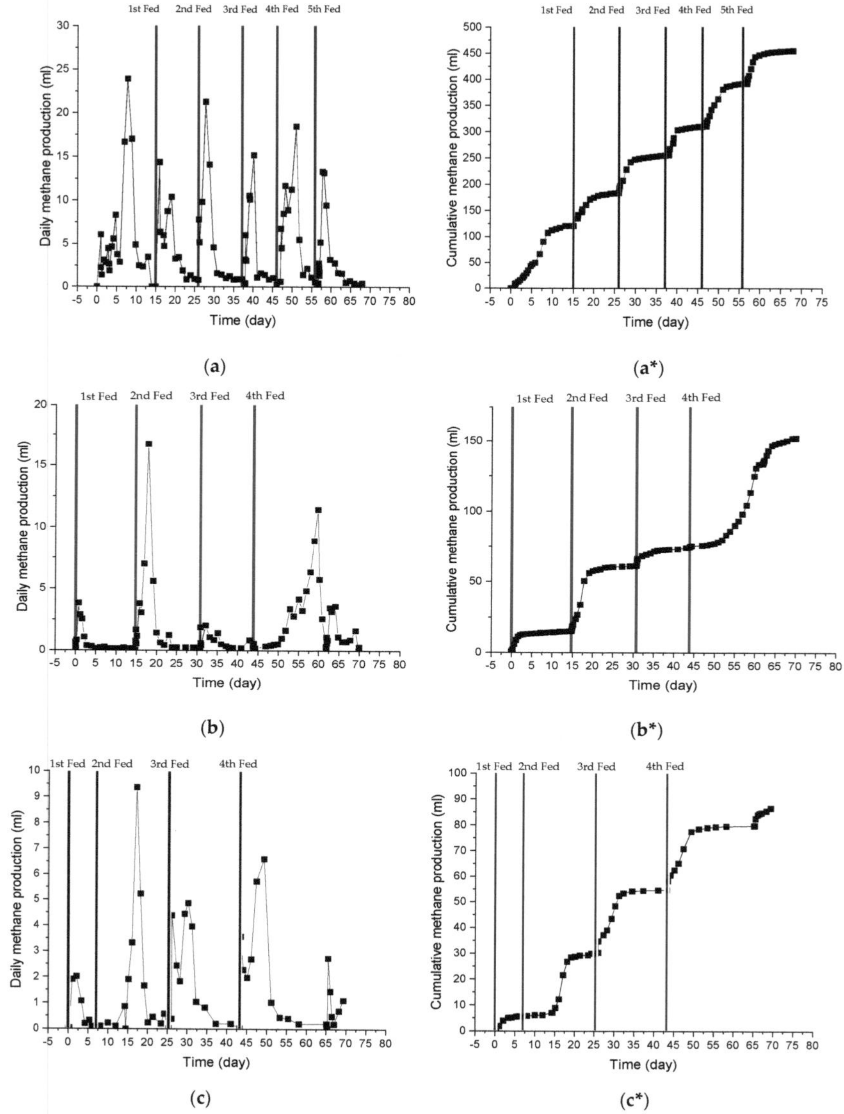

Figure 1. Daily (without *) and cumulative (with *) methane production of anaerobic sludge at (**a**,**a***) 42 °C, (**b**,**b***) shifted-up to 45 °C, and (**c**,**c***) shifted-up to 48 °C.

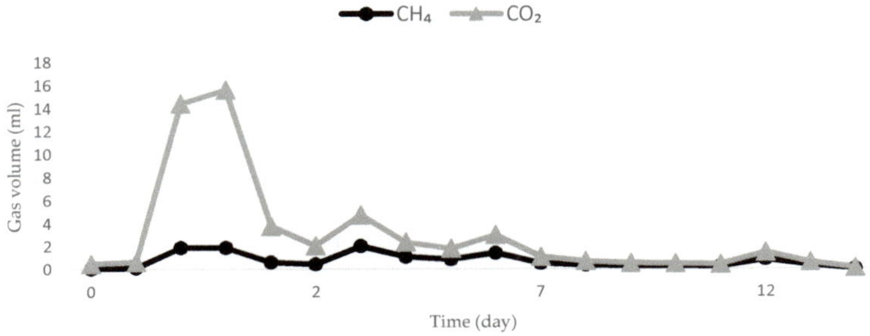

Figure 2. The production of CO_2 after the 3rd feed compared to CH_4 production in 12 consecutive days.

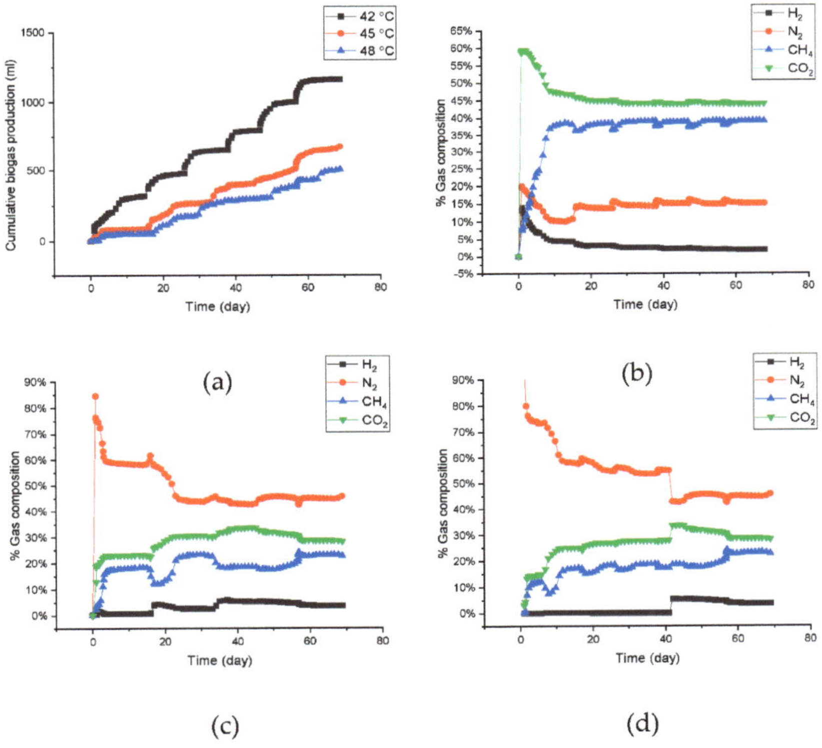

Figure 3. Total biogas production (H_2, N_2, CH_4, and CO_2) of anaerobic sludge during the shifted-up temperature conditions (**a**) and biogas composition at 42 °C (**b**), shifted-up to 45 °C (**c**), and shifted-up to 48 °C (**d**).

Table 2. Effluent quality in each temperature condition after incubation period.

Temperature (°C)	pH	Total Suspended Solids (TSS) in mg·L^{-1}	Volatile Suspended Solids (VSS) in mg·L^{-1}	VSS.TSS^{-1}	Chemical Oxygen Demand (COD) in mg·L^{-1}
42 °C	7.64	7300	4675	0.64	498.72
45 °C	8.20	6860	4105	0.60	1911.76
48 °C	8.33	6880	3465	0.50	3690.52

3.2. Alpha Diversity Analysis

To study the changes in microbial diversity at the upshift temperature, samples from the reactor were used for DNA isolation for bioinformatic analysis using NGS. For comparison, several samples from other fed-batch reactors with controlled temperatures of 42 °C, 45 °C, and 48 °C were examined for their microbial diversity. This action was intended to elucidate the differences among microbial communities and compositions that matured under controlled and shifted-up temperatures. The Shannon diversity index (SDI) and Simpson index were used to measure and compare the richness of the microbiota at a certain temperature, while the Shannon equitability index (SEI) was assigned to approximate the evenness of the microbiota diversity.

As illustrated by Figure 4a,b, the index value for both richness and evenness of the microbiota communities spread within the range 3.2–3.7 and 0.46–0.54, respectively. The Simpson index, as seen in Figure 4c, ranged from 0.89 to 0.96, indicating high diversity for all sample conditions. The deviation of the diversity index between the shifted-up temperature conditions and controlled temperature conditions was not significantly discerned, which signifies that each reactor has a close similarity of microbiota abundance and species to each other. However, compared to the controlled temperature conditions, the shifted-up temperature conditions showed a significant drop in microbiota diversity, with an increase in temperature. The diversity index value declined from 3.72 at 42 °C to 3.22 at 48 °C. This indicated that several bacteria communities were vanished during the temperature shift. This was also confirmed by the decrease in the equitability index value; however, since the equitability values were greater than 0.1, some microorganism colonies managed to acclimatize to this chaotic condition and experienced massive growth while the other colonies became extinct. The effects of temperature on diversity were confirmed using analysis of variance (ANOVA), with significance of $p < 0.05$. The probability value (p-value) for all diversity indices (SDI, SEI, and Simpson index) was $p < 0.001$, which signified that there were statistically significant differences in relative abundances and diversity indices between several temperature conditions.

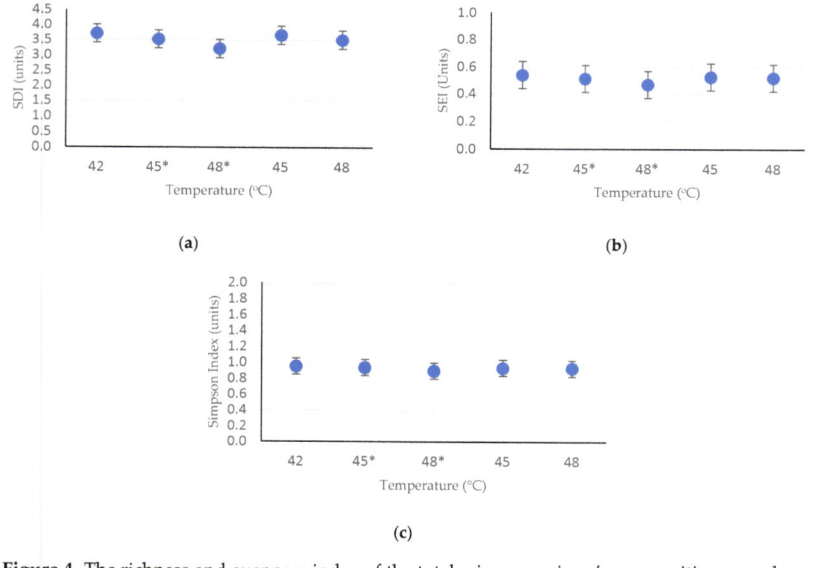

Figure 4. The richness and evenness index of the total microorganisms' communities on each reactor: (**a**) Shannon-Wiener Diversity Index; (**b**) Shannon Equitability Index; and (**c**) Simpson Diversity Index. **Note:** (*) signs the shifted-up temperature.

3.3. Microbial Community Structure

Overall, the number of methanogens decreased sharply when the temperature was shifted from 42 °C to 45 °C, as shown in Figure 5. *Methanosaeta* was the most dominant methanogenic archaea that existed during incubation at 42 °C, shifted up to 45 °C, subsequently shifted up to 48 °C, and also abundant during incubation atcontrolled temperatures of 45 °C and 48 °C. This result implied that *Methanosaeta* is a thermotolerant methanogen. The relative abundances of methanogens at the order level at various temperatures are shown in Figure 6a. The composition of methanogens in both shifted-up temperature and controlled temperature conditions was dominated by the orders *Methanobacteriales*, *Methanomicrobiales*, and *Methanosarcinales*. Among the three methanogens, *Methanosarcinales* was the most abundant (86.89% at 42 °C, 88.84% at shifted-up 45 °C, 59.14% at shifted-up 48 °C, 85.56% at controlled 45 °C, and 78.54% at controlled 48 °C).

At the family level, as illustrated in Figure 6b, the methanogen communities were composed of *Methanosaetaceae*, *Methanomicrobiaceae*, *Methanoregulaceae*, *Methanobacteriaceae*, *Methanosarcinaceae*, and *Methanospirillaceae*. The *Methanosaetaceae* family was abundant, with relative abundances of 85.01% at 42 °C, 88.05% at shifted-up 45 °C, 56.52% at shifted-up 48 °C, 84.76% at controlled 45 °C, and 51.14% at controlled 48 °C. In this study, *Methanosaeta* was the only descendant of the *Methanosaetaceae* family. At the genus level, *Methanoculleus*, *Methanolinea*, *Methanobacterium*, *Methanobrevibacter*, *Methanosarcina*, *Methanothermobacter*, *Methanofollis*, *Methanosalsum*, *Methanogenium*, and *Methanolobus* were detected at all temperatures (Figure 6c). However, during the incubation at a controlled temperature of 48 °C, the dominance of *Methanosaeta* was lower than under shifted-up temperature (84% of the total methanogens at shifted-up 48 °C and 51% at controlled 48 °C) while *Methanosarcina* genes were detected up to 27% of the total methanogens. These findings confirmed the results of Figeac et al. [44] who discovered that the family *Methanosarcinaceae* was the most abundant acetotrophic archaea in the initial thermophilic inoculum, whereas the *Methanosaetaceae* family was mostly found in the initial mesophilic inoculum. Therefore, the population of *Methanosaeta* with an initial temperature of 48 °C was considerably lower than that in the upshift condition that was initially acclimatized under mesophilic conditions (at 42 °C). Apart from acetotrophic methanogens, hydrogenotrophic methanogens, such as members of genera *Methanobacterium*, *Methanobrevibacter*, and *Methanothermobacter*, started to grow significantly after the reactor temperature was shifted up to 48 °C. Hydrogenotrophic methanogens favor thermophilic conditions, whereas acetotrophic methanogens cannot resist high temperatures [45].

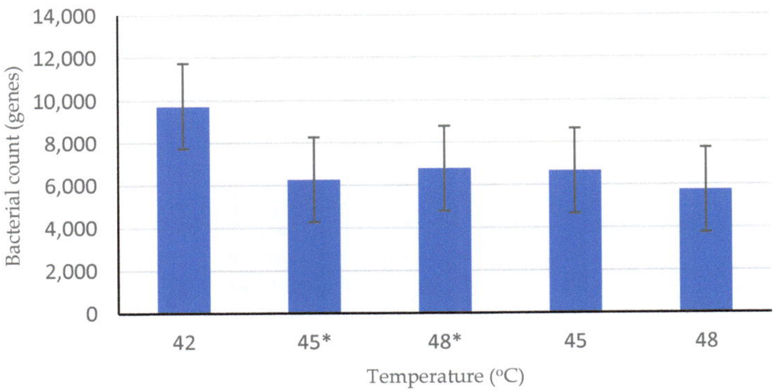

Figure 5. Methanogens total number of genes hit by NGS during anaerobic digestion process in several temperature conditions. **Note:** (*) signs the shifted-up temperature.

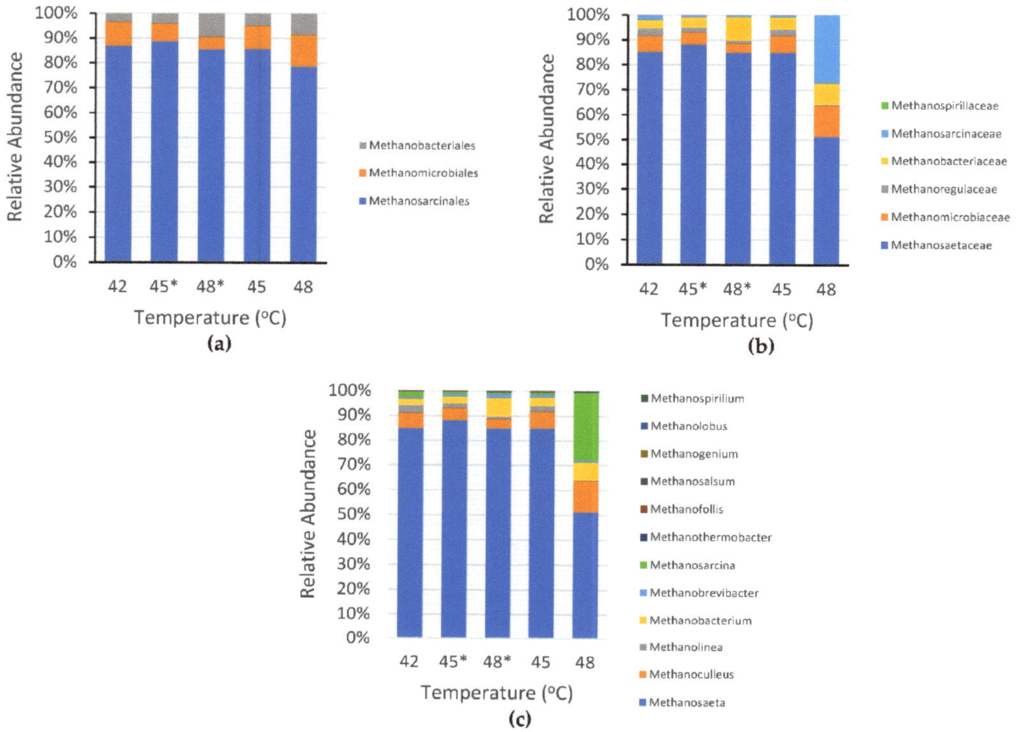

Figure 6. Methanogens distribution in shifted-up temperature (*) and controlled temperature: (**a**) order level; (**b**) family level; and (**c**) genus level.

Methanobacterium communities grew from 2.5% to 7.4% of the total methanogen population, and *Methanobrevibacter* population ranged from 0.76% to 1.64% at a shifted-up temperature of 42–48 °C. This finding contradicts previous research reporting that *Methanobacterium* genera were mostly found at lower mesophilic temperature (24–35% at 35 °C and 32–45% at 37 °C) and eradicated with increasing the temperature to 55 °C [44,46]. *Methanobrevibacter* genera were also found to be the dominant methanogens at 24 °C and 35 °C and vanished at 55 °C [44]. However, the researchers did not examine the existence of *Methanobrevibacter* genera at higher mesophilic temperatures (42–48 °C). Judging from the results of previous studies that showed *Methanobacterium* and *Methanobrevibacter* were abundant in the lower mesophilic conditions, our research reported a significant spike in the population of those methanogens after shifting up temperatures to higher mesophilic conditions, signifying that *Methanobacterium* and *Methanobrevibacter* genera potentially have heat-resistant characteristics that allow them to compromise the staggering increase in temperature conditions. Lastly, *Methanothermobacter* population increased from 0.17% to 0.24% relative abundance at 42–48 °C. This is not surprising as *Methanothermobacter* genera is a thermophilic methanogen that dominated methanogenesis at temperatures of 50 °C and higher [47–50].

The distribution of non-methanogenic bacteria is also an important factor for determining the influence of several bacteria on CH_4 production. As seen in Figure 7a, *Clostridia* and *Synergistia* were the most abundant bacteria at the order level, with respective relative abundances of 31.11% and 18.50% at 42 °C, 42.98% and 20.88% at shifted-up 45 °C, 24.38% and 34.83% at shifted-up 48 °C, 23.43% and 30.49% at 45 °C, and 27.88% and 28.17% at 48 °C. *Clostridia* was found to be dominant at both mesophilic (this research) and thermophilic temperatures (at 52 °C) [51], indicating that the *Clostridia* order belongs to thermotolerant bacteria. According to the experimental results, *Synergistia* were found at higher mesophilic temperatures, but in some cases, were also found abundantly at low temperatures of 20 °C [52]. This suggested that *Synergistia* was resistant to both low-and high-temperature environments, leading to the conclusion that temperature had a responsive connection with the microbial community structure. At the family level, *Anaerobaculaceae, Clostridiaceae,* and *Thermoanaerobacterceae* dominated the microbial communities, as shown in Figure 7b. At the genus level (see Figure 7c), *Anaerobaculum, Fervidobacterium, Tepidanaerobacter, Clostridium, Moorella, Aminiphilus, Carboxydocella*, and *Methanosaeta* are some microorganism genera that exhibited noteworthy growth during the temperature shift. *Anaerobaculum* and *Tepidanaerobacter* from the *Thermoanaerobacterales* family are syntrophic bacteria that have an essential role in converting short-chain fatty acids to methanogenic components such as acetate, H_2, and formate [53]. *Moorella* has been identified as an acidogenic bacterium [54], whereas *Clostridium* is a hydrogen-producing bacterium that plays a key role in the hydrolysis process [46].

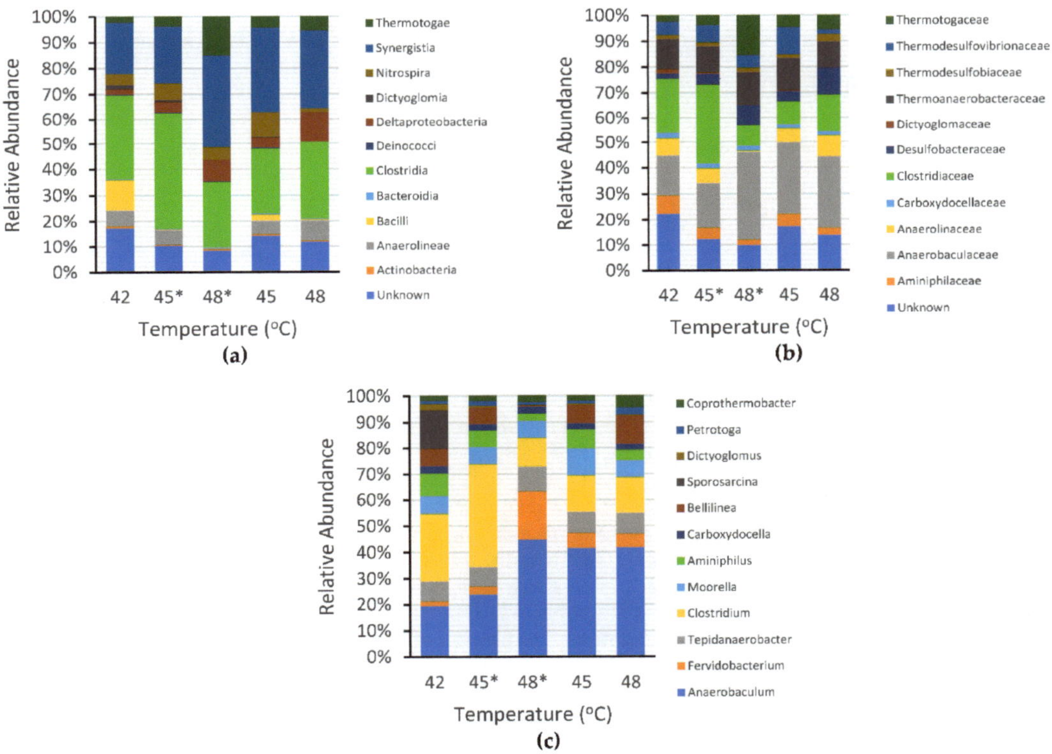

Figure 7. Bacteria (non-methanogens) community distribution in shifted-up temperature (*) and controlled temperature: (**a**) order level; (**b**) family level; and (**c**) genus level.

Compared to the other types of microorganisms, methanogenic archaea were shown to have the least portion of the population among the microbial communities. The decrease in the methanogen population accelerated faster in shifted-up temperature conditions than in controlled temperature conditions. However, the number of methanogen populations is unlikely to affect CH_4 production. The cumulative CH_4 production decreased along with the decline in the methanogen:total population ($M{\cdot}TP^{-1}$) ratio during the shifted-up temperature period. This result contradicts the outcome of the controlled temperature, in which the cumulative CH_4 production increased conspicuously despite the fluctuation in the $M{\cdot}TP^{-1}$ ratio (Figure 8a). To date, no particular ratio has been found to be effective in understanding the influence of the microbial ratio (the existence of a particular microorganism) on biogas production. The closest ratio was that of sulphate-reducing bacteria (SRB) to methanogens ($SRB{\cdot}M^{-1}$). Previous research has stated that the existence of SRBs in the AD process may inhibit CH_4 production as it would compete with methanogens for convenient H_2, acetate, propionate, and butyrate [55]. From the shifted-up temperature experiment, the $SRB{\cdot}M^{-1}$ ratio showed harmony with the statement of previous research. The higher the $SRB{\cdot}M^{-1}$ ratio, the lower the CH_4 production as the SRB emulated the methanogens in consuming available H_2 (Figure 8b).

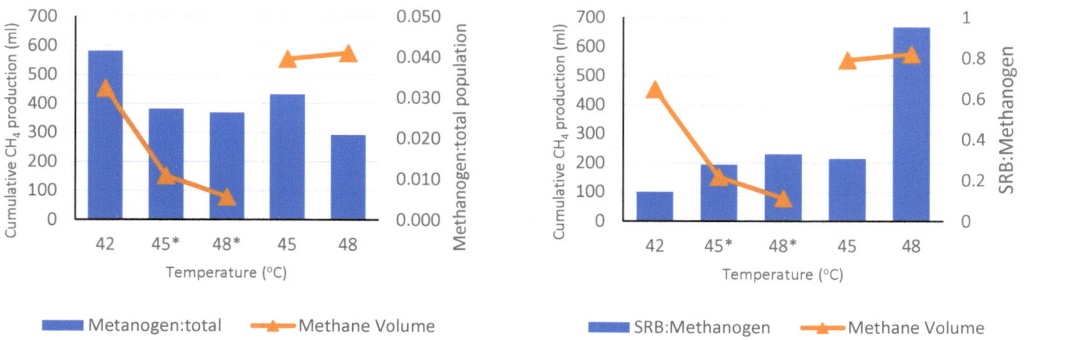

(a) **(b)**

Figure 8. Comparison between cumulative CH_4 production and $M{\cdot}TP^{-1}$ ratio (**a**) and sulphate-reducing bacteria (SRB) to methanogen ratio (**b**). **Note:** (*) signs the shifted-up temperature.

Nevertheless, the results from the controlled temperature experiments showed that the $SRB{\cdot}M^{-1}$ ratio was also ineffective in determining the relationship between the ratio and CH_4 production. At controlled 48 °C, the maximum CH_4 volume production was observed despite the low $M{\cdot}TP^{-1}$ ratio and high $SRB{\cdot}M^{-1}$ ratio. Under these conditions, the populations of *Methanosarcina* and *Methanoculleus* genera were the most abundant. *Methanosarcina* genera are known to be the major contributors to CH_4 production [56] and manage to perform all methanogenesis pathways (hydrogenotrophic, acetoclastic, and methylotrophic) that help them survive in food competition [57]. In contrast to other methanogens, *Methanosarcina* was capable of growing significantly under high concentrations of VFAs and ammonia, which are the foremost inhibitors in biogas production [58]. *Methanoculleus*, in contrast, has been reported to increase in abundance along with elevated sulfate concentration [59]. Thus, both *Methanosarcina* and *Methanoculleus* acclimatized well and were attributed to high CH_4 production under high inhibitor concentrations.

These findings may explain the phenomenon of increasing CH_4 production at a controlled temperature of 48 °C. The influence of *Methanosarcina* and *Methanoculleus* on CH_4 production was demonstrated by considering the decreasing volume of CH_4 production parallel to the decrease in *Methanosarcina* and *Methanoculleus* abundance during shifts in temperature. *Methanosarcina* abundances decreased from 2% to 0.75% and 0.59%, parallel to the *Methanoculleus* population that declined from 6.52% to 4.96% and 3.81% during the incubation at temperature of 42 °C, shifted up to 45 °C, and shifted up to 48 °C, respectively.

4. Discussion

In AD processes, temperature has a significant influence on biogas production and microbial ecology [60,61]. There has been a number of research that examined the effects of temperature in mesophilic and thermophilic conditions [18,62,63]. However, to the best of our knowledge, the assessment of biogas production (especially CH_4) and microbial community adaptation under multiple rising temperature conditions in a fed-batch reactor has not been widely studied. We expected instability in microbial communities (especially methanogens) and decreased CH_4 production, along with the temperature shift process owing to perturbations caused by sudden temperature changes. Among the three temperature conditions, there was a noticeable decrease in CH_4 production when the temperature was increased. The cumulative CH_4 production decreased from 454 mL at 42 °C to 152 mL after increasing the temperature to 45 °C and to 86.57 mL after increasing the temperature to 48 °C. This result is consistent with previous findings for the shifted-up temperature [51,64]. Each temperature shift was conducted after 70-day incubation periods in order to allow a period of acclimatization for methanogens and other bacteria. The total methanogen abundance in shifted-up 45 °C and 48 °C were close to that in controlled 45 °C and 48 °C after 70 days operation. Previous studies showed potential steadier operation condition in term of CH_4 production after acclimatization for 100–140 days, yet the risks of instability still exists [51,65].

Beale et al. [64] investigated the effect of upshift temperature shock from 37 °C to 42 °C on the biogas production volume of anaerobically digested sludge. Similarly, the biogas generated after the temperature was increased to 42 °C was persistently lower than that from the controlled digester at 37 °C during the first 32 days of operation. Identical results were also reported by Ziembinska-Buczynska et al. [66] who found a significant decrease in the biogas production rate from 70.5 $L \cdot day^{-1}$ to 28.6 $L \cdot day^{-1}$ along with an increase in temperature from 38 °C to 55 °C. Researchers also discovered that there was a decrease in microbiota diversity as the temperature of the digester influenced the evolution from mesophilic to thermophilic conditions. Some methanogens cannot survive at higher temperatures (heat unresistant), e.g., *Methanobrevibacter* (37–39 °C), *Methanogenium* (20–25 °C), or *Methanobacterium* (37–45 °C) [67]. However, our research contradicts the findings of Bouskova et al. [68] who discovered higher CH_4 production after the temperature was shifted from 42 °C to 47 °C, 51 °C, and 55 °C. A possible reason for the observed discrepancies was the characteristics and ratio of the inoculum and substrates. The researchers used an inoculum with TS of 31.24 $g \cdot L^{-1}$ and vs. of 14.48 $g \cdot L^{-1}$. Meanwhile in our study, the inoculum consisted of 8 $g \cdot L^{-1}$ TS and 3 $g \cdot L^{-1}$ VS. Previous researchers also utilized a mixture of primary sludge and waste activated sludge as substrates which also contained seeds of microorganisms that maintained the longevity of biogas production.

We also found that the presence of excessive CO_2 in the reactor may have led to lower methane production. Methanogenic bacteria require CO_2 and H_2 to produce CH_4, which indicates that if the CO_2 volume was greater than CH_4 after substrate feeding, this signifies the failure of these bacteria to consume sufficient quantities of CO_2 and H_2, which consequently would lead to the accumulation of VFAs, lower CH_4 yield, and low pH [69,70]. Low pH is a serious concern as it inhibits methanogenic bacteria due to the increase in the concentration of free acid molecules, which is harmful for microorganisms and impacts enzymatic activity [71]. It has been suggested that microbiota composition and methanogenic pathways are altered when encountering an immediate low pH and high acetate crisis (pH 5.5–6.5, completely hindered at pH 5.0) [72].

Temperature shifts also affected the microbial communities in the reactor, especially the methanogens. The number of methanogens decreased significantly after the temperature was shifted from 42 °C to 45 °C and then stabilized at 48 °C, as shown in Figure 4. This instability supported the study by Westerholm et al. [51] who found that immense perturbation occurred in the interval of 40–44 °C, signifying that the 40–44 °C temperature range had a significant impact on both mesophilic and thermophilic microbial populations. The only conceivable interpretation for this phenomenon is that the temperature range may be greater than the upper threshold for the growth of mesophiles but not sufficiently high for the growth of thermophiles [73]. Because our reactor was initially developed using mesophilic anaerobic sludge, this potentially limits the abundance of thermophilic microorganisms. Tian et al. [74] omitted the 40–44 °C area and still found a transitory decrease in total methanogen concentration after the temperature was shifted (37–55 °C), but then recovered quickly on day 11.

Among all the methanogens, *Methanosaeta* (*Methanosaetaceae* family) was the most abundant under all temperature conditions in this study. This finding contradicts the findings of Kim et al. [75] who reported that the *Methanosaetaceae* family started to dominate the microbial structure only at temperatures above 45 °C. A plausible explanation for this difference is that the researchers started cultivation at 35 °C, which was more favorable to the growth of *Methanomicrobiales* order than *Methanosarcinales* (the ancestor of *Methanosaetaceae*). The instability of methanogen populations in anaerobic digestion also led to an increase in some types of bacteria that contributed to H_2 and VFAs consumption, such as SRBs. As illustrated in Figure 7b, our research shows that the increasing SRB/methanogen ratio has a considerable influence on the decrease in CH_4 production under the shifted-up temperature conditions. This result supports the idea from previous studies that reported that sulfide generation by SRBs inhibits methanogenesis, with the latter being the leading rival of methanogens for electron donors and substrates [65,76]. In addition, Beale et al. [64] emphasized that even a small amount of SRBs was enough to inhibit biogas production, as methanogens were vulnerable to the toxicity caused by metabolic products of SRBs.

Although this study characterized the behavior of some microorganisms and their influence on CH_4 production during the shifted-up temperature, there are still several unidentified factors that can potentially affect the outcome of the AD process. Further studies are needed to determine and characterize the mechanism by which shifted temperature may affect the abundance of microorganisms, especially methanogenesis-related microbiota (e.g., methanogens, SRB, methanotrophs, hydrogenotrophs, acetotrophs, nitrogen-fixing bacteria, and sulfate-oxidizing bacteria) and the influence on CH_4 production. Other microbial communities with syntrophic and fermentative behaviors, as well as their metabolic networks, merit study. Minimizing the number of unknown microorganisms may also provide a clearer insight into the relationship between the abundance of microorganisms and CH_4 production, as in this study, we detected a large number of unknown bacteria.

5. Conclusions

Treating wastewater sludge using anaerobic digestion does not eliminate the risk of temperature instability. Consequently, the effects of shifting the temperature during anaerobic digestion of anaerobic sludge were investigated in this study. The results showed a considerable reduction in the CH_4 cumulative gas production, from 454 mL (4.55 $L \cdot g^{-1}$ COD) to 152 mL (1.52 $L \cdot g^{-1}$ COD) then to 86.57 mL (0.94 $L \cdot g^{-1}$ COD) when the temperature of the reactor was increased from 42 °C to 45 °C and subsequently to 48 °C, respectively. Several factors have been attributed to the decrease in CH_4 production under the shifted-up temperature, such as the decreasing methanogen population (expressed as the $M \cdot TP^{-1}$ ratio) due to intense food competition, increasing SRB populations over methanogens, and low abundance of major CH_4 producers (e.g., *Methanosarcina* and *Methanoculleus*). *Methanosaeta* was the most dominant methanogen in this study, while *Anaerobaculum* and *Tepidanaerobacter* were the most abundant syntrophic bacteria, and *Clostridium* which are known as hydrogen-producing bacteria. Overall, the diversity of the anaerobic microbial consortium observed in this study altered slightly during the shift in the thermal conditions. This indicated that the majority of the communities belonged to thermotolerant microorganisms.

Author Contributions: G.A.W.S.: Conceptualization, methodology, formal analysis, investigation, data curation, writing—original draft, and visualization. T.I.: conceptualization, methodology, resources, writing—review and editing, supervision, project administration, and funding acquisition. Y.-T.H.: conceptualization, methodology, writing—review and editing. All authors have read and agreed to the published version of the manuscript.

Funding: This research received no external funding.

Institutional Review Board Statement: Not applicable.

Informed Consent Statement: Not applicable.

Data Availability Statement: Not applicable.

Acknowledgments: The first author would like to thank the Japanese Ministry of Education, Culture, Sports, Science and Technology (Monbukagakusho) for the full support through the Doctoral Degree Scholarship Program.

Conflicts of Interest: The authors declare that they have no conflict of interest, no known competing financial interests or personal relationships that could have appeared to influence the work reported in this paper.

References

1. Nika, C.E.; Gusmaroli, L.; Ghafourian, M.; Atanasova, N.; Buttiglieri, G.; Katsou, E. Nature-Based Solutions as Enablers of Circularity in Water Systems: A Review on Assessment Methodologies, Tools and Indicators. *Water Res.* **2020**, *183*, 115988. [CrossRef] [PubMed]
2. Cayetano, R.D.A.; Kim, G.B.; Park, J.H.; Lee, M.J.; Kim, S.H. Anaerobic Digestion of Waste Activated Sludge Using Dynamic Membrane at Varying Substrate Concentration Reveals New Insight towards Methanogenic Pathway and Biofilm Formation. *Chem. Eng. J.* **2021**, *423*, 130249. [CrossRef]
3. Mateo-Sagasta, J.; Raschid-Sally, L.; Thebo, A. Global Wastewater and Sludge Production, Treatment and Use. In *Wastewater: Economic Asset in An Urbanizing World*; Springer: Berlin, Germany, 2015; ISBN 9789401795456.
4. Świerczek, L.; Cieślik, B.M.; Konieczka, P. The Potential of Raw Sewage Sludge in Construction Industry—A Review. *J. Clean. Prod.* **2018**, *200*, 342–356. [CrossRef]
5. Nazari, L.; Sarathy, S.; Santoro, D.; Ho, D.; Ray, M.B.; Xu, C.C. 3—Recent Advances in Energy Recovery from Wastewater Sludge. In *Direct Thermochemical Liquefaction for Energy Applications*; Woodhead Publishing: Thorston, UK, 2018; ISBN 9780081010259.
6. Singh, V.; Phuleria, H.C.; Chandel, M.K. Estimation of Energy Recovery Potential of Sewage Sludge in India: Waste to Watt Approach. *J. Clean. Prod.* **2020**, *276*, 122538. [CrossRef]
7. Yan, P.; Qin, R.C.; Guo, J.S.; Yu, Q.; Li, Z.; Chen, Y.P.; Shen, Y.; Fang, F. Net-Zero-Energy Model for Sustainable Wastewater Treatment. *Environ. Sci. Technol.* **2017**, *51*, 1017–1023. [CrossRef]
8. Haron, R.; Mat, R.; Tuan Abdullah, T.A.; Rahman, R.A. Overview on Utilization of Biodiesel By-Product for Biohydrogen Production. *J. Clean. Prod.* **2018**, *172*, 314–324. [CrossRef]
9. Young, M.N.; Marcus, A.K.; Rittmann, B.E. A Combined Activated Sludge Anaerobic Digestion Model (CASADM) to Understand the Role of Anaerobic Sludge Recycling in Wastewater Treatment Plant Performance. *Bioresour. Technol.* **2013**, *136*, 196–204. [CrossRef]
10. Campo, G.; Cerutti, A.; Zanetti, M.; Scibilia, G.; Lorenzi, E.; Ruffino, B. Enhancement of Waste Activated Sludge (WAS) Anaerobic Digestion by Means of Pre- and Intermediate Treatments. Technical and Economic Analysis at a Full-Scale WWTP. *J. Environ. Manag.* **2018**, *216*, 372–382. [CrossRef]
11. Hanum, F.; Yuan, L.C.; Kamahara, H.; Aziz, H.A.; Atsuta, Y.; Yamada, T.; Daimon, H. Treatment of Sewage Sludge Using Anaerobic Digestion in Malaysia: Current State and Challenges. *Front. Energy Res.* **2019**, *7*. [CrossRef]
12. Panepinto, D.; Riggio, V.A.; Campo, G.; Cerutti, A.; Comoglio, C.; Zanetti, M.C. Analysis of Two Treatment Technologies for Coffee Roasting Matrixes: Combustion and Anaerobic Digestion. *Clean Technol. Environ. Policy* **2019**, *21*, 685–694. [CrossRef]
13. Pramanik, S.K.; Suja, F.B.; Zain, S.M.; Pramanik, B.K. The Anaerobic Digestion Process of Biogas Production from Food Waste: Prospects and Constraints. *Bioresour. Technol. Rep.* **2019**, *8*, 100310. [CrossRef]
14. Silva, F.M.S.; Mahler, C.F.; Oliveira, L.B.; Bassin, J.P. Hydrogen and Methane Production in a Two-Stage Anaerobic Digestion System by Co-Digestion of Food Waste, Sewage Sludge and Glycerol. *Waste Manag.* **2018**, *76*, 339–349. [CrossRef] [PubMed]
15. Maragkaki, A.E.; Fountoulakis, M.; Kyriakou, A.; Lasaridi, K.; Manios, T. Boosting Biogas Production from Sewage Sludge by Adding Small Amount of Agro-Industrial by-Products and Food Waste Residues. *Waste Manag.* **2018**, *71*, 605–611. [CrossRef] [PubMed]
16. Goswami, R.; Chattopadhyay, P.; Shome, A.; Banerjee, S.N.; Chakraborty, A.K.; Mathew, A.K.; Chaudhury, S. An Overview of Physico-Chemical Mechanisms of Biogas Production by Microbial Communities: A Step towards Sustainable Waste Management. *3 Biotech* **2016**, *6*, 1–12. [CrossRef] [PubMed]
17. Mirmasoumi, S.; Ebrahimi, S.; Saray, R.K. Enhancement of Biogas Production from Sewage Sludge in a Wastewater Treatment Plant: Evaluation of Pretreatment Techniques and Co-Digestion under Mesophilic and Thermophilic Conditions. *Energy* **2018**, *157*, 707–717. [CrossRef]
18. Kasinski, S. Mesophilic and Thermophilic Anaerobic Digestion of Organic Fraction Separated during Mechanical Heat Treatment of Municipal Waste. *Appl. Sci.* **2020**, *10*, 2412. [CrossRef]
19. Matsushita, K.; Azuma, Y.; Kosaka, T.; Yakushi, T.; Hoshida, H.; Akada, R.; Yamada, M. Genomic Analyses of Thermotolerant Microorganisms Used for High-Temperature Fermentations. *Biosci. Biotechnol. Biochem.* **2016**, *80*, 655–668. [CrossRef]
20. Su, Z.; Ge, X.; Zhang, W.; Wang, L.; Yu, Z.; Li, Y. Methanol Production from Biogas with a Thermotolerant Methanotrophic Consortium Isolated from an Anaerobic Digestion System. *Energy Fuels* **2017**, *31*, 2970–2975. [CrossRef]

21. Moonmangmee, D.; Adachi, O.; Ano, Y.; Shinagawa, E.; Toyama, H.; Theeragool, G.; Lotong, N.; Matsushita, K. Isolation and Characterization of Thermotolerant Gluconobacter Strains Catalyzing Oxidative Fermentation at Higher Temperatures. *Biosci. Biotechnol. Biochem.* **2000**, *64*, 2306–2315. [CrossRef]
22. Limtong, S.; Sringiew, C.; Yongmanitchai, W. Production of Fuel Ethanol at High Temperature from Sugar Cane Juice by a Newly Isolated Kluyveromyces Marxianus. *Bioresour. Technol.* **2007**, *98*, 3367–3374. [CrossRef]
23. Manaia, C.M.; Moore, E.R.B. Pseudomonas Thermotolerans Sp. Nov., a Thermotolerant Species of the Genus Pseudomonas Sensu Stricto. *Int. J. Syst. Evol. Microbiol.* **2002**, *52*, 2203–2209. [CrossRef] [PubMed]
24. Ndoye, B.; Lebecque, S.; Dubois-Dauphin, R.; Tounkara, L.; Guiro, A.T.; Kere, C.; Diawara, B.; Thonart, P. Thermoresistant Properties of Acetic Acids Bacteria Isolated from Tropical Products of Sub-Saharan Africa and Destined to Industrial Vinegar. *Enzyme Microb. Technol.* **2006**, *39*, 916–923. [CrossRef]
25. Suksong, W.; Kongjan, P.; Prasertsan, P.; O-Thong, S. Thermotolerant Cellulolytic Clostridiaceae and Lachnospiraceae Rich Consortium Enhanced Biogas Production from Oil Palm Empty Fruit Bunches by Solid-State Anaerobic Digestion. *Bioresour. Technol.* **2019**, *291*, 121851. [CrossRef] [PubMed]
26. Łukajtis, R.; Hołowacz, I.; Kucharska, K.; Glinka, M.; Rybarczyk, P.; Przyjazny, A.; Kamiński, M. Hydrogen Production from Biomass Using Dark Fermentation. *Renew. Sustain. Energy Rev.* **2018**, *91*, 665–694. [CrossRef]
27. Scherhag, P.; Ackermann, J.U. Removal of Sugars in Wastewater from Food Production through Heterotrophic Growth of Galdieria Sulphuraria. *Eng. Life Sci.* **2021**, *21*, 233–241. [CrossRef]
28. Yoon, Y.M.; Kim, S.H.; Shin, K.S.; Kim, C.H. Effects of Substrate to Inoculum Ratio on the Biochemical Methane Potential of Piggery Slaughterhouse Wastes. *Asian-Australas. J. Anim. Sci.* **2014**, *27*, 600–607. [CrossRef]
29. Filer, J.; Ding, H.H.; Chang, S. Biochemical Methane Potential (BMP) Assay Method for Anaerobic Digestion Research. *Water* **2019**, *11*, 921. [CrossRef]
30. Raposo, F.; De La Rubia, M.A.; Fernández-Cegrí, V.; Borja, R. Anaerobic Digestion of Solid Organic Substrates in Batch Mode: An Overview Relating to Methane Yields and Experimental Procedures. *Renew. Sustain. Energy Rev.* **2012**, *16*, 861–877. [CrossRef]
31. Raposo, F.; Banks, C.J.; Siegert, I.; Heaven, S.; Borja, R. Influence of Inoculum to Substrate Ratio on the Biochemical Methane Potential of Maize in Batch Tests. *Process Biochem.* **2006**, *41*, 1444–1450. [CrossRef]
32. Liu, Y. Bioenergetic Interpretation on the S0/X0 Ratio in Substrate-Sufficient Batch Culture. *Water Res.* **1996**, *30*, 2766–2770. [CrossRef]
33. Wang, B. *Factors That Influence the Biochemical Methane Potential (BMP) Test: Steps towards the Standardisation of BMP Test*; Lund University: Lund, Sweden, 2016; ISBN 9789174224368.
34. Ma, X.; Jiang, T.; Chang, J.; Tang, Q.; Luo, T.; Cui, Z. Effect of Substrate to Inoculum Ratio on Biogas Production and Microbial Community during Hemi-Solid-State Batch Anaerobic Co-Digestion of Rape Straw and Dairy Manure. *Appl. Biochem. Biotechnol.* **2019**, *189*, 884–902. [CrossRef] [PubMed]
35. Dixon, P.J.; Ergas, S.J.; Mihelcic, J.R.; Hobbs, S.R. Effect of Substrate to Inoculum Ratio on Bioenergy Recovery from Food Waste, Yard Waste, and Biosolids by High Solids Anaerobic Digestion. *Environ. Eng. Sci.* **2019**, *36*, 1459–1465. [CrossRef]
36. Bolzonella, D.; Pavan, P.; Battistoni, P.; Cecchi, F. Mesophilic Anaerobic Digestion of Waste Activated Sludge: Influence of the Solid Retention Time in the Wastewater Treatment Process. *Process Biochem.* **2005**, *40*, 1453–1460. [CrossRef]
37. Ramdial, P.K.; Bastian, B.C.; Goodlad, J.; McGrath, J.A.; Lazar, A. Specialized Techniques in Dermatopathology. In *McKee's Pathology of the Skin*; Elsevier: Amsterdam, The Netherlands, 2012.
38. Tucker, C.M.; Cadotte, M.W.; Carvalho, S.B.; Jonathan Davies, T.; Ferrier, S.; Fritz, S.A.; Grenyer, R.; Helmus, M.R.; Jin, L.S.; Mooers, A.O.; et al. A Guide to Phylogenetic Metrics for Conservation, Community Ecology and Macroecology. *Biol. Rev.* **2017**, *92*, 698–715. [CrossRef]
39. Wei, C.H.; Harb, M.; Amy, G.; Hong, P.Y.; Leiknes, T.O. Sustainable Organic Loading Rate and Energy Recovery Potential of Mesophilic Anaerobic Membrane Bioreactor for Municipal Wastewater Treatment. *Bioresour. Technol.* **2014**, *166*, 326–334. [CrossRef]
40. Tetteh, E.K.; Rathilal, S. Biophotocatalytic Reduction of CO2 in Anaerobic Biogas Produced from Wastewater Treatment Using an Integrated System. *Catalysts* **2022**, *12*, 76. [CrossRef]
41. El-Gendy, N.S.; Nassar, H.N. Biosynthesized Magnetite Nanoparticles as an Environmental Opulence and Sustainable Wastewater Treatment. *Sci. Total Environ.* **2021**, *774*, 145610. [CrossRef]
42. González, J.; Sánchez, M.; Gómez, X. Enhancing Anaerobic Digestion: The Effect of Carbon Conductive Materials. *C* **2018**, *4*, 59. [CrossRef]
43. Kouzi, A.I.; Puranen, M.; Kontro, M.H. Evaluation of the Factors Limiting Biogas Production in Full-Scale Processes and Increasing the Biogas Production Efficiency. *Environ. Sci. Pollut. Res.* **2020**, *27*, 28155–28168. [CrossRef]
44. Figeac, N.; Trably, E.; Bernet, N.; Delgenès, J.P.; Escudié, R. Temperature and Inoculum Origin Influence the Performance of Ex-Situ Biological Hydrogen Methanation. *Molecules* **2020**, *25*, 5665. [CrossRef]
45. Zabranska, J.; Pokorna, D. Bioconversion of Carbon Dioxide to Methane Using Hydrogen and Hydrogenotrophic Methanogens. *Biotechnol. Adv.* **2018**, *36*, 707–720. [CrossRef] [PubMed]
46. Tang, Q.; Xu, J.; Liu, Z.; Huang, Z.; Zhao, M.; Shi, W.; Ruan, W. Optimal the Ex-Situ Biogas Biological Upgrading to Biomethane and Its Combined Application with the Anaerobic Digestion Stage. *Energy Sources Part A Recover. Util. Environ. Eff.* **2021**, *43*, 2147–2159. [CrossRef]

47. Guneratnam, A.J.; Ahern, E.; FitzGerald, J.A.; Jackson, S.A.; Xia, A.; Dobson, A.D.W.; Murphy, J.D. Study of the Performance of a Thermophilic Biological Methanation System. *Bioresour. Technol.* **2017**, *225*, 308–315. [CrossRef] [PubMed]
48. Martin, M.R.; Fornero, J.J.; Stark, R.; Mets, L.; Angenent, L.T. A Single-Culture Bioprocess of Methanothermobacter Thermautotrophicus to Upgrade Digester Biogas by CO_2-to-CH_4 Conversion with H2. *Archaea* **2013**, *2013*, 1–11. [CrossRef]
49. Bassani, I.; Kougias, P.G.; Treu, L.; Porté, H.; Campanaro, S.; Angelidaki, I. Optimization of Hydrogen Dispersion in Thermophilic Up-Flow Reactors for Ex Situ Biogas Upgrading. *Bioresour. Technol.* **2017**, *234*, 310–319. [CrossRef]
50. Porté, H.; Kougias, P.G.; Alfaro, N.; Treu, L.; Campanaro, S.; Angelidaki, I. Process Performance and Microbial Community Structure in Thermophilic Trickling Biofilter Reactors for Biogas Upgrading. *Sci. Total Environ.* **2019**, *655*, 529–538. [CrossRef]
51. Westerholm, M.; Isaksson, S.; Karlsson Lindsjö, O.; Schnürer, A. Microbial Community Adaptability to Altered Temperature Conditions Determines the Potential for Process Optimisation in Biogas Production. *Appl. Energy* **2018**, *226*, 838–848. [CrossRef]
52. Wang, S.; Ma, F.; Ma, W.; Wang, P.; Zhao, G.; Lu, X. Influence of Temperature on Biogas Production Efficiency and Microbial Community in a Two-Phase Anaerobic Digestion System. *Water* **2019**, *11*, 133. [CrossRef]
53. Hao, L.; Michaelsen, T.Y.; Singleton, C.M.; Dottorini, G.; Kirkegaard, R.H.; Albertsen, M.; Nielsen, P.H.; Dueholm, M.S. Novel Syntrophic Bacteria in Full-Scale Anaerobic Digesters Revealed by Genome-Centric Metatranscriptomics. *ISME J.* **2020**, *14*, 906–918. [CrossRef]
54. Alves, J.I.; van Gelder, A.H.; Alves, M.M.; Sousa, D.Z.; Plugge, C.M. Moorella Stamsii Sp. Nov., a New Anaerobic Thermophilic Hydrogenogenic Carboxydotroph Isolated from Digester Sludge. *Int. J. Syst. Evol. Microbiol.* **2013**, *63*, 4072–4076. [CrossRef]
55. McCartney, D.M.; Oleszkiewicz, J.A. Competition between Methanogens and Sulfate Reducers: Effect of COD:Sulfate Ratio and Acclimation. *Water Environ. Res.* **1993**, *65*, 655–664. [CrossRef]
56. Conrad, R. Contribution of Hydrogen to Methane Production and Control of Hydrogen Concentrations in Methanogenic Soils and Sediments. *FEMS Microbiol. Ecol.* **2006**, *28*, 193–202. [CrossRef]
57. Habtewold, J.; Gordon, R.; Sokolov, V.; VanderZaag, A.; Wagner-Riddle, C.; Dunfield, K. Reduction in Methane Emissions from Acidified Dairy Slurry Is Related to Inhibition of Methanosarcina Species. *Front. Microbiol.* **2018**, *9*, 2806. [CrossRef] [PubMed]
58. Demirel, B.; Scherer, P. The Roles of Acetotrophic and Hydrogenotrophic Methanogens during Anaerobic Conversion of Biomass to Methane: A Review. *Rev. Environ. Sci. Biotechnol.* **2008**, *7*, 173–190. [CrossRef]
59. Ma, T.T.; Liu, L.Y.; Rui, J.P.; Yuan, Q.; Feng, D.S.; Zhou, Z.; Dai, L.R.; Zeng, W.Q.; Zhang, H.; Cheng, L. Coexistence and Competition of Sulfate-Reducing and Methanogenic Populations in an Anaerobic Hexadecane-Degrading Culture. *Biotechnol. Biofuels* **2017**, *10*, 207. [CrossRef]
60. Mei, R.; Nobu, M.K.; Narihiro, T.; Kuroda, K.; Muñoz Sierra, J.; Wu, Z.; Ye, L.; Lee, P.K.H.; Lee, P.H.; van Lier, J.B.; et al. Operation-Driven Heterogeneity and Overlooked Feed-Associated Populations in Global Anaerobic Digester Microbiome. *Water Res.* **2017**, *124*, 77–84. [CrossRef]
61. Guo, X.; Wang, C.; Sun, F.; Zhu, W.; Wu, W. A Comparison of Microbial Characteristics between the Thermophilic and Mesophilic Anaerobic Digesters Exposed to Elevated Food Waste Loadings. *Bioresour. Technol.* **2014**, *152*, 420–428. [CrossRef]
62. Odirile, P.T.; Marumoloa, P.M.; Manali, A.; Gikas, P. Anaerobic Digestion for Biogas Production from Municipal Sewage Sludge: A Comparative Study between Fine Mesh Sieved Primary Sludge and Sedimented Primary Sludge. *Water* **2021**, *13*, 3532. [CrossRef]
63. Shu, C.H.; Jaiswal, R.; Kuo, M.-d.; Yu, B.H. Enhancing Methane Production in a Two-Stage Anaerobic Digestion of Spent Mushroom Substrate and Chicken Manure via Activation of Sludge, Optimization of Temperature, and C/N Ratio. *Front. Environ. Sci.* **2022**, *9*, 1–11. [CrossRef]
64. Beale, D.J.; Karpe, A.V.; McLeod, J.D.; Gondalia, S.V.; Muster, T.H.; Othman, M.Z.; Palombo, E.A.; Joshi, D. An "omics" Approach towards the Characterisation of Laboratory Scale Anaerobic Digesters Treating Municipal Sewage Sludge. *Water Res.* **2016**, *88*, 346–357. [CrossRef]
65. Moestedt, J.; Nilsson Påledal, S.; Schnürer, A. The Effect of Substrate and Operational Parameters on the Abundance of Sulphate-Reducing Bacteria in Industrial Anaerobic Biogas Digesters. *Bioresour. Technol.* **2013**, *132*, 327–332. [CrossRef] [PubMed]
66. Ziembińska-Buczyńska, A.; Banach, A.; Bacza, T.; Pieczykolan, M. Diversity and Variability of Methanogens during the Shift from Mesophilic to Thermohilic Conditions While Biogas Production. *World J. Microbiol. Biotechnol.* **2014**, *30*, 3047–3053. [CrossRef] [PubMed]
67. Ahring, B.K. Methanogenesis in Thermophilic Biogas Reactors. *Antonie Van Leeuwenhoek* **1995**, *67*, 91–102. [CrossRef]
68. Boušková, A.; Dohányos, M.; Schmidt, J.E.; Angelidaki, I. Strategies for Changing Temperature from Mesophilic to Thermophilic Conditions in Anaerobic CSTR Reactors Treating Sewage Sludge. *Water Res.* **2005**, *39*, 1481–1488. [CrossRef] [PubMed]
69. Metcalf, W.; Eddy, C. *Wastewater Engineering Treatment and Reuse*; McGraw Hill: New York, NY, USA, 2003.
70. Al-mashhadani, M.K.H.; Wilkinson, S.J.; Zimmerman, W.B. Carbon Dioxide Rich Microbubble Acceleration of Biogas Production in Anaerobic Digestion. *Chem. Eng. Sci.* **2016**, *156*, 24–35. [CrossRef]
71. Han, W.; He, P.; Lin, Y.; Shao, L.; Lü, F. A Methanogenic Consortium Was Active and Exhibited Long-Term Survival in an Extremely Acidified Thermophilic Bioreactor. *Front. Microbiol.* **2019**, *10*, 2757. [CrossRef]
72. Hao, L.P.; Lü, F.; Li, L.; Shao, L.M.; He, P.J. Shift of Pathways during Initiation of Thermophilic Methanogenesis at Different Initial PH. *Bioresour. Technol.* **2012**, *126*, 418–424. [CrossRef]
73. Tezel, U.; Tandukar, M.; Hajaya, M.G.; Pavlostathis, S.G. Transition of Municipal Sludge Anaerobic Digestion from Mesophilic to Thermophilic and Long-Term Performance Evaluation. *Bioresour. Technol.* **2014**, *170*, 385–394. [CrossRef]

74. Tian, Z.; Zhang, Y.; Li, Y.; Chi, Y.; Yang, M. Rapid Establishment of Thermophilic Anaerobic Microbial Community during the One-Step Startup of Thermophilic Anaerobic Digestion from a Mesophilic Digester. *Water Res.* **2015**, *69*, 9–19. [CrossRef]
75. Kim, J.; Lee, C. Response of a Continuous Anaerobic Digester to Temperature Transitions: A Critical Range for Restructuring the Microbial Community Structure and Function. *Water Res.* **2016**, *89*, 241–251. [CrossRef]
76. van den Brand, T.P.H.; Roest, K.; Brdjanovic, D.; Chen, G.H.; van Loosdrecht, M.C.M. Influence of Acetate and Propionate on Sulphate-Reducing Bacteria Activity. *J. Appl. Microbiol.* **2014**, *117*, 1839–1847. [CrossRef] [PubMed]

International Journal of
Environmental Research and Public Health

MDPI

Article

Influence of Particle Size and Zeta Potential in Treating Highly Coloured Old Landfill Leachate by Tin Tetrachloride and Rubber Seed

Siti Fatihah Ramli [1], Hamidi Abdul Aziz [1,2,*], Fatehah Mohd Omar [1], Mohd Suffian Yusoff [1,2], Herni Halim [1], Mohamad Anuar Kamaruddin [3], Kamar Shah Ariffin [4] and Yung-Tse Hung [5]

1 School of Civil Engineering, Engineering Campus, Universiti Sains Malaysia, Seberang Perai 14300, Malaysia; ctfatihah88@gmail.com (S.F.R.); cefatehah@usm.my (F.M.O.); suffian@usm.my (M.S.Y.); ceherni@usm.my (H.H.)
2 Solid Waste Management Cluster, Science and Technology Research Centre, Engineering Campus, Universiti Sains Malaysia, Seberang Perai 14300, Malaysia
3 School of Industrial Technology, Universiti Sains Malaysia, Gelugor 11800, Malaysia; anuarkamaruddin@usm.my
4 School of Materials and Mineral Resources Engineering, Engineering Campus, Universiti Sains Malaysia, Seberang Perai 14300, Malaysia; kamarsha@usm.my
5 Department of Civil and Environmental Engineering, Cleveland State University, Cleveland, OH 44115, USA; yungtsehung@gmail.com
* Correspondence: cehamidi@usm.my

Abstract: Old leachate normally has a low organic compound content, poor biodegradability and is hard to biologically treat. The efficacy of tetravalent metal salts as a coagulant and the application of a natural coagulant as a flocculant in landfill leachate treatment is still inconclusive. Hence, this study aimed to evaluate the potential application of tin tetrachloride ($SnCl_4$) as the main coagulant and the rubber seed (*Hevea brasiliensis*) (RS) as the natural coagulant aid as the sole treatment in eradicating highly coloured and turbid stabilised landfill leachate present at one of the old local landfills in Malaysia. The standard jar test conducted revealed that $SnCl_4$ was able to eliminate 99% and 97.3% of suspended solids (SS) and colour, respectively, at pH8, with 10,000 mg/L dosages, an average particle size of 2419 d·nm, and a zeta potential (ZP) of −0.4 mV. However, RS was found to be ineffective as the main coagulant and could only remove 46.7% of SS and 76.5% of colour at pH3 with 6000 mg/L dosages, and also exhibited smaller particles (933 d·nm) with ZP values of −6.3 mV. When used as a coagulant aid, the polymer bridging mechanism in RS helped in reducing the $SnCl_4$ concentration from 10,000 mg/L to 8000 mg/L by maintaining the same performances. The presence of 1000 mg/L RS as a coagulant aid was able to remove 100% of SS and 97.6% of colour. The study concluded that RS has the potential to be used together with $SnCl_4$ in treating concentrated leachate with SS and colour.

Keywords: solid waste; landfill; leachate; tin tetrachloride; *Hevea brasiliensis*

check for updates

Citation: Ramli, S.F.; Aziz, H.A.; Omar, F.M.; Yusoff, M.S.; Halim, H.; Kamaruddin, M.A.; Ariffin, K.S.; Hung, Y.-T. Influence of Particle Size and Zeta Potential in Treating Highly Coloured Old Landfill Leachate by Tin Tetrachloride and Rubber Seed. *Int. J. Environ. Res. Public Health* **2022**, *19*, 3016. https://doi.org/10.3390/ijerph19053016

Academic Editor: Paul B. Tchounwou

Received: 14 January 2022
Accepted: 2 March 2022
Published: 4 March 2022

1. Introduction

Among the available methods of waste disposal, landfilling is still one of the most commonly used methods worldwide. This is because compared to the other waste disposal methods, landfilling provides a more economical and simple procedure to be applied. However, without proper management, it may lead to serious environmental problems, as landfilling tends to produce toxic gases and liquids, such as leachate [1,2]. Leachate is high-strength wastewater and is black in colour, containing numerous toxic pollutants that can pose a danger to the surroundings. Old leachate is categorised as stabilised and matured leachate, which is in the methanogenic phase. Stabilized leachate usually contains a lower concentration of organic compounds than younger leachate [3]. However, this leachate has low biodegradability due to its high content of non-biodegradable organic compounds [4–6]. Physical-chemical processes such as the coagulation-flocculation technique are appropriate and recommended for stabilised leachate to treat chemical oxygen demand (COD), suspended solids (SS), colour, and heavy metals [7,8]. Leachate is very difficult to deal with and to treat due to its complex composition. It normally requires multiple treatment methods, which depend on the type and age of the landfill.

The process of removing the pollutants from the leachate can be achieved by disrupting the stability of its surface particles. A coagulation and flocculation procedure causes the surface charge of the particles to be reduced or destabilised by the addition of chemicals, known as a coagulant. Meanwhile, the flocculation process involves the agglomeration of particles, forming into a bigger size [9–11]. By applying the coagulation and flocculation treatments, concentrated heavy metals, organics, colour, chemical oxygen demand (COD) and total suspended solids are expected to be removed effectively [12].

Primary coagulants and coagulant aids are used in this process. Usually, primary coagulants consist of metal salts, and coagulant aids are mostly derived from polymers/ polyelectrolytes. The coagulant aid acts to improve the roughness of the floc, hence increasing its density, which then minimises the breakup of the floc during the mixing and settling process, as well as reducing the flocculation and settling period [13,14]. The amount of primary coagulant (metal salt coagulant) in the treatment process can be decreased by the coagulant aid. At the same time, the sludge also contains a lower level of chemical residues.

The amount of electrolyte needed to coagulate a certain quantity of a colloidal solution relies on the valency of the ion having a counter charge to that of the colloidal particles. As stated by the Schulze-Hardy rule, stability decreases with the increase in counter-ion valency [15]. The degree of destabilisation is different depending on the coagulant. Coagulant with a higher valence of the counter-ion provides a greater destabilisation effect, which in turn lowers the dose required in the process [16]. This is due to the attractive electrostatic forces between the ions. The ionic strength of a metal ion (of a salt) used in the destabilisation of the colloidal particles is determined by the valence. As the electrostatic forces of repulsion are decreased, the attraction forces of van der Waals are increased, which induce the ions' kinetic movement through the diffuse layer [17]. Hence, for the coagulation of negatively charged particles, tetravalent cations are more efficient than trivalent cations, which are more efficient than divalent ions, which are far more efficient than monovalent cations.

Several works have been performed to identify the possibility of four valence state coagulants through the coagulation-flocculation process. Zhao et al., (2013) [18] used the poly titanium tetrachloride (PTC) in treating synthetic water, which removed 89.2% of UV_{254} and 61.6% of dissolved organic carbon (DOC). Aziz et al., (2015) [19] studied leachate treatment by titanium tetrachloride ($TiCl_4$) obtained 81.4%, 86.7%, and 76.5% reductions in colour, SS, and ammoniacal nitrogen, respectively. At the same time, Aziz and Ramli (2014) [20] found that zirconium tetrachloride ($ZrCl_4$) was able to remove 93.4% of SS, 94.3% of colour and 97% of UV_{254} from the leachate. We used tin (IV) chloride ($SnCl_4$) as a sole coagulant in our previous works in treating COD in the same leachate as in the current study. The optimum conditions occurred at pH7.17 and $SnCl_4$ dosage of 15,012 mg/L, which exhibited 67.6% reductions in COD [21].

Tin salts have been applied in most other water/wastewater treatments. For example, a study conducted by Mathews et al., (2015) [22] involves the use of tin (II) chloride in the removal of Hg (mercury) in stream water by air stripping method. From the study, it was found that the use of $SnCl_2$ as the coagulant was able to remove more than 90% of Hg (mercury) from the stream water. The removal achieved was relatively higher. In addition, another study conducted by Kennedy et al., (2020) [23] found that the use of $SnCl_2$ together with rapid sand filtration in a pilot scale in the treatment of groundwater achieved a consistent removal of chromium (Cr) from the groundwater. Moreover, Zepeda et al., (2018) [24] applied the usage of tin oxide (SnO_2) synthesised from $SnCl_4$, finding that it allowed the significant removal of both nickel (Ni^{2+}) and copper (Cu^{2+}) ions from an aqueous solution. Therefore, theoretically, it was expected that $SnCl_4$, as the tetravalent coagulant with four valencies, is able to remove and act as one of the best coagulants in removing pollutants from the leachate.

Hence, the current work was undertaken to examine the possibility of employing tin (IV) chloride and RS as an alternative coagulant and coagulant aid (flocculant), respectively. These substances have not yet been tested together in treating water, wastewater, and landfill leachate as a sole treatment. The target pollutant parameters were colour and SS, which are among the hardest parameters to treat in stabilised leachate [25]. The goal of this research was to assess the possibility of reducing the primary coagulant quantity of tin (IV) chloride ($SnCl_4$) by adding RS as a flocculant.

2. Materials and Methods

2.1. Characterisation of Leachate

The case study leachate was sampled from the Alor Pongsu Landfill Site (APLS), located at 5°04′ N, 100°35′ E in Alor Pongsu, Perak, Malaysia. The monitoring of the leachate characteristics was conducted for 12 months starting from November 2018 to October 2019, and the results are outlined in Table 1.

Table 1. Raw compositions of Alor Pongsu Landfill leachate.

Parameter.	Min	Max	Average	Discharge Limit [1]	Method Number	Instrument
pH			8.12	6.0–9.0		
BOD_5 (mg/L)	45	112	85	20	5210 B	DO meter
COD (mg/L)	1390	5078	2937	400	5220 D	DR Spectrophotometer
BOD_5/COD	0.02	0.07	0.03	-		
Suspended Solids (mg/L)	258	547	411	50	HACH Methd 8006	DR Spectrophotometer
Ammoniacal nitrogen (mg/L NH_3-N)	1160	2820	1448	5	HACH Method 8038	DR Spectrophotometer
Colour (Pt.Co)	9480	22,970	15,062	-	HACH Method 8025	DR Spectrophotometer
Turbidity (NTU)	9.68	44.59	22.0		APHA 2130 B	Turbiditimeter
Zeta Potential	−18.6	−22.4	−20.5			Zetasizer

[1] According to Environmental Quality (Control of Pollution from Solid Waste Transfer Station and Landfill) Regulations 2009 (PU (A) 433).

The BOD_5/COD ratio of Alor Pongsu leachate is 0.03, which is <0.1; hence, it is in a stable form and is difficult to biologically treat. It is necessary for this type of leachate to be pretreated so that the organic compound becomes much simpler and can easily be broken down by the bacteria. The colour level is also considered extremely high (maximum value almost reached 23,000 PtCo), and this could be associated with the high organic content (humic and fulvic acids) present in the leachate.

2.2. Preparation of Tin Tetrachloride ($SnCl_4$) Stock Solution

Tin tetrachloride ($SnCl_4$) was provided by BG OilChem Sdn. Bhd. (Permatang Pauh, Pulau Pinang, Malaysia). The stock solution of 40 g/L was prepared, and the appropriate working concentrations were obtained based on the dilution principle. The working concentrations of $SnCl_4$ varied from 8000 mg/L to 18,000 mg/L.

2.3. Preparation of Rubber Seed (*Hevea brasiliensis*) Stock Solution

The raw rubber seeds used in this study were obtained from a local rubber tree farm located in Sg. Petani, Kedah, Malaysia. The seed coats were removed, and the seeds were dried at 105 °C in an oven for 30–60 min to remove the moisture. Next, a heavy-duty grinder was used for crushing the dried rubber seeds into powder. The stock solution of rubber seed was prepared by mixing the rubber seed powder with distilled water, and the mixture was agitated at room temperature for 30 min. Then, the blend was filtered thrice using a filter cloth. The solution was then diluted to the desired concentration needed for the experiment. The working concentrations used for RS were varied from 1000 mg/L to 10,000 mg/L. These were prepared by adding an appropriate amount of RS into distilled water. One gram of RS added into 1 L of distilled water is equivalent to a 1000 mg/L solution. The non-diluted solution was made daily and used on the day it was prepared.

2.4. Zeta Potential and Particle Size Measurement

The zeta potential was evaluated using a 25 °C Malvern Zetasizer Nano ZS (Malvern Instruments, Malvern, UK) instrument. The measurement was conducted for the raw leachate and also for the sample obtained immediately after the coagulation and flocculation process. All samples were tested in triplicate. For each test, the capillary cell was injected with 1 mL of supernatant by means of a syringe [26]. The particle electrophoretic mobility was measured by means of the light dispersion/dynamic light scattering method. This method is sometimes also referred to as photon correlation spectroscopy or quasi-elastic light scattering, and is a technique used for measuring the size or particles especially in the sub-micron region. The sample was tested following the standard procedures provided in the Malvern Zetasizer manual. In an applied electric field, the motion of the particles leads to a minor frequency change in the dispersed light, and laser Doppler electrophoresis typically detects this slight movement [27,28].

2.5. Coagulation and Flocculation Test

The coagulation and flocculation tests were conducted in the laboratory by means of a standard jar test instrument equipped with 6 paddles (VELP Scientifica Srl, Usmate, Italy). The leachate samples used in the experiment was 500 mL for each beaker. Therefore, a total of 3 L was required for the 6 beakers. To prevent the settling of solids, the leachate sample was completely agitated, and the sample was allowed to condition at the ambient temperature for about 30 min. Using 3 M hydrochloric acid (HCl) or 3 M sodium hydroxide, the pH of the samples was modified to the appropriate pH value (pH3–11) (NaOH). The acid and base used in this study were also used by other researchers, for example [29,30]. Then, using the pre-determined value of the operational condition, the leachate was agitated simultaneously. For the determination of $SnCl_4$ and RS as the main coagulant, the doses were varied from 8000 mg/L to 18,000 mg/L and from 1000 mg/L to 10,000 mg/L, respectively. Meanwhile, for the determination of RS as the coagulant aid, the doses were varied in the range of 100 mg/L to 250 mg/L, 500 mg/L, 750 mg/L, 1000 mg/L, 2000 mg/L, 4000 mg/L, 6000 mg/L, 8000 mg/L and 10,000 mg/L. The tests were conducted at an optimum pH8. The jar test was run at 220 rpm for 5 min (rapid mixing) and 60 rpm for 30 min (slow mixing). The samples were allowed to settle for 40 min. The samples were then tested for zeta potential, colour and suspended solids. The removal percentages were calculated based on the SS and colour concentrations before (raw) and after the experiments in relation to their raw concentrations.

$$\text{Percentage Removal (\%)} = [(C_i - C_f)/C_i] \times 100 \quad (1)$$

where:

C_i = initial concentrations.

C_f = final of suspended solids and colour.

3. Results and Discussion

3.1. Influence of Coagulant Dosages

Figures 1 and 2 illustrate the effect of tin tetrachloride ($SnCl_4$) and rubber seed (RS) dosage on the removal performance. The tests were conducted by varying the coagulant dosage of $SnCl_4$ and RS at leachate original pH, which is 8.24. Based on the preliminary experiments, for $SnCl_4$, the dosages were varied from 8000 mg/L to 18,000 mg/L, while for RS, the dosages of coagulants ranged from 1000 mg/L to 10,000 mg/L.

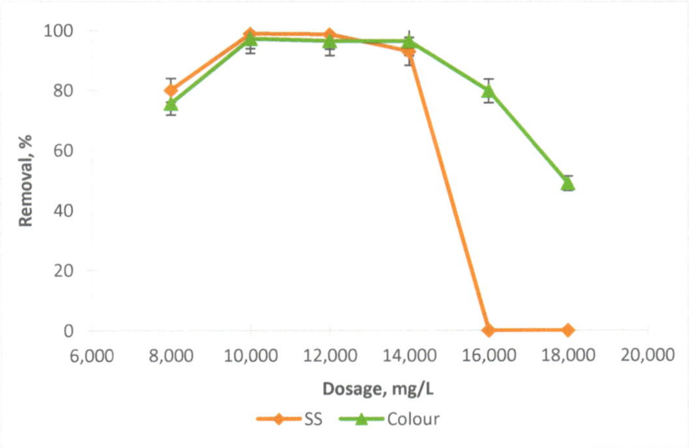

Figure 1. Impact of coagulant dosage on SS and colour removals with $SnCl_4$.

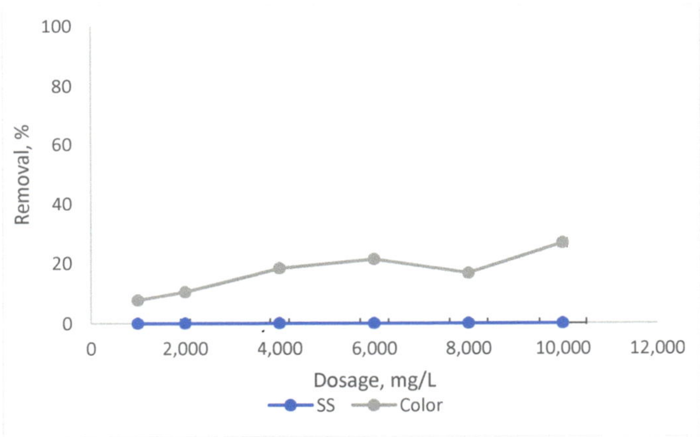

Figure 2. Impact of coagulant dosage on SS and colour removals with RS.

It can be noted from Figure 1 that when the amount of $SnCl_4$ was increased from 8000 mg/L to 10,000 mg/L, the elimination for SS and colour also increased from 80.1% to 99% and 75.6% to 97.3%, respectively. Beyond 10,000 mg/L, the removals for both dropped, gradually for colour and significantly for SS. This was due to the overdosing of coagulant dosage, which leads to the destabilisation of the particles (partly associated as organics in colour) and hence gives a lower removal compared to at the optimum point [31]. However, RS was found not as effective as a coagulant, as shown in Figure 2. At an optimal dosage of 6000 mg/L, the removals for both were less than 30%, a little bit better for colour (about 23%) compared to SS. This is because the addition of RS tends to increase the organic matter content in leachate, which could not be effectively settled even after the treatment.

3.2. Influence of pH

The impact of pH on performance is shown in Figure 3. The pH values were varied from pH3 to pH11. We excluded pH2 and pH12 because these values are too extreme. The optimum pre-determined dosages for both $SnCl_4$ (10,000 mg/L) and RS (6000 mg/L) were applied in this investigation.

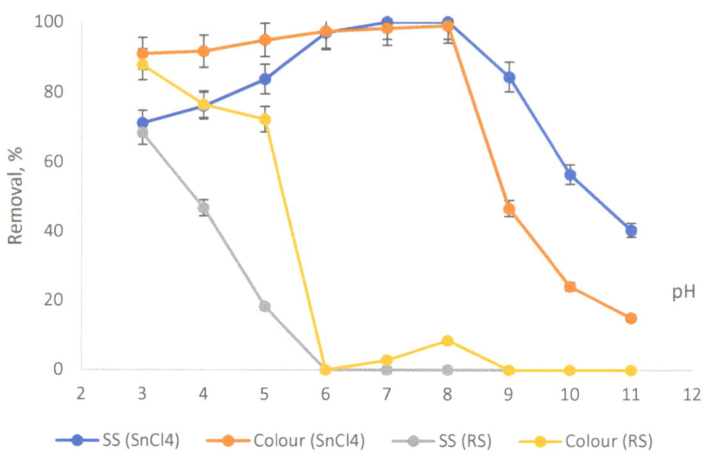

Figure 3. Impact of pH on SS and colour removals by $SnCl_4$ and RSS.

Based on Figure 3, the removal of SS and colour increased with the increase in pH from 3 to pH8, when $SnCl_4$ was applied as the main coagulant. The elimination was increased from 71.2% for SS and 90.9% for colour at pH3 to almost complete removal (100%) at pH8 for both. Beyond pH8 as the optimum, the removals gradually dropped until pH11, where the removals for SS and colour were 40.3% for colour and 15% for SS. A number of soluble hydrolysis species were then produced when $SnCl_4$ (a metal salt coagulant) was applied to the sample. The species may be charged either positively or negatively, based on the pH of the sample. The formation of a positively charged surface could destabilise the negatively charged colloidal particles. The destabilisation of the particles will lead to the aggregation of particles and the establishment of flocs. The condition where positively charged surface species adsorb onto the surface of the negatively charged particles is called a charged neutralisation mechanism [32]. When the amount of the coagulant exceeds the optimum level, this overdosing condition leads to the destabilisation of the particles, which hinders the settlement. As the SS also constitutes some organics that may also contribute to the colour level, a non-efficient settlement of particles would also exhibit poor removal of colour.

As opposed to the $SnCl_4$, when RS was applied as the main coagulant, the SS and colour reductions decreased with the increase in the pH values. The highest removal was at pH3 with 68.4% and 87.8% removals for SS and colour, respectively. Then, as the pH increased from pH4 to 11, the reduction in SS and colour started to gradually drop until it reached the lowest removal at pH 6 with less than 5% reductions for both. Better removals at pH3 are due to the acidic effect from the acid used and not due to the RS applied. A few studies on natural coagulant have also reported good performances at acidic pH. Kristianto (2017) [33] used *Leucaena leucocephala* seed's extract as a natural coagulant to eradicate dye from synthetic wastewater and managed to reduce 99% of colour at pH3. In addition, Zonoozi et al., (2011) [34] removed 83% of colour from dye-containing solutions when chitosan was employed as the main natural coagulant. The high performance obtained at acidic conditions is due to the protonation of the extracted protein from the natural coagulant. Basically, under acidic conditions, a higher H^+ concentration from the protonation will neutralise the negative charges of the wastewater surface particles. Therefore, more removal was obtained under this condition when natural coagulant was used as the main coagulant [20,33,34].

3.3. Rubber Seed as Coagulant Aid

We examined the potential of RS as a coagulant aid in reducing the optimum $SnCl_4$ dose (10,000 mg/L) as determined in the previous experiment. We ranged the $SnCl_4$ dosages from 4000 mg/L to 8000 mg/L. Our previous experiment indicated that the $SnCl_4$ dosage could not be reduced below 8000 mg/L, as the removal for both SS and colour were fairly low (less than 50%), as also shown in Figure 1. This is because too much or not enough coagulant added into water/wastewater will make the removal less efficient. This is due to the overdosing or insufficient surface charge to affect the stability of the colloid's particles [35]. For this, we run the experiment at a fixed $SnCl_4$ of 8000 mg/L, and we vary the coagulant aid dosages (RS) from 100 mg/L to 250 mg/L, 500 mg/L, 750 mg/L, 1000 mg/L, 2000 mg/L, 4000 mg/L, 6000 mg/L, 8000 mg/L and 10,000 mg/L. The tests were undertaken at an optimum pH8. The results are shown in Figure 4.

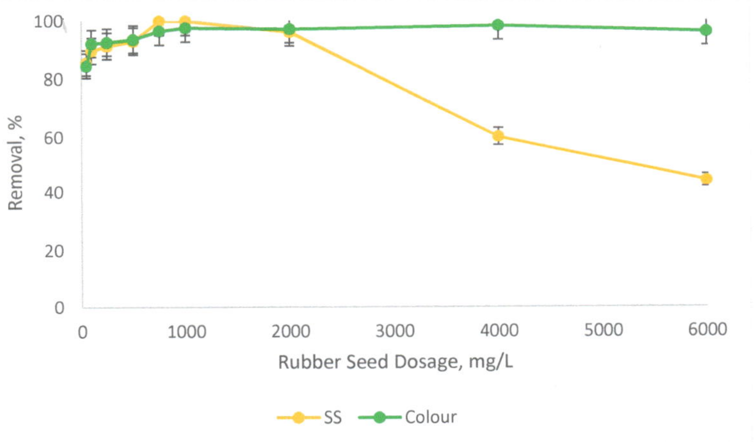

Figure 4. Percentage reductions in SS and colour at 8000 mg/L $SnCl_4$ with different ranges of RS as the coagulant aid at pH8.

Figure 4 indicates that the removal of SS and colour increased as the RS dosages increased. When the RS dose rose from 100 mg/L to 1000 mg/L, the removal for SS was improved from about 90% to almost 100% and from 92% to about 97% for colour. Beyond 1000 mg/L, the reductions in the performance for colour were not significant and were maintained above 90% at different RS dosages. However, there is a significant drop in the SS removals beyond 1000 mg/L of RS. This is due to the overdosing phenomenon, as explained before. Even though the concentration of $SnCl_4$ was reduced from 10,000 mg/L to 8000 mg/L, in the presence of RS as a coagulant aid, the removals for SS and colour were on par with the performance of $SnCl_4$ alone, as presented in Figure 1. According to Aygun and Yilmaz (2010) [32], the use of natural polyelectrolyte as coagulant aid can assist in reducing the concentration of metal coagulant without affecting the removal of the parameters. Basically, there are four types of mechanisms involved in the coagulation and flocculation process. In the case of natural coagulant aid, the possible mechanism is charge neutralisation or polymer bridging. However, compared to charge neutralisation, the polymer bridging mechanism is more dominant in plant-based coagulant aid. During the polymer bridging mechanism, the surface of the colloidal particles will attach to the polymer chains because of the existence of affinity between them. In addition, for a bridging mechanism to be effective, the addition of enough polymer dosage is an important criterion, as it helps to provide a sufficient site for particle attachment. It will provide sufficient unoccupied surface for particle attachment and the bridging span, which should be at a distance so that the interparticle repulsion between the particles is hindered. Therefore, adequate natural coagulant dosage will provide sufficient bridging links and surfaces for the removal of pollutants [36]. In polymer bridging, the particle surface of the colloid is able to attach to the loop of RS polymer chain. This polymer bridging mechanism helps in reducing the amount of chemical dosage ($SnCl_4$), through the charge neutralisation mechanism. The combination of these two mechanisms is able to remove a high percentage of SS and colour.

Organic matter (some in insoluble forms), measured as turbidity, and SS contribute mainly to colour in landfill leachates [13]. $SnCl_4$ is a cationic coagulant with positive electric charges. It helps to decrease the negative charge of hydroxide-forming colloids when dissolved in water. This hydroxide is hydrophobic in nature, meaning it can adsorb to the surface of organic anionic particles and become insoluble [37]. According to (Awang et al., 2012) [38], 'bridging' is the main mechanism that governs the agglomeration of the constituents for the coagulants that present as a negative surface charge. The addition of RS could help to form bigger flocs and enhance particle settlement, thus improving the removal rates. According to the findings, it was found that RS could be used as a coagulant aid to minimise the amount of metal coagulant used.

3.4. Comparisons of Zeta Potential (ZP) and Particle Size of $SnCl_4$ and Rubber Seed

The effect of pH on the ZP and particle size on the reduction in SS and colour using $SnCl_4$ and RS are illustrated in Figures 5 and 6. The particle size measurement was carried out by using the Malvern Zetasizer Nano ZS by applying the dynamic light scattering method (DLS). First, a laser is employed to provide a light source for the sample particles within a cell. Then, the intensity of the dispersed light is measured using a detector. Because a particle scatters light in all directions, the detector can theoretically be placed anywhere and yet detect the scattering. The detector's scattering intensity signal is sent to a correlator, which is a digital signal processor. The information from the correlator is then sent to a computer, where the Zetasizer software analyses the data and calculates the size. The determination of the sample particle size in this study is important, as it provides the effectiveness of the coagulant when added into the treatment. This is because, depending on the surface charge of the coagulant, the agglomeration will be induced to form a bigger size of particles/colloid, which then settles down during the settlement process. The relationship of ZP and particle size in removing the SS and colour was conducted in the existence of 10,000 mg/L of $SnCl_4$, which was the optimum dosage pre-determined before and in the pH range of 3–11.

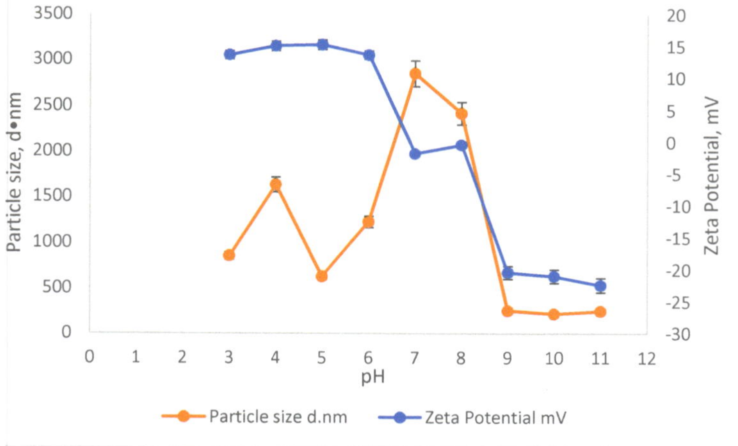

Figure 5. The influence of pH on the Zeta potential and particle size of $SnCl_4$.

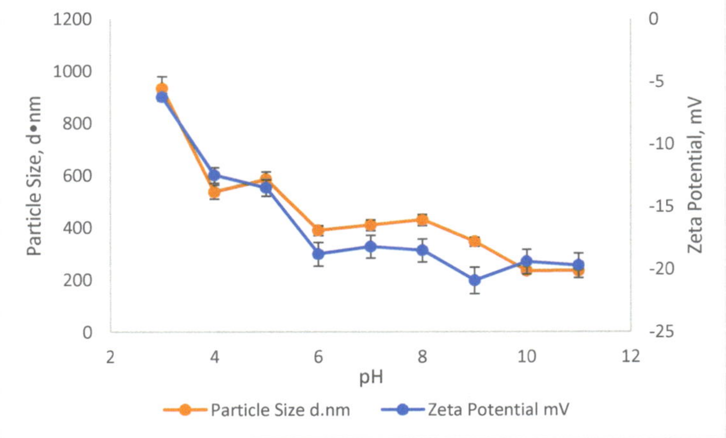

Figure 6. The influence of pH on the Zeta potential and Particle Size of Rubber Seed.

The ZP during the removal of SS and colour showed a gradual decrease throughout the pH values. Starting from pH3, the ZP increased slightly from 13.6 mV to 15.2 mV at pH5. Then, it slowly started to become less positive as the pH moved from pH6 until it approached a negative ZP value. The ZP values in the range of pH7 to pH8 were −1.8 mv to −0.4 mV, respectively, approaching the zero value. The point where the ZP becomes zero is known as the isoelectric point (IEP). Therefore, it can be said that the IEP value for $SnCl_4$ to remove the pollutants from leachate lies in this pH range (pH7–8). Starting from pH7 onwards, the ZP of the sample became more negative, from −1.8 mV at pH7 to −22.4 mV at pH11.

This can be correlated from the results in Figure 3 showing that the removals were obtained at a pH range between 7 and 8 with an associated particle size of 2419 d·nm and ZP of −0.4 mV. This enhances the settlement of the flocs. The removal for colour and SS at this pH range was about 98% for both. The particle size at approaching zero ZP will have a greater size compared to the particles with a higher (more negative or positive) ZP, which occurred at acidic (pH < 6) and basic (pH > 8) conditions, as shown in Figure 5. This situation happens because the particle surface charge in the IEP range (pH7–8) is less compared to the more negatively or too positively charged ZP. The thickness of the diffusion layer in the particles is reduced when the surface charge of the particle is lower. The reduction in the thickness of the layer of particles hence will decrease the repulsive force between the particles. This led to the agglomeration and induced them to come closer and form a much larger particle [39,40].

Meanwhile, the effect of pH on the ZP and particle size in removing the SS and colour from the leachate using RS as the main coagulant is shown in Figure 6. Compared to ZP and particle size when $SnCl_4$ was used as the coagulant, there was not much effect of RS as a coagulant on the ZP and the size of the particles. As the pH moved from pH3 to pH11, the ZP of the sample became more and more negatively charged. It started with −6.3 mV at pH3 to −19.7 mV at pH11. Since the ZP of the sample became more negatively charged, the size of the particle also became much smaller, which contributed to low removals for pollutants. The highest removal obtained at pH3 was 46.7% for SS and 76.5% for colour with an associated ZP of −6.3 mV and an average particle size of 933 d·nm. As the pH increased, the removal of SS and colour started to decrease. Smaller particle sizes were recorded with an increase in pH. For example, the particles at pH4 were 535 d·nm and reduced to 234 d·nm at pH11. These conditions did not favour the settlement of flocs. This is because the negatively charged particle (pH6–11) appear to repel each other and prevent the particles from coming closer to form much bigger particles.

3.5. *Comparisons of $SnCl_4$ with Different Types of Metal Coagulants*

Table 2 compares the performance of different metal salts and their valencies in landfill leachate treatment.

Table 2. Comparisons of optimal dosage on the removal of SS and colour using different types of coagulant from landfill leachate.

Types of Coagulant	Optimal Dosage (mg/L)	Optimal pH	Raw Leachate Concentration		Efficiency Removal (%)		References
			Colour (PtCo)	SS (mg/L)	Colour	SS	
Aluminium Sulfate	9000	7	4372	351	84	96	[41]
PAC	5000	6	5517.5	745	98	99.5	[25]
	3000	7	4372	351	92	89	[41]
$FeCl_3$	3600	6	5318	297	95.5	-	[20]
	1500	6	3199	407	95	94	[42]
$ZrCl_4$	1500	4	5000	441	94.3	93.4	
$TiCl_4$	1200	4	17,075	397	82	92	[29]
	600	6	4253	330	81.4	86.7	[19]
$SnCl_4$	10,000	8	22,970	548	97.3	99	This Study
$SnCl_4$ + RS	8000 $SnCl_4$ + 1000 RS	8	22,970	548	97.6	100	This Study

Compared with the previous works, the raw leachate that we examined in the current study was generally the highest initial concentration of the raw leachate (colour 22,970 PtCo and SS 548 mg/L). A higher concentration of leachate pollutants usually requires a higher coagulant dosage. Our conclusion (8000 $SnCl_4$ and 1000 RS) in the current work is considered in line with the results of previous workers, especially by Rui et al., (2015) [41] (9000 mg/L Alum). However, they used less concentrated leachate. Even though lower dosages of metal salts were found by other researchers, most of them used fairly low to moderate levels of leachate pollutants.

3.6. *Operating Cost Calculation*

The estimation of the coagulation treatment costs in removing suspended solids and colour removal is shown in Table 3. However, our calculation was derived from laboratory work only. The reagent cost involved the three types of chemical reagents: hydrochloric acid (HCl) and sodium hydroxide (NaOH) for pH alteration, and tin tetrachloride ($SnCl_4$). The energy cost was estimated according to the electricity unit cost based on the electricity tariff by our local energy provider, i.e., Tenaga Nasional Berhad (TNB), Malaysia.

Table 3. The quantity and cost of coagulant required for SS and colour treatment at lab scale (per 3 L of leachate).

Item	Estimated Usage	Cost (per kg or L or W)	Total	Reference
Reagents Cost				
HCl MERK 37%	25 mL	RM 550/2.5 L	RM 5.50	[43,44]
NaOH 97% Pellets	25 mL	RM 335/500 g	RM 16.80	
$SnCl_4$	75 mL (for 6 beakers)	RM 269/100 g	RM 201.8	
Rubber Seed	120 mL (for 6 beakers)	RM 1.00/kg	RM 120.00	
Energy Cost				
Jar test (VELP-Scientifica, Model: JLT6)	19 W	RM 0.73/kW	RM 14.00	[45]
		Total	RM 358.10	

4. Conclusions

- At pH8 and with 10,000 mg/L concentration, $SnCl_4$ as the main coagulant was able to remove 99% and 97.3% of SS and colour, respectively, with an associated particle size of 2850.7 d·nm and ZP values of −0.4 mV.
- RS was ineffective as a sole coagulant with only 46.7% and 76.5% reductions in SS and colour, respectively, at pH3, a 6000 mg/L dosage, 933 d·nm, and a ZP value of −6.3 mV.
- In the presence of 1000 mg/L of RS, the dosage of $SnCl_4$ could be reduced from 10,000 mg/L to 8000 mg/L by the bridging mechanism, with almost complete removal of 100% of SS and colour.
- As a sole landfill leachate treatment system, this study concludes that RS showed better potential as a coagulant aid compared to when it is used as a main coagulant.

Author Contributions: Data curation, investigation, writing original draft preparation, S.F.R.; conceptualization, methodology, supervision, H.A.A.; supervision, F.M.O.; supervision, M.S.Y.; funding acquisition, H.H.; writing—review and editing, M.A.K.; project administration, K.S.A.; validation, Y.-T.H. All authors have read and agreed to the published version of the manuscript.

Funding: The authors would like to express the highest gratitude to Universiti Sains Malaysia (USM) for the provision of facilities required for this study. This work was also partly funded by the FRGS grant No. FRGS/1/2018/TK01/USM/01/1 and USM RUI grant No. 1001/PAWAM/8014081.

Institutional Review Board Statement: Not applicable.

Informed Consent Statement: Not applicable.

Data Availability Statement: Not applicable.

Acknowledgments: Thanks is also extended to the Majlis Daerah Kerian (M.D.K.), Perak, for their cooperation during the study and also to the laboratory technical staff who also contributed to the leachate sampling and laboratory works.

Conflicts of Interest: The authors declare no conflict of interest.

References

1. Rahim, N.; Apendi, S.R.M.; Farook, F.; Ismail, A. Environmental attitudes inventory (EAI) of UiTM Penang hospitality students. *Herit. Cult. Soc.* **2016**, 613–616.
2. Zin, N.S.M.; Aziz, H.A.; Adlan, M.N.; Ariffin, A.; Yusoff, M.S.; Dahlan, I. Application of a pre-hydrolysed iron coagulant on partially stabilised leachate. *Desalination Water Treat.* **2015**, *54*, 2951–2958. [CrossRef]
3. Kulikowska, D.; Zielińska, M.; Konopka, K. Treatment of stabilised landfill leachate in an integrated adsorption–fine-ultrafiltration system. *Int. J. Environ. Sci. Technol.* **2019**, *16*, 423–430. [CrossRef]
4. Adhikari, B.; Khanal, S.N. Qualitative study of landfill leachate from different ages of landfill sites of various countries including Nepal. *IOSR J. Environ. Sci. Toxicol. Food Technol.* **2015**, *9*, 23–36.
5. Bolyard, S.C.; Reinhart, D.R. Application of landfill treatment approaches for stabilisation of municipal solid waste. *Waste Manag.* **2016**, *55*, 22–30. [CrossRef] [PubMed]
6. Shadi, A.; Niza, N.M.; Ijanu, M.E. Characterization of stabilised leachate and evaluation of LPI from sanitary landfill in Penang, Malaysia. *Desalination Water Treat.* **2020**, *189*, 152–164. [CrossRef]
7. Aziz, H.A.; Rahim, N.A.; Ramli, S.F.; Alazaiza, M.Y.D.; Omar, F.M.; Hung, Y.T. Potential use of Dimocarpus longan seeds as a flocculant in landfill leachate treatment. *Water* **2018**, *10*, 1672. [CrossRef]
8. Torretta, V.; Ferronato, N.; Katsoyiannis, I.A.; Tolkou, A.K.; Airoldi, M. Novel and conventional technologies for landfill leachates treatment: A review. *Sustainability* **2017**, *9*, 9. [CrossRef]
9. Wang, W.L.; Mohammad, A.W.; Benamor, A.; Hilal, N. Chitosan as natural coagulant in hybrid coagulation-nanofiltration membrane process for water treatment. *J. Environ. Chem. Eng.* **2016**, *4*, 4857–4862.
10. Talib, Y.; Idris, A.; Aslina, S. A tannin-based agent for coagulation and flocculation of municipal wastewater: Chemical composition, performance assessment compared to polyaluminum chloride, and application in a pilot plant. *J. Environ. Manag.* **2016**, *184*, 494–503.
11. Rajasulochana, P.; Preethy, V. Comparison on efficiency of various techniques in treatment of waste and sewage water—A comprehensive review. *Resour. Effic. Technol.* **2016**, *2*, 175–184. [CrossRef]
12. Altaher, H.; Khalil, T.; Abubeah, R. An agricultural waste as a novel coagulant aid to treat high turbid water containing humic acids. *Glob. Nest J.* **2016**, *18*, 279–290.

13. Hamidi, A.A.; Alias, S.; Assari, F.; Adlan, M.N. The use of alum, ferric chloride and ferrous sulphate as coagulants in removing suspended solids, colour and COD from semi-aerobic landfill leachate at controlled pH. *Waste Manag. Res.* **2007**, *25*, 556–565. [CrossRef]
14. Lichtfouse, E.; Navarrete, M.; Hamelin, M.; Debaeke, P. *Sustainable Agriculture*; Springer: New York, NY, USA, 2011; Volume 2, pp. 1–895.
15. Lyklema, J. Coagulation by multivalent counterions and the Schulze-Hardy rule. *J. Colloid Interface Sci.* **2013**, *392*, 102–104. [CrossRef]
16. Patel, H.; Vashi, R.T. Removal of Congo Red dye from its aqueous solution using natural coagulants. *J. Saudi Chem. Soc.* **2012**, *16*, 131–136. [CrossRef]
17. Ntwampe, I.O.; Waanders, F.B.; Bunt, J.R. Destabilization dynamics of clay and acid-free polymers of ferric and magnesium salts in AMD without pH adjustment. *Water Sci. Technol.* **2016**, *1*, 861–875. [CrossRef] [PubMed]
18. Zhao, Y.X.; Phuntsho, S.; Gao, B.Y.; Huang, X.; Qi, Q.B.; Yue, Q.Y.; Shon, H.K. Preparation and characterisation of novel polytitanium tetrachloride coagulant for water purification. *Environ. Sci. Technol.* **2013**, *19*, 12966–12975. [CrossRef]
19. Aziz, H.A.; Amr, S.A.; Hussain, S. Potential use of Titanium Tetrachloride as coagulant to treat semi aerobic leachate treatment potential use of Titanium Tetrachloride as coagulant to treat semi aerobic leachate treatment. *Aust. J. Basic Appl. Sci.* **2015**, *9*, 37–44.
20. Aziz, H.A.; Ramli, S.F. Removal of COD and colour from landfill leachate using ferric chloride by coagulation-flocculation treatment. *Adv. Environ. Biol.* **2014**, *8*, 83–90.
21. Hamidi, A.A.; Zainal, S.F.F.S.; Motasem, Y.D.A. Optimization of coagulation-flocculation process of landfill leachate by Tin (IV) Chloride using response surface methodology. *J. Hamadan Univ. Med. Sci.* **2019**, *6*, 41–48. [CrossRef]
22. Mathews, T.J.; Looney, B.B.; Bryan, A.L.; Smith, J.G.; Miller, C.L.; Southworth, G.R.; Peterson, M.J. The effects of a stannous chloride-based water treatment system in a mercury contaminated stream. *Chemosphere* **2015**, *138*, 190–196. [CrossRef] [PubMed]
23. Kennedy, A.; Croft, R.; Flint, L.; Arias-Paić, M. Stannous chloride reduction—Filtration for hexavalent and total chromium removal from groundwater. *AWWA Water Sci.* **2020**, *2*, e1174. [CrossRef]
24. Zepeda, A.M.; Gonzalez, D.; Heredia, L.G.; Marquez, K.; Perez, C.; Pena, E.; Cantu, J. Removal of Cu^{2+} and Ni^{2+} from aqueous solution using SnO_2 nanomaterial effect of: PH, time, temperature, interfering cations. *Microchem. J.* **2018**, *141*, 188–196. [CrossRef] [PubMed]
25. Aziz, H.A.; Yii, Y.C.; Syed Zainal, S.F.F.; Ramli, S.F.; Akinbile, C.O. Effects of using Tamarindus indica seeds as a natural coagulant aid in landfill leachate treatment. *Glob. Nest J.* **2018**, *20*, 373–380.
26. Kaszuba, M.; Corbett, J.; Watson, F.M.; Jones, A. High-concentration zeta potential measurements using light-scattering techniques. *Philos. Trans. R. Soc. Lond.* **2010**, *368*, 4439–4451. [CrossRef] [PubMed]
27. Delgado, A.V.; Koopal, L.K. Measurement and Interpretation of electrokinetic phenomena (IUPAC). *Pure Appl. Chem.* **2005**, *77*, 1753–1805. [CrossRef]
28. Hunter, R.J. *Zeta Potential in Colloid Science: Principles and Applications*; Academic Press: London, UK, 1988.
29. Wern, T.Y.; Ardani, M.R.; Ramli, S.F.; Rezan, S.A.; Aziz, H.A.; Ibrahim, I. Coagulation Performance of Titanium Tetrachloride for Alor Pongsu Wastewater Treatment. In Proceedings of the 3rd International Postgraduate Conference on Materials, Minerals & Polymer (MAMIP), Seberang Perai, Malaysia, 31 October–1 November 2019.
30. Bogacki, J.; Marcinowski, P.; El-Khozondar, B. Treatment of landfill leachates with combined acidification/coagulation and the FeO/H_2O_2 process. *Water* **2019**, *11*, 194. [CrossRef]
31. Amuda, O.S.; Amoo, I.A.; Ajayi, O.O. Performance optimization of coagulant/flocculant in the treatment of wastewater from a beverage industry. *J. Hazard. Mater.* **2006**, *129*, 69–72. [CrossRef]
32. Aygun, A.; Yilmaz, T. Improvement of coagulation-flocculation process for treatment of detergent wastewaters using coagulant aids. *Int. J. Chem. Environ. Eng.* **2010**, *1*, 97–101.
33. Kristianto, H. The potency of Indonesia native plants as natural coagulant: A mini review. *Water Conserv. Sci. Eng.* **2017**, *2*, 51–60. [CrossRef]
34. Zonoozi, M.H.; Alavi Moghaddam, M.R.; Arami, M. Study on the removal of acid dyes using chitosan as a natural coagulant/coagulant aid. *Water Sci. Technol.* **2011**, *63*, 403–409. [CrossRef] [PubMed]
35. Ernest, E.; Onyeka, O.; David, N.; Blessing, O. Effects of pH, dosage, temperature and mixing speed on the efficiency of watermelon seed in removing the turbidity and colour of Atabong River, Awka-Ibom State, Nigeria. *Int. J. Adv. Eng. Manag. Sci.* **2017**, *3*, 427–434.
36. Amran, A.H.; Zaidi, N.S.; Muda, K.; Loan, L.W. Effectiveness of natural coagulant in coagulation process: A review. *Int. J. Eng. Technol.* **2018**, *7*, 34–37. [CrossRef]
37. Irfan, M.; Butt, T.; Imtiaz, N.; Abbas, N. The removal of COD, TSS and colour of black liquor by coagulation—Flocculation process at optimized pH, settling and dosing rate. *Arab. J. Chem.* **2014**, *48*, S2307–S2318. [CrossRef]
38. Awang, N.A.; Aziz, H.A.; Tunku, U.; Rahman, A.; Umar, M. Comparative removal of suspended solids from landfill leachate by Hibiscus rosa-sinensis leaf extract and alum. *Desalination Water Treat.* **2012**, *51*, 2005–2013. [CrossRef]
39. Fathiah, M.Z.; Edyvean, R.G. The role of ionic strength and mineral size to zeta potential for the adhesion of putida to mineral surfaces. *Int. J. Biotechnol. Bioeng.* **2015**, *9*, 805–810.

40. Xiaodi, S.; Shuntang, G. Effect of diluent type on analysis of zeta potential of colloid particles of soymilk protein. *Trans. Chin. Soc. Agric. Eng.* **2016**, *32*, 270–275.
41. Rui, L.M.; Daud, Z.; Aziz, A.; Latif, A. Treatment of leachate by coagulation-flocculation using different coagulants and polymer: A Review. *Int. J. Adv. Sci. Eng. Inf. Technol.* **2015**, *2*, 1–4. [CrossRef]
42. Zin, N.S.M.; Aziz, H.A.; Adlan, N.M.; Ariffin, A.; Suffian, M. Removal of color, suspended solids, COD and ammonia from partially stabilize landfill leachate by using iron chloride through coagulation process. *Int. J. Eng. Technol.* **2016**, *5*, 736–739. [CrossRef]
43. Ebrahimi, A.; Mahdavi, M.; Meghdad, P.; Alimohammadi, F.; Mahvi, A.H. Dataset on the cost estimation for spent filter backwash water (SFBW) treatment. *Data Brief.* **2017**, *15*, 1043–1047. [CrossRef]
44. MERK. Chemical Price. Retrieved November 4, 1BC. 2020. Available online: https://www.sigmaaldrich.com/catalog/search?term=NaOH&interface=All&N=0&mode=matchpartialmax&lang=en&region=MY&focus=product (accessed on 11 November 2020).
45. TNB. Current Rate According to Tariff Category. 2020. Available online: https://www.tnb.com.my/residential/pricing-tariffs (accessed on 11 November 2020).

International Journal of
Environmental Research and Public Health

MDPI

Article

Heavy Metals in Harvested Rainwater Used for Domestic Purposes in Rural Areas: Yatta Area, Palestine as a Case Study

Fathi Anabtawi [1], Nidal Mahmoud [2], Issam A. Al-Khatib [2,*] and Yung-Tse Hung [3]

1 Faculty of Graduate Studies, Birzeit University, Birzeit P.O. Box 14, West Bank, Palestine; fathianabtawi@gmail.com
2 Institute of Environmental and Water Studies, Birzeit University, Birzeit P.O. Box 14, West Bank, Palestine; nmahmoud@birzeit.edu
3 Department of Civil and Environmental Engineering, Cleveland State University, Cleveland, OH 44115, USA; yungtsehung@gmail.com
* Correspondence: ikhatib@birzeit.edu; Tel.: +970-2298-2120

Abstract: Rainwater harvesting is considered one of the most important water resources in the Palestinian countryside. In this research, the study area chosen for the study was Yatta town in Hebron city. 75 water samples were collected from 74 cisterns in a number of neighborhoods in Yatta, and a structured household survey was conducted with the same households where the water samples were collected. Statistical analysis was made using the SPSS software. An analysis for the samples was made using ICP-MS to test the existence of a number of heavy metals, namely Pb, Cr, Mn, Co, Ni, Cu, Zn and Cd. The results were compared with the WHO and Palestinian limits for drinking water quality. Considering the metals Mn, Co, Cu and Cd, neither of the samples exceeded any of the two limits. For the metals, Pb, Cr, and Ni, two samples exceeded both limits. For the metal, Zn, one sample exceeded the WHO limit only. Sources of pollution by heavy metals of the harvested rainwater were identified by means of a questionnaire distributed to the households. The results showed that except for nickel and the water collection surface of the cistern factor, there is no direct relationship between the factors and activities that may contribute to contaminate harvested rainwater with heavy metals and the existence of heavy metals beyond local and international limits. Based on the questionnaire and literature: Possible sources of lead and zinc are the roof, storage tanks, distribution systems and plumbing; possible sources of chromium are road dust, asbestos brakes and anthropogenic activities occurring around the house; possible source of nickel is leaching from metals in contact with harvested rainwater such as pipes and fittings which are used to collect the harvested rainwater. In addition, an assessment of the potential health risks due to contamination of the harvested rainwater by heavy metals was made for all the samples that exceeded either WHO limit or the Palestinian limit or both. The Chronic Daily Intake (CDI) and the Health Risk Index (HRI) were calculated. The assessment was made for both adults and children. The results showed that all the samples are considered safe (HRI < 1), which means that there are no potential health risks for consumers.

Keywords: heavy metals; harvested rainwater; health risk; developing countries

check for updates

Citation: Anabtawi, F.; Mahmoud, N.; Al-Khatib, I.A.; Hung, Y.-T. Heavy Metals in Harvested Rainwater Used for Domestic Purposes in Rural Areas: Yatta Area, Palestine as a Case Study. *Int. J. Environ. Res. Public Health* **2022**, *19*, 2683. https://doi.org/10.3390/ijerph19052683

Academic Editor: Miklas Scholz

Received: 13 January 2022
Accepted: 23 February 2022
Published: 25 February 2022

1. Introduction

Supply of water security that meets the quality parameters specified in applicable standards, is now the basis for the functioning of most societies [1]. Worldwide, one in three people do not have access to safe drinking water, and so billions of people still lack access to clean drinking water [2]. Ensuring access to water and sanitation for all through sustainable management of water resources is the sixth of the 17 goals of the Sustainable Development Goals, developed by the UN General Assembly in 2015 [1]. The climate change impacts on the water resources quantity and quality, besides the increasing frequency of extreme events, e.g., floods and droughts, are becoming an important problem [3]. Water is the most important resource; it is utilized largely in agricultural production and is fundamental to ensuring global food security. The overutilization of water resources creates complex problems, such as waterlogging and salinization, and results in the depletion of groundwater resources [4]. Water is a vital substance in the environment, and its contamination with heavy metals is considered a worldwide environmental problem [5–7]. Heavy metals are important pollutants to groundwater, surface water, and harvested rainwater [8–11]. Common heavy metals that human beings are exposed to include: Aluminum (Al), Cadmium (Cd), Chromium (Cr), Lead (Pb), Mercury (Hg), Copper (Cu), Zinc (Zn), Iron (Fe), Nickel (Ni) and Cobalt (Co).

Among the various pollutants of harvested rainwater, heavy metals are always of big concerns due to their severe toxicities, and so heavy metals are regulated in drinking water by the United States Environmental Protection Agency (U.S. EPA, USA) [12]. When heavy metals enter into the human body, they could easily bind to vital cellular components and accumulate in organisms, resulting in a series of diseases and disorders, e.g., cancers, osteomalacia, kidney malfunction, etc. [13], depending on the type and amount of the metal involved [14,15]. Their toxicity is made by forming complexes with proteins where they contain carboxylic acid ($-COOH$), amine ($-NH_2$) and thiol ($-SH$) groups.

Worldwide, cancer is considered a significant health care problem, and it is considered the second most cause of death. According to WHO, environmental factors are responsible for more than 70% of cancer cases. Many heavy metals are considered carcinogenic according to the classification of WHO and the International Agency for Research on Cancer (IARC), e.g., cobalt, mercury, lead, arsenic, nickel, cadmium, beryllium, chromium and others [16]. For example, many studies showed that people who drink water containing high levels of arsenic have higher risks of bladder, kidney, lung, colon, liver and skin cancer [17,18]. In addition, cadmium is known to cause kidney, prostate and lung cancer, aluminum can cause lung and bladder cancer [16].

As a first step of water pollution prevention, accurate and rapid monitoring of the heavy metals is vital. Ideally monitoring methods are expected to identify point sources of pollutants and the variation of non-point sources of pollutants in the environment. One of the main sources of heavy metals in harvested rainwater collected from roof surfaces is the roofing material. Storm water runoff, collected from both roof surfaces and ground, collects a variety of pollutants, e.g., excess nutrients, metals, hydrocarbons and pesticides that may leach from traditional roofing materials or may be introduced onto roofs through wet and dry depositions [19,20]. The first flush of runoff water occurring at the beginning of the storm contains a high proportion of the pollutants loads, including heavy metals [21]. Other sources of heavy metals are mineral particles from ground surface, components originating from industrial emissions, vehicles emissions and fuel combustion products emitted to the atmosphere [22,23].

Rainwater harvesting is a common practice in the West Bank of Palestine, especially in the southern part where water is very scarce [24]. In these areas, during winter season, rainwater is collected from the rooftops of the houses and stored in cisterns. There is a high probability that this water might be contaminated with heavy metals coming from, for instance, from dust, roof materials, etc. [25]. Problems related to rainwater quality and its contamination with heavy metals are significant from the point of view of health. While scientific knowledge is important for knowledge-based policy development, combining science and local knowledge from stakeholders is necessary for developing more inclusive approaches and locally targeted solutions [26].

Harvested rainwater, used for domestic purposes including human consumption, cooking, toilets and showers etc. in the southern part of the West Bank, was the main focus of this research. Rural areas in Hebron, specifically different neighborhoods of Yatta area, were selected as the study area, since many of them are not served with water distribution networks, experiencing a shortage of water, and thus resorting to rainwater harvesting techniques [27]. This research brings new cognitive and practical conclusions from the

Palestine to science to ensure the safe use of the limited water resources. It concerns both rainwater and surface and groundwater. The research addresses the important issue of evaluating the safety of harvested rainwater. This issue is timely as climate change is impacting freshwater resources and an understanding of the factors that impact contaminant levels is urgently needed. this paper adds to the knowledge pool of water quality, especially rainwater, in water stressed countries of arid to semi-arid climates. The main objectives of this research were to investigate the occurrence of different heavy metals (Pb, Cr, Mn, Co, Ni, Cu, Zn and Cd) in harvested rainwater collected in cisterns; to identify sources of heavy metals pollution of the harvested rainwater; and to assess the potential health risks due to contamination of harvested rainwater used for drinking by heavy metals.

2. Methodology

2.1. Study Area

The focus of this research is on Yatta area. Yatta town is approximately 8 km south of the city of Hebron. Hebron is a Palestinian city, located in the southern West Bank, 30 km south of Jerusalem. It lies 930 m above sea level. It is the largest city in the West Bank and the second largest in the Palestinian Territories after Gaza. Hebron is attached to cities of Adh Dhahiriya, Dura, Yatta, and the surrounding villages with no borders. In 2020, its population reached 68,094 capita [28]. Figure 1 shows the Hebron District Map.

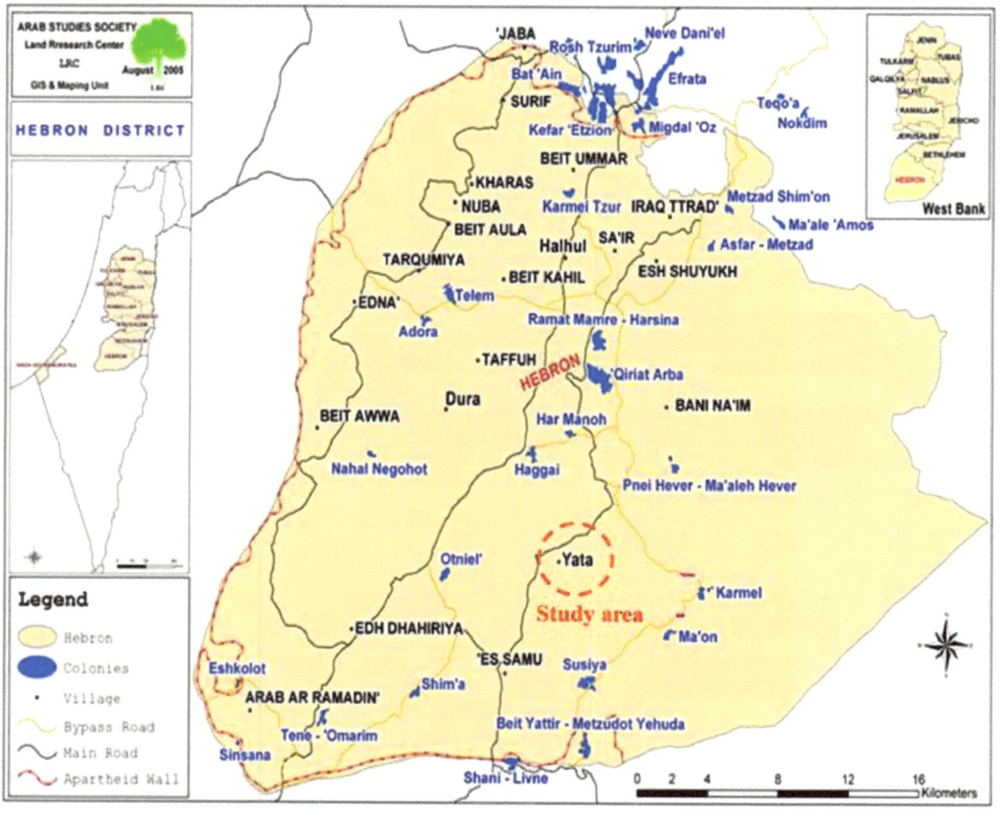

Figure 1. Hebron district including the study area, (PCBS, 2020).

Hebron District climate ranges from arid to semiarid with an increase in aridity towards the Negev Desert in the south and the Jordan Valley in the east. The monthly average temperature ranges from 7.5 to 10 °C in winter to 22 °C in summer. The minimum temperature is −3 °C in January and the maximum is 40 °C in August. Most of the rainfalls are during December through February, although there may be rain from mid-October to the end of April. The number of rainfalls per month ranges between 400 mm during the rainfall season and 0 mm during the dry season. Hebron District is facing amplified water scarcity due to the arid and semiarid climatic conditions [29].

2.2. Water Sampling

In this case, 74 water samples were randomly collected from five different neighborhoods in Yatta area, with estimated 75,000 inhabitants having around 4500 cistern. The samples were collected during the rainy season in February 2016 over a period of 5 days. One-time water sample was collected from each of the 74 cistern, and the water samples distribution was as follows: 12 samples from Al-Hila village, 21 from Yatta town, 13 from Khallet Saleh village, 13 samples from Khallet al Maiyya village, and 15 from Al-Hadidya village. First, the water samples were collected in a 1-L high density polyethylene bottles; pre-cleaned with 10% nitric acid followed by repeated rinsing with bi-distilled water, stabilized with ultrapure nitric acid (0.5% HNO_3), preserved in a cooler of about 4 °C and transported to the lab of Al-Quds University for analysis. The samples were analyzed for heavy metal content (Pb, Cr, Mn, Co, Ni, Cu, Zn and Cd) using Inductively Coupled Plasma Mass Spectrometry (ICP-MS). Preparation of samples was made by diluting 1.0 mL of the water samples to 10.0 mL with 0.3% ultrapure nitric acid. After that, the samples were analyzed by ICP-MS. ICP-MS (Agilent 7500) with an onboard peristaltic pump, a nebulizer (MicroMist nebulizer), an ICP argon plasma torch, two pumps for evacuation, a quadruple mass analyzer, an octapole reaction system (ORS), and an electron multiplier detector was used for the analysis of the heavy metals in this study.

ICP-MS is a type of mass spectrometry that is capable of detecting several trace metals and non-metals at concentrations as low as one part in 10^{15} (part per quadrillion, ppq) on non-interfered low-background isotopes. The detection is achieved by ionizing the sample with inductively coupled plasma first and then using a mass spectrometer to separate and quantify the ions [30]. The results of heavy metal concentrations in the analyzed samples were compared with the Palestinian standards and WHO guidelines for drinking water, as well as the values mentioned in some other developing countries that use harvested rainwater for domestic purposes.

2.3. Human Health Risk Assessment

As the harvested rainwater, in the study area, is to be used for drinking and other domestic purposes, it is important to make sure that this water is safe to be used by consumers.

Two approaches were used to test the safety of the harvested rainwater [4]:

1. Chronic daily intakes of metals (CDIs) and,
2. Health risk indexes of metals (HRIs).

The health risk assessment was made for all the samples that exceeded either the WHO or the Palestinian standards for drinking water quality or both.

2.3.1. Chronic Daily Intakes of Metals (CDIs)

Heavy metals enter the human body through several pathways including: food intake, dermal contact and inhalation. In comparison with oral intake, however, all other pathways are considered negligible [31]. The *CDI* (μg/(kg.day)) of a heavy metal through water ingestion was calculated by Equation (1) [31,32].

$$CDI = \frac{C_m}{W_b} \times I_w \tag{1}$$

where, C_m (μg/L) is the heavy metal concentration in water, I_w (L/day) is the average daily intake of water (assumed to be 2 L/day for adult and 1 L/day for child), and W_b (kg) is the average body weight (assumed to be 72 kg for adult and 32.7 kg for child), respectively [31].

2.3.2. Health Risk Indexes of Metals (HRIs)

To estimate the chronic health risks, HRIs were calculated by Equation (2) [31,32].

$$HRI = \frac{CDI}{RfD} \tag{2}$$

where, the oral toxicity reference dose (*RfD*, μg/(kg.day)) values for Cd, Cr, Cu, Mn, Ni, Pb, Zn and Co are 5×10^{-1}, 1.5×10^{1}, 3.7×10^{1}, 1.4×10^{2}, 2×10^{1}, 3.6×10^{1}, 3×10^{2} and 3×10^{1}, respectively [31,32]. The *HRI* value less than 1 is considered to be safe for the consumers [33].

2.4. Household Survey

A structured household survey was conducted with the same households where the water samples were collected. A questionnaire of 19 questions in multiple choice format was designed and distributed for this purpose (Appendix A). The questionnaire included questions about the source of water in the cistern, the frequency of cleaning the roof and the cistern, age of the cistern, shape of the cistern, existence of impurities or algae on the surface and sides of cistern and many others. The main aim of the questionnaire was to relate the results of the sampling analysis by the sources of pollution of the samples to heavy metals. In order to connect them together, statistical analysis was made using the Statistical Science Software Program (SPSS) software version 20 [34]. All the sample results exceeding the WHO and Palestinian limits were considered polluted by heavy metals. Cross tabulation method was used in the statistical analysis in order to determine whether there is a statistical significance between the questions, at a significance level of 0.05; which are considered as possible sources of pollution by heavy metals, and the sample results.

3. Results and Discussion

3.1. Main Characteristics and Usage of Rainwater Harvesting Cisterns in Yatta Area

The main characteristics of rain fed cisterns in Yatta area are discussed in this section. Considering water collection surface, 77.3% of the households depend on rainwater collected from the roof of the house as a major source of domestic water. During winter, they harvest rainwater in cisterns laid in the backyard of the houses. In addition, 1.3% collect water from the garden or the backyard of the house, and 2.7% collect rainwater from the streets.

The age of cistern varies between the households, 56.4% of the households indicated that their cistern's age is less than or equal 20 years, while 20.5% had older cisterns with an age of more than 30 years. The age of the cistern can be linked to the quality of the water collected inside cistern. The older the cistern, the higher the probability that it contains impurities and accumulates heavy metals, and therefore the harvested rainwater might become contaminated inside the cistern. The volume of the rainfed cisterns of the households varies between a few to 200 cubic meters, but 34.8% of the households have a cistern of 40 cubic meters or less. 81.3% of the cisterns are made from concrete, while only 14.7% are made from rock.

The acidic components of rainwater react with the alkaline components of concrete cisterns or cement mortar, dissolving mineral salts (mainly calcium carbonate). Therefore, using harvested rainwater in concrete cisterns might affect its quality. Regarding the cistern top, most of the cisterns (98.6%) have a cover. Closing the cisterns totally guarantee that no or little impurities and heavy metals may enter into them. 61.3% of the households have a cuboid cistern shape; while 38.7% have a peer-shaped cistern. 61.3% of respondents use the roof of the house for hanging the laundry.

Most of the households (98.7%) use the harvested rainwater for drinking, while 62.7% and 61.3% of them use it for irrigation and cleaning and laundry. Since harvested rainwater is used for many purposes, it is crucial to guarantee that the harvested rainwater is safe for the consumers, i.e., in case of the presence of heavy metals, they should be in minimal concentrations or concentrations below local and international limits, also, the CDI and HRI must be within the international limits to consider the water safe for daily intake.

With respect to the environmental conditions regarding the rainwater harvesting cisterns, the distance between the cesspit and the cistern in most of households (86%) was 15 m or less. The greater the distance between the cesspit and the cistern, the lower the probability of wastewater leakage from the cesspit into the cistern [35]. 48.5% of the households have the level of cesspit below the cistern level; while 32.4% have the level of cesspit the same level as cistern. It is better that the level of cesspit be lower than the cistern level to prevent wastewater leakage from the cesspit into the cistern. 93.8% of the households emptied the cesspits within the last year. 4.8% of the households emptied cesspits within 3 years and 1.6% within 15 years. Emptying the cesspits periodically guarantees that the wastewater leakage from these cesspits into the cisterns is kept as minimum, therefore making sure that the quality of the harvested rainwater is kept at its maximum level. Most of the households (74.7%) do not have trees close to the cistern, while 25.3% have trees around the house and close to the cistern. Heavy metals, coming from streets through the air, can adsorb to these trees and then might desorb and enter the cisterns polluting the harvested rainwater.

3.2. Cisterns Owners' Awareness

The indicators of cisterns owners' awareness to the factors that contribute to pollution prevention in rainwater harvesting cisterns are presented in this section. In this case, 96% clean the roof and 81.3% remove the first rain before start collecting rainwater in the cistern. Cleaning the roofs guarantees that no or little amounts of heavy metals can present in the harvested rainwater and that is the case in this study. 75.3% of the households raise animals and birds around the house. Raising animals or birds in the backyard of the house increases the opportunity that these animals pick the heavy metals from streets and therefore these heavy metals may enter the harvested rainwater affecting its quality. 70% of the households clean the cisterns periodically before harvesting the rainwater. Cleaning the cisterns from time to time guarantees that no or very little amounts of heavy metals will be stuck on their walls and this keeps the low concentrations of the heavy metals analyzed in the samples. 77.3% of the households allow solid wastes accumulate in the yard of the house. Solid wastes include many components that may contain heavy metals, and so collecting the waste in the backyard of the house increases the opportunity of contamination of the harvested rainwater with a number of heavy metals such as Pb and Cr coming, for example, from batteries.

3.3. Heavy Metals' Concentrations

The results of the investigated heavy metals concentration in the harvested rainwater are presented in Table 1. Out of the eight tested heavy metals, Cr, Ni and Cu were detected in all samples. Pb, Mn, Co, Zn and Cd were detected in 71, 60, 40, 68 and 46 out of 74 samples, respectively. The results of the current study are compared against international and local studies on the occurrence of heavy metals in harvested rainwater (Table 2). The heavy metal presented in Tables 1 and 2 are discussed separately hereafter.

Table 1. Concentrations of the heavy metals for all of the samples analyzed in harvested rainwater, Yatta area.

Element	Average (STD) (μg/L) (SD)	Samples Where Each Heavy Metal Was Detected, *n* (%)	Range (μg/L)	WHO Limits (μg/L)	Palestinian Limits (μg/L)	Samples Exceeding WHO Limits, *n* (%)	Samples Exceeding Palestinian Limits, *n* (%)
Pb	1.80 (1.49)	71 (96)	n.d.–24.00	10	10	2 (2.70)	2 (2.70)
Cr	4.31 (3.69)	74 (100)	0.11–101.00	50	50	2 (2.70)	2 (2.70)
Mn	4.32 (2.99)	60 (81)	n.d.–58.00	500	100	0 (0)	0 (0)
Co	0.24 (0.66)	40 (54)	n.d.–3.00	10	-	0 (0)	0 (0)
Ni	19.06 (8.01)	74 (100)	0.60–518.00	70	50	2 (2.70)	2 (2.70)
Cu	8.18 (3.99)	74 (100)	0.54–123.00	2000	1000	0 (0)	0 (0)
Zn	201.34 (21.04)	68 (92)	n.d.–3453.00	3000	5000	1 (1.35)	0 (0)
Cd	0.10 (0.56)	46 (62)	n.d.–2.00	3	3	0 (0)	0 (0)

n.d.: not detected.

Cr. The water samples are considered safe in terms of the occurrence of the heavy metal Cr, except two samples that exceeded the WHO and Palestinian limits (Tables 1 and 2). Possible sources of Cr are road dust, asbestos brakes or anthropogenic activities occurring around the house [36,37]. In addition, Cr is emitted from solid waste, fossil fuel combustion and steel industry [38,39]. Indeed, several Chromium leather tanning factories exist in Hebron [40]. However, no heavy industrial or nuclear activities are occurring in the study area, and this explains the low value of this metal in the harvested rainwater samples of this research.

Mn. The harvested rainwater samples in the all the studies are considered safe in terms of the occurrence of Mn. In the study area, none of the samples exceeded the WHO and Palestinian limits for this metal. Based on the literature, a possible source of Mn is dust transported through wind to the roofs of the houses [41].

Co. The heavy metal Co occurred in very low amounts in the study areas that were lower than the acceptable limits for drinking water set by the WHO and the Palestinian standards institution (Tables 1 and 2). Possible sources of Co in harvested rainwater are uncontrolled incineration of solid wastes in illegal dumping sites, vehicles' exhausts and leakage from engines, pesticides, and sand, soil, silt and others [41].

Ni. The occurrence of the metal Ni in the majority of the samples was far below the WHO and the local Palestinian limits (Tables 1 and 2). Possible source of Ni is leaching from metals in contact with harvested rainwater such as pipes and fittings which are used to collect the harvested rainwater [36,42]. For this research, the statistical analysis showed that there is a relationship between the source of the water in the cistern and the high level of Ni in two samples. The chemical reactions occurring between the roofing material or pipe with the harvested rainwater may leach out many chemicals including nickel [21,43].

Cu. None of the harvested rainwater samples in this study contained the heavy metal Cu in concentration above the WHO or the Palestinian limits. Possible sources of Cu in harvested rainwater are vehicles' exhausts, pesticides and industrial activities [39,41,44].

Zn. The metal Zn concentrations, in all the tested samples, were below the Palestinian limits, but only one sample exceeded the WHO limit (Tables 1 and 2). Based on the questionnaire results and literature, possible sources of Zn are the roof and storage tanks. Rainwater can dissolve the heavy metal Zn and other impurities from materials of the catchment and storage tank. Other source of Zn is the distribution system (pipes) and plumbing, as the pipes are used to collect harvested rainwater from the roofs [36].

Cd. All of the tested harvested rainwater samples were considered safe in terms of the heavy metal Cd. Possible sources of this metal in harvested rainwater are vehicles' exhausts, leakage from engines, ashes and dust containing Cd transported through wind to roof of houses, pesticides and others [41].

Pb. The water samples are considered safe in terms of the occurrence of the heavy metal Pb, except two samples that exceeded the WHO and Palestinian limits (Tables 1 and 2). Based on the questionnaire results and literature, possible sources of Pb are the roof and storage tanks. Elevated levels of Pb could be from leaching from metallic roofs and storage tanks or from atmospheric precipitation. In addition, municipal solid waste incinerators are a major source of Pb [45], because of the use of their metal oxides as pigments, stabilizers and catalysts in plastic processing [44].

Table 2. Concentration of the heavy metals in harvested rainwater in several countries, as well as the WHO and Palestinian limits.

Element	1 Palestine	2 Palestine	3 Palestine	4 Turkey	5 Pakistan	6 Iran	7 WHO Limits	8 Palestinian Limits
Cr	4.31	56.1	0.008	4.40	3.61	1.74	50	50
Mn	4.32	112.6	0.003	3.98	14.01	n.a.	500	100
Co	0.24	3.16	n.d.	n.d.	0.52	n.a.	10	n.a.
Ni	19.06	26.7	0.003	3.82	4.25	7.14	70	50
Cu	8.18	143.6	0.003	6.01	65.8	21.4	2000	1000
Zn	201.34	111.8	0.05	6.12	34.2	80.93	3000	5000
Cd	0.10	1.17	n.d.	n.a.	0.53	0.67	3	3
Pb	1.80	45.8	n.d.	n.a.	5.03	69.7	10	10

1, This study; 2, [42]; 3, [46]; 4, [47]; 5, [31]; 6, [48]; 7, [49]; 8, [50]. All concentrations are in μg/L; n.d.: not detected; n.a: not available.

3.4. Factors and Activities Contribute to Harvested Rainwater Contamination with Heavy Metals

To study the effect of the various factors surrounding the well and the activities undertaken by well owners to reduce or prevent water pollution with heavy elements, cross tabulations between these factors and activities, and the four heavy elements polluting water (Table 1): Pb, Cr, Ni and Zn was conducted. Table 3 shows a summary of these cross tabulations. It can be noticed that there are only two statistically significant relationship ($p < 0.05$) between various factors surrounding the well and the activities undertaken by well owners to reduce or prevent water pollution with four heavy elements: the factor "Type of water collection surface" with the metal Ni and the factor "Actions taken before collecting rainwater" with the metals Pb and Cr. Possible source of Ni is leaching from metal pipes and fittings which are used to collect the harvested rainwater [36,42]. Pb could be from metallic roofs and storage tanks. In addition, municipal solid waste incinerators are a major source of Pb [44,45], because of the use of their metal oxides as pigments, stabilizers and catalysts in plastic processing [43,44]. Chromium could had been originated from leather tanning factories [39,40] (see Section 3.3).

3.5. Health Risk Assessment

The CDI and HRI values calculated for the meals exceeded the WHO and the Palestinian standards (Table 4) show that the harvested rainwater is safe for consumers (Table 5). This makes sense, as the samples exceeded the limits by small amounts only.

Table 3. Cross tabulation between the factors that may lead to contamination with heavy metals and the four heavy elements polluting water: Pb, Cr, Ni and Zn.

Factor	Significance *p*-Value			
	Zn	Ni	Cr	Pb
Type of water collection surface	1.000	0.000	0.996	0.996
Take actions before collecting rainwater (Cleaning the roof of the house, Getting rid of first rain water, etc.).	0.972	0.925	0.000	0.000
For what purposes is the water in the cistern used	0.977	0.996	0.983	0.983
When the cistern was cleaned last time?	1.000	0.997	0.997	0.997
Walls material of cistern	0.905	0.817	0.817	0.817
Situation of the cistern top	0.906	0.867	0.867	0.867
Shape of the cistern	0.424	0.739	0.255	0.255
Distance between cistern and cesspit	1.000	1.000	1.000	1.000
Level of cesspit with respect to cistern	0.192	0.230	0.230	0.230
When the cesspit was discharged last time	0.999	0.996	0.996	0.996
Is there any animals or birds raised in the house	0.443	0.714	0.714	0.714
Is there any trees close to the cistern	0.558	0.404	0.404	0.404
Do you use the roof of the house for hanging the laundry	0.424	0.255	0.255	0.255
Do you allow solid waste to accumulate in cistern vicinity	0.586	0.438	0.349	0.349

Table 4. Summary of the health risk assessment calculations (for adults).

Heavy Metal	Number of Sampled that Exceeded the Standard Drinking Water	CDI Value (µg/kg.day)	HRI Value	Result
Pb	1	0.668	0.019	Safe
	2	0.382	0.011	Safe
Cr	3	2.82	0.188	Safe
	4	1.70	0.113	Safe
Ni	5	5.97	0.30	Safe
	6	14.39	0.72	Safe
Zn	7	95.90	0.32	Safe

Table 5. Summary of the health risk assessment calculations (for children).

Heavy Metal	Sample No.	CDI Value (µg/kg.day)	HRI Value	Result
Pb	1	0.735	0.020	Safe
	2	0.421	0.012	Safe
Cr	3	3.10	0.207	Safe
	4	1.87	0.125	Safe
Ni	5	6.57	0.33	Safe
	6	15.84	0.79	Safe
Zn	7	105.60	0.35	Safe

The results of this study show that heavy metals in the harvested rainwater in the study area were not a big issue, as their concentrations were usually below the local and international limits. Most of the studies of in the international literature focus on testing many parameters in drinking water including physiochemical parameters (pH, electrical conductivity, turbidity, alkalinity, hardness, calcium, magnesium and many others) and microbiological parameters (Fecal coliform and Total coliform). For example, a study in the area of Tulkarm in the West Bank, which considered 12 rural areas, focused on these parameters, but the heavy metals were not considered. In general, rural areas do not have heavy industrial nor anthropogenic activities that may produce high amounts of heavy metals such as Pb, Cr and Zn. On the contrary, in the study carried in Tulkarem area, high percentage of the samples were contaminated with Total coliform and Fecal coliform of 86% and 80%, respectively [46]. Therefore, it can be concluded that more attention must be given to the other water quality parameters, especially microbial quality of harvested and drinking waters, especially in rural areas in Palestine. In addition, a one major limitation of this study is that only one sample was collected from each study site which means that the samples collected might not be representative of the true heavy metal levels, and so more samples are recommended to be analyzed. The study did not include analysis of rainwater before it was collected, but this could be considered in a further research.

4. Conclusions

This study only presents results for heavy metals but there are several other important contaminants that must be considered for drinking water safety. So, concerning heavy metals contamination, the results of this research showed that most of the water samples analyzed were safe. For the metals, Pb, Cr and Ni; 97.3% of the samples were below the WHO and Palestinian limits, for the metals, Mn, Co, Cu and Cd; none of the samples exceeded the WHO nor the Palestinian limits. For the metal, Zn, 98.7% of the samples were below the WHO limits and none of them exceeded the Palestinian limits.

Statistical analysis showed that there was no relationship between the factors that may lead to contamination with heavy metals and the laboratory results except for the metal Ni and the "type of water collection surface" factor and for the metals Pb and Cr and "Actions taken before collecting rainwater" factor.

Health risk assessment was made for all samples that exceeded the WHO and Palestinian limits. However, and as the metals' concentrations, which exceeded the limits, exceeded them by small amounts, the samples were considered all safe. This means that the harvested rainwater in the Palestinian rural areas imposes no health risks on consumers, whether they were adults or children, and so can be safely consumed for dinking and other domestic use regarding heavy metals.

Author Contributions: Conceptualization, I.A.A.-K. and N.M.; methodology, I.A.A.-K.; validation, F.A., I.A.A.-K. and N.M.; formal analysis, F.A.; investigation, F.A.; resources, I.A.A.-K.; data curation, F.A.; writing—original draft preparation, F.A.; writing—review and editing, F.A., N.M. and Y.-T.H.; visualization, N.M.; supervision, N.M.; project administration, I.A.A.-K.; funding acquisition, I.A.A.-K. All authors have read and agreed to the published version of the manuscript.

Funding: This study was funded by the Partnerships for Enhanced Engagement in Research (PEER) program, implemented by the U.S. National Academy of Sciences—Sponsor Grant No.: AID-OAA-A-11-00012 and USAID-USGS Grant No. G17AS00001.

Institutional Review Board Statement: Not applicable.

Informed Consent Statement: Not applicable.

Data Availability Statement: The datasets generated during and/or analyzed during the current study are available and can be obtained by contacting the corresponding author.

Acknowledgments: This work was carried out as part of the 'Rainwater Harvesting Analysis using Water Harvesting Evaluation Tool (WHEAT)' project supported by the USAID-funded Partnerships for Enhanced Engagement in Research (PEER) program, implemented by the U.S. National Academy of Sciences—Sponsor Grant No.: AID-OAA-A-11-00012.

Conflicts of Interest: The authors declare no conflict of interest.

Appendix A Questionnaire

Q 1	• What is the source of water in cistern? 1. Roof of the house 2. Garden or the backyard of the house 3. Street 4. Otherwise (indicate)
Q 2	• Do you take any actions before collecting the rainwater? 1. Yes 2. No
Q 3	• If the answer of Q 2 is Yes, what are these actions? 1. Cleaning the roof of the house 2. Getting rid of first rain water 3. Both 1 and 2 4. Otherwise (indicate)
Q 4	• When the cistern was cleaned last time? Before:
Q 5	• For what purposes is the water in the cistern used? (The answer may be more than one choice) 1. Drinking 2. Animals 3. Irrigation 4. Cooking 5. Cleaning the house 6. Otherwise (indicate)
Q 6	• Indicate approximately the age of cistern in years
Q 7	• What is the volume of the cistern
Q 8	• The walls of the cistern are from: 1. Concrete 2. Rock 3. Otherwise (indicate)
Q 9	• Is the cover of the cistern: 1. Open 2. Closed 3. Perforated
Q 10	• The shape of cistern: 1. Cuboid 2. Peer-shaped 3. Otherwise (indicate)
Q 11	• The distance between cistern and cesspit (in meters)
Q 12	• The level of cesspit: 1. Above cistern level 2. Below cistern level 3. Same level as cistern
Q 13	• When the cistern was discharged last time? Before
Q 14	• Do you raise animals or birds around the house? 1. Yes, always 2. Sometimes 3. No

Q 15	• Is there any trees close to the house? 1. Yes 2. No
Q 16	• Do you notice any impurities on the surface of the water in the cistern? 1. Yes 2. No
Q 17	• Do you notice algae on the sides of the cistern? 1. Yes 2. No
Q 18	• Do you use the roof of the house in winter for hanging the laundry? 1. Yes 2. No
Q 19	• Do you collect solid waste in the backyard of the house? 1. Yes 2. No

References

1. Ober, J.; Karwot, J. Tap Water Quality: Seasonal User Surveys in Poland. *Energies* **2021**, *14*, 3841. [CrossRef]
2. UN. Sustainable Development Goals. Goal 6: Ensure Access to Water and Sanitation for All. Available online: https://www.un.org/sustainabledevelopment/water-and-sanitation/ (accessed on 17 February 2022).
3. WHO; UNICEF. *Progress on Drinking Water, Sanitation and Hygiene: 2017 Update and SDG Baselines*; World Health Organization (WHO) and the United Nations Children's Fund (UNICEF): Geneva, Switzerland, 2017; Available online: https://www.un.org/africarenewal/sites/www.un.org.africarenewal/files/JMP-2017-report-launch-version_0.pdf (accessed on 17 February 2022).
4. Yaradua, A.I.; Alhassan, A.J.; Nasir, A.; Matazu, K.I.; Usman, A.; Yau, S.; Idi, A.; Muhammad, I.U.; Yaro, S.A.; Mashi, J.A. Health Risk Assessment for Carcinogenic and Non Carcinogenic Heavy Metal Exposure from Hibiscus Leaf Cultivated in Katsina State, North West Nigeria. *Asian J. Agric. Hortic. Res.* **2019**, *3*, 1–12. [CrossRef]
5. Pejakov, T.; Tepic, I.; Vidikant, P.; Kulundzic, R. Reducing harmful effects of metals in water. *Metalurgija* **2014**, *53*, 677–680.
6. Oyewo, O.A.; Amos Adeniyi, A.; Bopape, M.F.; Onyango, M.S. Chapter 4—Heavy metal mobility in surface water and soil, climate change, and soil interactions. In *Climate Change and Soil Interactions*; Elsevier: Pretoria, South Africa, 2020; pp. 51–88. [CrossRef]
7. Ahmed, S.F.; Kumar, P.S.; Rozbu, M.R.; Chowdhury, A.T.; Nuzhat, S.; Rafa, N.; Mahlia, T.M.I.; Ong, H.C.; Mofijur, M. Heavy metal toxicity, sources, and remediation techniques for contaminated water and soil. *Environ. Technol. Innov.* **2022**, *25*, 102114. [CrossRef]
8. Batayneh, A.T. Heavy metals in water springs of the Yarmouk Basin, North Jordan and their potentiality in health risk assessment. *Int. J. Phys. Sci.* **2010**, *5*, 997–1003.
9. Kaplan, O.; Yildirim, N.C.; Yildirim, N.; Tayhan, N. Assessment of some heavy metals in drinking water samples of Tunceli, Turkey. *E-J. Chem.* **2011**, *8*, 276–280. [CrossRef]
10. Pirsaheb, M.; Khosravi, T.; Sharafi, K.; Babajani, L.; Rezaei, M. Measurement of Heavy Metals Concentration in Drinking Water from Source to Consumption Site in Kermanshah-Iran. *J. World Appl. Sci.* **2013**, *21*, 416–423.
11. Shaheed, R.; Mohtar, W.H.M.W.; El-Shafie, A. Ensuring water security by utilizing roof-harvested rainwater and lake watertreated with a low-cost integrated adsorption-filtration system. *Water Sci. Eng.* **2017**, *10*, 115–124. [CrossRef]
12. Jomova, K.; Valko, M. Advances in metal-induced oxidative stress and human disease. *Toxicology* **2011**, *283*, 65–87. [CrossRef]
13. Boyd, R.S. Heavy metal pollutants and chemical ecology: Exploring new frontiers. *J. Chem. Ecol.* **2010**, *36*, 46–58. [CrossRef]
14. Adepoju-Bello, A.A.; Alabi, O.M. Heavy metals: A review. *Nig. J. Pharm.* **2005**, *37*, 41–45.
15. Rai, P.K.; Lee, S.S.; Zhang, M.; Tsang, Y.F.; Kim, K.-H. Heavy metals in food crops: Health risks, fate, mechanisms, and management. *Environ. Int.* **2019**, *125*, 365–385. [CrossRef]
16. International Agency for Research on Cancer (IARC). IARC Monographs on the Evaluation of Carcinogenic Risks to Humans, Agents Classified by the IARC Monographs, World Health Organization (WHO) 2012. Available online: https://monographs.iarc.who.int/agents-classified-by-the-iarc/ (accessed on 17 February 2022).
17. American Cancer Society. Cancer Facts & Figures. 2013. Available online: https://www.cancer.org/research/cancer-facts-statistics/all-cancer-facts-figures/cancer-facts-figures-2013.html (accessed on 29 June 2019).
18. Alarcón-Herrera, M.T.; Martin-Alarcon, D.A.; Gutiérrez, M.; Reynoso-Cuevas, L.; Martín-Domínguez, A.; Olmos-Márquez, M.A.; Bundschuh, J. Co-occurrence, possible origin, and health-risk assessment of arsenic and fluoride in drinking water sources in Mexico: Geographical data visualization. *Sci. Total. Environ.* **2020**, *698*, 134168. [CrossRef] [PubMed]

19. Berndtsson, J.C.; Bengtsson, L.; Jinno, K. Runoff water quality from intensive and extensive vegetated roofs. *Ecol. Eng.* **2009**, *35*, 369–380. [CrossRef]
20. Mao, J.; Xia, B.; Zhou, Y.; Bi, F.; Xia, S. Effect of roof materials and weather patterns on the quality of harvested rainwater in Shanghai, China. *J. Clean. Prod.* **2021**, *279*, 123419. [CrossRef]
21. Spinks, A.T.; Coombes, O.; Dunstan, R.H.; Kuczera, G. Water quality treatment processes in domestic RWH systems. In *Proceedings of the 28th International Hydrology and Water Resources Symposium 2003, Wollongong, Australia, 10–14 November 2003*; The Institute of Engineers: Wollongong, Australia, 2003; p. 8.
22. Walna, B.; Kurzyca, I. Tendencies of changes in the chemical composition of precipitation in the Wielkopolski National Park. *J. Water Land Dev.* **2009**, *13*, 53–69. [CrossRef]
23. Ki, S.J.; Kang, J.H.; Lee, S.W.; Lee, Y.S.; Cho, K.H.; An, K.G.; Kim, J.H. Advancing assessment and design of stormwater monitoring programs using a self-organizing map: Characterization of trace metal concentration profiles in stormwater runoff. *Water Res.* **2011**, *45*, 4183–4197. [CrossRef]
24. Shadeed, S.; Judeh, T.; Riksen, M. Rainwater Harvesting for Sustainable Agriculture in High Water-Poor Areas in the West Bank, Palestine. *Water* **2020**, *12*, 380. [CrossRef]
25. Mosley, L. SOPAC water quality officer, water quality of rainwater harvesting systems. *SOPAC Misc. Rep.* **2005**, 579.
26. Nesheim, I.; Sundnes, F.; Enge, C.; Graversgaard, M.; van den Brink, C.; Farrow, L.; Glavan, M.; Hansen, B.; Leitão, I.A.; Rowbottom, J.; et al. Multi-Actor Platforms in the Water–Agriculture Nexus: Synergies and Long-Term Meaningful Engagement. *Water* **2021**, *13*, 3204. [CrossRef]
27. PWA. *Annual Water Status Report, 2011*; Palestinian Water Authority: Ramallah, Palestine, 2011.
28. Palestinian Central Bureau of Statistics (PCBS). *Projected Mid-Year Population for Hebron Governorate by Locality 2017–2021*; PCBS: Ramallah, Palestine, 2020.
29. Applied Research Institute, Jerusalem (ARIJ). *GIS Database (2006–2009): Hebron District. Hebron Governorate*; ARIJ: Bethlehem, Palestine, 2009.
30. Corman Center for Mass Spectrometry. (LA-) ICP-MS. Available online: http://ees2.geo.rpi.edu/icpms (accessed on 20 June 2019).
31. Muhammad, S.; Shah, M.T.; Khan, S. Health risk assessment of heavy metals and their source apportionment in drinking water of Kohistan region, northern Pakistan. *Microchem. J.* **2011**, *98*, 334–343. [CrossRef]
32. Shah, M.T.; Ara, J.; Muhammad, S.; Khan, S.; Tariq, S. Health risk assessment via surface water and sub-surface water consumption in the mafic and ultramafic terrain, Mohmand agency, northern Pakistan. *J. Geochem. Explor.* **2012**, *118*, 60–70. [CrossRef]
33. United States Environmental Protection Agency (US EPA, 2016). *Health Effects Assessment Summary Tables (HEAST)*; FY 2016 Annual Update; United States Environmental Protection Agency: Washington, DC, USA, 2016.
34. IBM Corp. *SPSS Statistics for Windows, Version 20.0*; IBM Corp.: Armonk, NY, USA, 2011.
35. PWA. *General Technical Specifications for the Construction of Harvesting Cisterns (Pear-Shaped and Rectangular)*; Palestinian Water Authority: Ramallah, Palestine, 2003.
36. Mendez, C.B.; Klenzendorf, J.B.; Afshar, B.R.; Simmons, M.T.; Barrett, M.E.; Kinney, K.A.; Kirisits, M.J. The effect of roofing material on the quality of harvested rainwater. *Water Res.* **2011**, *45*, 2049–2059. [CrossRef] [PubMed]
37. Agency for Toxic Substances and Disease Registry (ATSDR). Case Studies in Environmental Medicine (CSEM) Chromium Toxicity. 2018. Available online: https://www.atsdr.cdc.gov/csem/chromium/docs/chromium.pdf (accessed on 24 September 2020).
38. Wise Sr, J.P.; Payne, R.; Wise, S.S.; LaCerte, C.; Wise, J.; Gianios, C., Jr.; Thompson, W.D.; Perkins, C.; Zheng, T.; Zhu, C.; et al. A global assessment of chromium pollution using sperm whales (*Physeter macrocephalus*) as an indicator species. *Chemosphere* **2009**, *75*, 1461–1467. [CrossRef]
39. Cong, Z.; Kang, S.; Zhang, Y.; Li, X. Atmospheric wet deposition of trace elements to central Tibetan Plateau. *Appl. Geochem.* **2010**, *25*, 1415–1421. [CrossRef]
40. Al-Jabari, M.; Sawalha, H.; Pugazhendhi, A.; Rene, E.R. Cleaner production and resource recovery opportunities in leather tanneries: Technological applications and perspectives. *Bioresour. Technol. Rep.* **2021**, *16*, 100815. [CrossRef]
41. Malassa, H.; Al-Rimawi, F.; Al-Khatib, M.; Al-Qutob, M. Determination of trace heavy metals in harvested rainwater used for drinking in Hebron (south West Bank, Palestine) by ICP-MS. *Environ. Monit. Assess.* **2014**, *186*, 6985–6992. [CrossRef]
42. WHO. WHO Guidelines for Drinking Water Quality. In *Nickel in Drinking Water*; WHO: Geneva, Switzerland, 2005.
43. Che, W.; Wang, H.; Li, J.; Liu, H.; Meng, G. The quality and major influencing factors of runoff in Beijing's urban area. In Proceedings of the 10th International Rainwater Catchment Systems Conference 2001, Weikersheim, Germany, 1 October 2001; p. 4.
44. Hu, G.-P.; Balasubramanian, R. Wet deposition of trace metals in Singapore. *Water Air Soil Pollut.* **2003**, *144*, 285–300. [CrossRef]
45. Özsoy, T.; Örnektekin, S. Trace elements in urban and suburban rainfall, Mersin, Northeastern Mediterranean. *Atmos. Res.* **2009**, *94*, 203–219. [CrossRef]
46. Almur, S.A. Assessing Water Quality of Cisterns in Sha'rawiya Area "Tulkarm Governorate" for Drinking Purposes. Master's Thesis, An-Najah National University, Nablus, Palestine, 2016.
47. Tuzen, M.; Soylak, M. Evaluation of Metal Levels of Drinking Waters from the Tokat–Black Sea Region of Turkey. *Polish J. Environ. Stud.* **2006**, *15*, 915–919.

48. Kamani, H.; Hoseini, M.; Safari, G.; Jaafari, J.; Mahvi, A. Study of trace elements in wet atmospheric precipitation in Tehran, Iran. *Environ. Monit. Assess.* **2014**, *186*, 5059–5067. [CrossRef] [PubMed]
49. WHO. *Guidelines for Drinking Water Quality*, 4th ed.; WHO: Geneva, Switzerland, 2011.
50. Palestinian Standard Institution (PSI). *Standards Manual for Drinking Water Quality*; Palestinian Standard Institution: Ramallah, Palestine, 2010.

International Journal of
Environmental Research and Public Health

MDPI

Article

The Impact of Treated Wastewater Irrigation on the Metabolism of Barley Grown in Arid and Semi-Arid Regions

Alan Alvarez-Holguin [1], Gabriel Sosa-Perez [1], Omar Castor Ponce-Garcia [1], Carlos Rene Lara-Macias [1], Federico Villarreal-Guerrero [2], Carlos Gustavo Monzon-Burgos [2] and Jesus Manuel Ochoa-Rivero [1,*]

1 Instituto Nacional de Investigaciones Forestales, Agricolas y Pecuarias (INIFAP), Campo Experimental La Campana, Aldama, Chihuahua 32910, Mexico; alvarez.alan@inifap.gob.mx (A.A.-H.); sosa.gabriel@inifap.gob.mx (G.S.-P.); ponce.omar@inifap.gob.mx (O.C.P.-G.); lara.rene@inifap.gob.mx (C.R.L.-M.)

2 College of Animal Production and Ecology, Autonomous University of Chihuahua, Chihuahua 31453, Mexico; fvillarreal@uach.mx (F.V.-G.); carlosmonzon19@gmail.com (C.G.M.-B.)

* Correspondence: ochoa.jesus@inifap.gob.mx; Tel.: +52-01(55)-3871-8770

Abstract: The use of treated wastewater (TWW) for irrigation has gained global attention since it reduces pressure on groundwater (GW) and surface water. This study aimed to evaluate the effect of TWW on agronomic, photosynthetic, stomatal, and nutritional characteristics of barley plants. The experiment with barley was established on two bands: one band was irrigated with GW and the other with TWW. The evaluation was performed 25, 40, 60, 90, and 115 days after sowing (DAS). Results showed that irrigation with TWW increased ($p < 0.01$) grain yield by 54.3% and forage yield by 39.4% compared to GW irrigation. In addition, it increased plant height (PH) ($p = 0.013$), chlorophyll concentration index (CCI) ($p = 0.006$), and leaf area index (LAI) ($p = 0.002$). TWW also produced a positive effect ($p < 0.05$) in all the photosynthetic efficiency parameters evaluated. Barley plants irrigated with TWW had lower stomatal density (SD) and area (SA) ($p < 0.001$) than plants irrigated with GW. Plants irrigated with TWW had a higher P concentration ($p < 0.05$) in stems and roots and K concentration in leaves than plants irrigated with GW. We concluded that the use of TWW induced important biochemical, physiological, and agronomic changes in barley plants. Hence, the use of TWW may be a sustainable alternative for barley production in arid and semi-arid regions. This study was part of a government project, which aimed to develop a new metropolitan irrigation district with TWW. This study may contribute to the sustainability of water resources and agricultural practices in northern Mexico.

Keywords: agriculture; nutritional composition; photosynthetic efficiency; stomata; wastewater reuse

Citation: Alvarez-Holguin, A.; Sosa-Perez, G.; Ponce-Garcia, O.C.; Lara-Macias, C.R.; Villarreal-Guerrero, F.; Monzon-Burgos, C.G.; Ochoa-Rivero, J.M. The Impact of Treated Wastewater Irrigation on the Metabolism of Barley Grown in Arid and Semi-Arid Regions. *Int. J. Environ. Res. Public Health* **2022**, *19*, 2345. https://doi.org/10.3390/ijerph19042345

Academic Editors: Yung-Tse Hung, Hamidi Abdul Aziz, Issam A. Al-Khatib and Paul B. Tchounwou

Received: 6 January 2022
Accepted: 14 February 2022
Published: 18 February 2022

1. Introduction

Anthropic activities, population growth, and climate change have caused a decrease in water quality and quantity, generating huge amounts of wastewater (WW). Around 380 billion m^3 of WW are generated globally each year, and this amount could increase 24% by 2030 and 51% by 2050 [1]. Therefore, there is a need of alternatives for the use of WW on a sustainable basis.

Agriculture is the activity with the highest water demand on Earth, and this demand is exacerbated in arid and semi-arid lands, where water availability is limited [2]. According to the FAO, 2.75 million km^2 are irrigated with WW around the world [3]; however, more than 80% of WW produced worldwide is released into the environment without any treatment. Thus, the use of treated wastewater (TWW) for agricultural irrigation has gained global attention since water consumption has significantly increased due to the growing demand for food [4]. Currently, around 20 million hectares are irrigated with TWW in 50 countries, contributing to 40% of the food produced [5].

The use of TWW in agriculture can generate numerous benefits. For instance, the use of TWW for agricultural irrigation reduces pressure on groundwater (GW) and surface water, which could be directed towards domestic use [6–8]. Moreover, the use of TWW is cheaper than pumping GW [9,10].

From an ecological perspective, irrigation with TWW decreases the eutrophication rate of aquatic ecosystems [1]. It has been shown that TWW can considerably increase the concentration of macro (N, P, and K) and micronutrients (Fe, Mn, Zn, and Cu), which increases soil fertility [11,12]. Qadir et al. [1] reported that worldwide, TWW carries 16.6 Tg (Tg, million metric ton) of N, 3.0 Tg of K, and 6.3 Tg of P annually. This amount represents 13.4% of the global demand for nutrients in agriculture. In addition, the increase in nutrient concentration decreases fertilization needs and reduces costs [13]. Previous research has found that TWW decreases the use of fertilizers from 50% to 100% in forage species, such as alfalfa and wheat [14,15]. In this sense, the use of TWW contributes to reduce the use of agrochemicals, generating positive ecological effects [1,7]. Hence, there is a great opportunity for using TWW in agriculture.

Several studies have revealed that the use of TWW can improve physicochemical properties and enhance the productivity of many economically important crops [5,16,17]. For example, the use of TWW enhanced the chlorophyll content, photosynthetic efficiency, and forage yield in alfalfa (*Medicago sativa*) [18]. Likewise, the irrigation with TWW improved the forage yield of kikuyu grass (*Pennisetum clandestinum*) in the absence of chemical fertilizers compared to the irrigation with GW [19]. Furthermore, the use of this source of water increased the chlorophyll fluorescence, stomatal conductance, and the photosynthetic rate and decreased stress metabolites, such as leaf phenolic compounds and carbohydrates in olive trees (*Olea europaea*) [20]. Such benefits were induced by the high concentration of nutrients in the TWW.

Previous studies have shown that TWW irrigation also has beneficial effects on barley (*Hordeum vulgare* L.) productivity [14,21,22]. This crop is the fourth most important cereal in the world, just behind wheat, corn, and rice [21]. Around 150 million tons year^{-1} of barley are globally produced. In 2013, barley exports and imports were valued at USD 8.5 and USD 9.4 trillion, respectively [22]. However, studies on the physiological and biochemical effects on this species due to TWW irrigation are limited. Hence, the objective of this study was to evaluate the effect of TWW on agronomic, photosynthetic, stomatal, and nutritional characteristics of barley plants. This information will build a deeper understanding of the use of TWW in agriculture. It may also contribute to reducing the city's freshwater requirements. Thus, it may contribute to the sustainability of water resources and agricultural practices, consequently improving the availability of this natural resource.

2. Materials and Methods

2.1. Study Area and Experimental Setup

The experiment was conducted during 2020 on a plot identified as "Los Alamitos", located in the municipality of Aldama, in the state of Chihuahua, Mexico (Figure 1). The plot is located at 28°44′44.28″ N and 105°57′28.52″ W. The climate is dry semi-warm, with a mean maximum temperature of 23.5 °C during summer and a mean minimum temperature of 5.8 °C during winter. The mean annual precipitation is 298 mm, which mainly occurs during July–September [23,24]. The plot under study had 19 stripes or bands 10-m wide and 150-m long, with aisles of 1.0 m in between bands (Figure 1). These bands have conventionally been irrigated for 30 years with two water sources: one section of three bands with GW and the rest with TWW. The experiment with barley was established on two bands: one band was irrigated with GW and the other with TWW. The distance between these two experimental bands was 150 m. The remaining bands were sown with oats. The TWW used for irrigation was from a wastewater treatment plant (WWTP) located in the south of Chihuahua City. The WWTP has a secondary treatment with an effluent of 1100 L seg^{-1} from domestic, industrial, and storm sewages.

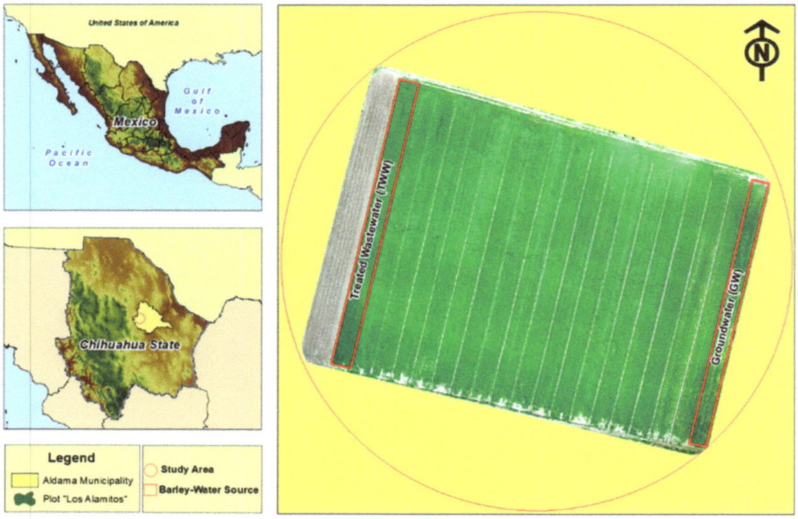

Figure 1. Location of the plot under study (municipality of Aldama) and an aerial image showing its characteristics.

The soil preparation for sowing was carried out from 10–25 January by performing the following practices: fallow, tracing, subsoiling, and furrowing. The soil characteristics were determined before sowing by analyzing a composite soil sample, which was extracted from five subsamples collected in the upper soil profile (0–30 cm). The composite sample was analyzed according to Mexican Standard NOM-021-RECNAT-2000 at the National Laboratory of Water, Soil, Plant, and Atmosphere of INIFAP, located in Gomez Palacio, Durango, Mexico. The physicochemical characteristics of the soil under study can be seen in Table 1.

Table 1. Physicochemical characteristics of the soils irrigated with groundwater (GW) and treated wastewater (TWW).

Variable	Units	GW	TWW
pH		8.51	8.44
Electrical Conductivity (EC)	dS m^{-1}	1.17	1.01
Texture		Clay loam	Clay loam
Cation Exchange Capacity (CEC)	meq 100 g^{-1}	36.01	32.71
* Organic matter (OM)	%	1.3	1.5
Nitrogen (N-NO_3−)		29.11	9.49
[1] Phosphosrus available (P)		8.24	12.76
[2] Removable potassium (K)		693.98	895.39
[3] Removable calcium (Ca)		4102.53	4702.02
[4] Removable magnesium (Mg)	mg kg^{-1}	449.26	388.48
* Copper (Cu)		1.59	1.20
* Iron (Fe)		0.6	1.76
* Manganese (Mn)		2.66	2.30
* Zinc (Zn)		2.01	3.31

[1,2,3] Units are kg ha^{-1} and reported as P_2O_5, K_2O, and CaO, respectively; [4] reported MgO; * units are mg Kg^{-1}.

The experimental design consisted of randomized blocks. The band of each treatment or water source was divided into six blocks 25-m long. Seeds of barley, var. "Alina", were sown on 1 February 2020, using a seeding rate of 120 kg ha^{-1}. Five days after sowing (DAS), 100 kg ha^{-1} of urea ($CO(NH_2)_2$) were applied to the soil surface by using a tractor-driven fertilizing machine. Furrow irrigation events were carried out on 7 February, 21 February, 6 March, 23 March, 3 April, and 17 April, with 120 mm per irrigation. To dismiss differences in the results due to high soil moisture content in some of the treatments, the soil moisture (volumetric water content) was measured with a FieldScout TDR 300 portable meter (Time-Domain Reflectometer- Spectrum Technologies Inc., Aurora, IL, USA) one or two days before each watering, beginning in March. The moisture was measured using 20-cm rods at randomly selected points on one band irrigated with TWW and one band with GW. Soil moisture was similar between the treatments, and the general mean of the volumetric water content, before each irrigation date, was 18.7% (s.d. = 0.9%), 19.3% (s.d. = 1.3%), 20.9% (s.d. = 1.6%), and 21.7% (s.d. = 2.6%), respectively.

Water quality was analyzed for each water source (GW and TWW). For that, two water samples from each water source were collected at each irrigation event, beginning on 21 February. A total of 10 water samples was collected from each source by the time the experiment concluded. The on-site measurements included pH, electrical conductivity (EC), and temperature. Concentrations of nitrates (NO_3^- as N), phosphates (PO_4^{3-}), and sulfates (SO_4^{2-}) were analyzed at the College of Chemical Sciences of the Autonomous University of Chihuahua. The concentrations of heavy metals and metalloids were determined at the Mexican Geological Service (SGM), located in Chihuahua City. The analyses were based on the Mexican Regulation NOM-001-SEMARNAT-1996 and [25]. The results are presented in Table 2. In general, the values of the water quality parameters from the two sources were within acceptable levels according to the Mexican Regulations for irrigation [26].

Table 2. Physicochemical characteristics of the water used for the irrigation of barley (*Hordeum vulgare*).

			GW			TWW		
Variable	**Units**	**Mean**	**Min**	**Max**	**Mean**	**Min**	**Max**	**Quality Standards**
pH		7.67	7.18	8.89	7.78	7.25	8.75	6.0–9.0 *
Electrical Conductivity (EC)	mS cm^{-1}	0.94	0.25	1.12	0.94	0.76	1.18	NA
Temperature (°C)	°C	22.34	17.10	31.10	21.46	14.30	30.40	NA
Nitrate–nitrogen (NO_3 as N)	mg L^{-1}	33.63	0.00	42.49	31.85	0.00	60.76	NA
Phosphate (PO_4^{3-})	mg L^{-1}	0.00	0.00	0.00	6.80	0.00	8.75	NA
Sulfate (SO_4^{2-})	mg L^{-1}	137.80	95.89	168.36	97.85	0.00	123.86	250.00 *
Copper (Cu)	mg L^{-1}	0.00	0.00	0.00	0.00	0.00	0.00	0.20 *
Iron (Fe)	mg L^{-1}	0.02	0.00	0.06	0.00	0.00	0.01	5.00 *
Manganese (Mn)	mg L^{-1}	0.78	0.36	1.25	0.60	0.38	0.72	0.20 *
Molybdenum (Mo)	mg L^{-1}	0.005	0.001	0.015	0.015	0.004	0.100	NA
Nickel (Ni)	mg L^{-1}	0.00	0.00	0.00	0.00	0.00	0.00	0.20 *
Zinc (Zn)	mg L^{-1}	0.00	0.00	0.00	0.00	0.00	0.00	2.00 *
K	mg L^{-1}	0.16	0.06	0.24	0.24	0.22	0.26	0–2 **
Ca	meq L^{-1}	4.06	2.16	5.10	4.93	4.93	5.73	0–20 **
Mg	meq L^{-1}	0.78	0.36	1.25	0.60	0.38	0.72	0–5 **

Quality standard of water for agriculture as indicated by the * Mexican Federal Law [26] and ** FAO guideline for the quality of water used for irrigation [27].

2.2. Agronomic Attributes

Plant height (PH), chlorophyll concentration index (CCI), and leaf area index (LAI) were evaluated 25, 40, 60, and 90 DAS. PH was measured on three randomly selected plants from each block. Measurements of CCI (CCM-200 device, Opti-Sciences, Inc., Hudson,

NH, USA) were made on the second leaf of three plants from each block. Likewise, measurements of LAI were performed with a ceptometer (LP-80, Decagon Devices, Inc., Pullman, WA, USA) in areas of 0.25 m^2 (0.4 × 0.62 m), with one sample per block (n = 6). The sensor bar of the ceptometer was positioned at 0.05 m above the ground level, and two measurements were taken every 0.1 m. Meanwhile, the external PAR (photosynthetically active radiation) sensor of the ceptometer was positioned at 1.5-m height, at an angle of 90° with respect to the ground level, in a shadow-free location.

Forage yield was also determined on the aforementioned sampling dates. For that, forage cuts were made at 0.05 m from the ground level in six quadrants (n = 6) of 0.25 m^2 per treatment or water source. The samples were then stored in paper bags and dried at 65 °C during 72 h. An additional sampling was performed at 115 DAS to calculate grain yield. For that, the seed was separated from the forage and weighed to estimate grain yield.

2.3. Photosynthetic Efficiency

The photosynthetic efficiency was evaluated through the fluorescence parameters of chlorophylls: maximum quantum yield of photosystem II (Fv/Fm), photochemical efficiency of photosystem II (ΦPSII), quantum efficiency of unregulated energy dissipation in PSII (YNO), quantum efficiency of regulated energy dissipation in the PSII (YNPQ), and electron transfer rate (ETR). These parameters were evaluated using a portable photosynthesis yield analyzer (Mini-PAM II; Walz, Effeltrich, Germany).

Photosynthetic efficiency was evaluated 80 DAS, when the plants were at their flowering stage. This evaluation was performed in three plants per block (i.e., 18 plants per treatment) and consisted of randomly measuring three healthy leaves from each plant. The measurements were done on both dark-acclimated (one hour after dark) and light-exposed (11:30 a.m.–12:30 p.m.) leaves. At night, the plants were subjected to a photon light pulse of approximately 0.5 µmol m^{-2} s^{-1}, with a frequency of 600 Hz, to define the minimum fluorescence signal (F0). A photon saturation pulse of approximately 6000 µmol m^{-2} s^{-1} was then applied during 0.8 s to find the maximum fluorescence signal (Fm). The parameters of F0 and Fm served to calculate the maximum quantum yield of photosystem II through the equation proposed by Kitajima and Butler [28]:

$$Fv/Fm = (Fm - F0)/Fm$$

These steps were repeated at noon to find the fluorescence level (Ft) and the maximum fluorescence level (F'm) of light-exposed leaves.

The values obtained from the aforementioned measurements were used to calculate the photochemical efficiency of photosystem II by using the equation proposed by Genty et al. [29]:

$$\Phi PSII = (F'm - Ft)/F'm$$

In addition, this equation was used to calculate the quantum efficiency of unregulated energy dissipation in the PSII:

$$YNO = Ft/Fm$$

The quantum efficiency of the regulated dissipation of energy in the PSII was obtained by the following equation:

$$YNPQ = (Ft/F'm) - (Ft/Fm).$$

Finally, the ETR was calculated using this equation:

$$ETR = [(\Phi PSII) \times (PAR\ radiation\ received) \times (0.84)]$$

Data were collected and processed in the WinControl-3 software, version 3.23.

2.4. Stomatal Characteristics

Stomata of experimental plants were characterized by assessing stomatal density (SD) and stomatal area (SA). For that, leaf blade imprints were extracted from 18 plants per treatment, three plants per block, and three healthy leaves per plant. Imprints were taken from approximately 1 cm^2 of adaxial and abaxial leaf surfaces. The imprints were then analyzed under a phase contrast microscope (Model Axio imager 2, Carl Zeiss, Jena, Germany) at a magnification of 200×. Five fields of view per imprint were photographed with an AxioCam MRc5 camera (Carl Zeiss). The number of stomata was counted, and data were normalized to 1 mm^2. The SA was calculated from three randomly selected stomata per optic field by using the Zen 2 core software. A total of 810 stomata per treatment were analyzed from 270 optical fields.

2.5. Analysis of Nutritional Components

The nutritional components of the plants were analyzed at the National Laboratory of Water, Soil, Plant, and Atmosphere of INIFAP located in Durango, Mexico. A total of 24 plants (12 plants per treatment) were randomly selected and split into roots, stems, leaves, and grain. The samples were dried and homogenized. Then, nitrogen (N), phosphorus (P), potassium (K), calcium (Ca), magnesium (Mg), sodium (Na), and iron (Fe) were quantified using acid digestion of samples with 0.5 g of plant tissue. The digested samples were analyzed in the atomic absorption equipment AAnalyst700-Perkin Elmer. All the analyses were determined according to standard methods [30–32].

2.6. Statistical Analysis

Data were analyzed using a multi-factor analysis of variance. The variables of PH, CCI, and LAI were analyzed using a four-factor model utilizing water source, sampling date, block, and plant as factors. The plant was included in the analysis to avoid pseudo-replication. The forage yield analysis was adjusted to a three-factor model (water source, sampling date, and block) since it was evaluated by area and not by plants. Grain yield, photosynthetic parameters, stomatal characteristics, and nutritional components were analyzed from plants of a single sampling date; thus, data were analyzed using three factors; water source, block, and plant. Regarding the physicochemical characteristics, water from the two sources, from each sampling date, were compared using Dunnett's comparison test ($\alpha = 0.05$). Data of soil moisture were analyzed using repeated measurements with two factors (water source and watering date), where the subjects were the bands. Data were analyzed using the R software version 4.0.5.

3. Results and Discussion

3.1. Treated Wastewater Improve Agronomic Traits on Barley

Irrigation with TWW increased grain yield by 54.3% ($p = 0.007$) compared to GW irrigation (Figure 2a). Samarah et al. [21] found similar results in different barley cultivars, where the use of TWW increased grain yield. The increase in grain yield is attributed to the enriched mix of nutrients contained in the TWW, which benefits crops' growth and productivity [1,33,34].

Grain yield with TWW was 6.42 t ha^{-1}, using 46 N units (100 kg ha^{-1} of urea). Ramírez-Novoa et al. [35] obtained a yield of 7.36 t ha^{-1} of the same variety of barley "Alina"; however, the fertilizer applied in that study was 45-60-00 (urea and triple calcium superphosphate) at the time of sowing and 45-00-00 during the first irrigation. The results prove the feasibility of producing barley with low fertilization costs using TWW. This is also supported by other studies, where the nutritional contribution of TWW to crops was reported [1].

The maximum forage yield using TWW was 39.4% higher ($p = 0.006$) than using GW, and the difference was consistent during all the evaluations (Figure 2b). Previous studies have also found that irrigation with TWW can generate an increase in forage yield [33]. For instance, Elfanssi et al. [18] found that irrigating with TWW increased productivity in alfalfa (*Medicago sativa*), and Aghtape et al. [34] reported an increase in yield and forage quality in foxtail millet (*Setaria italica*) when irrigated with TWW. In this study, forage yield,

PH (p = 0.013), CCI (p = 0.006), and LAI (p = 0.002) were higher using TWW instead of GW (Figure 3). These findings are explained by the chlorophyll since it is the pigment responsible for capturing solar radiation for photosynthesis, and it occurs mainly on the leaves. Tambussi et al. [36] found that barley cultivars with a larger leaf area are more efficient in terms of photosynthesis and produce a greater amount of forage and grain.

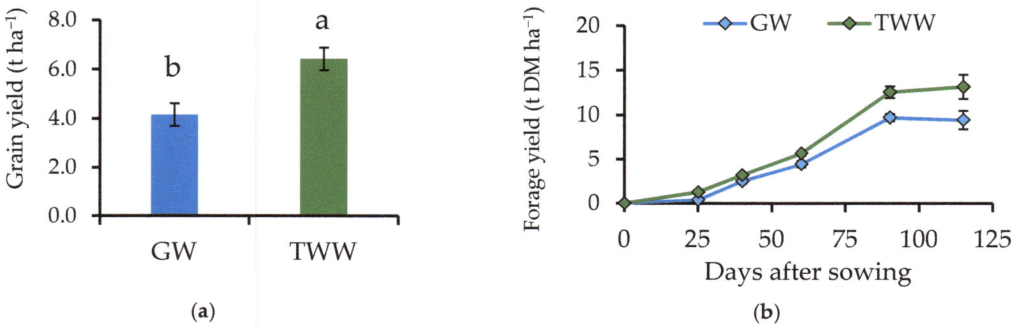

Figure 2. (**a**) Grain and forage yield in barley plants (*Hordeum vulgare*) irrigated with treated wastewater (TWW) and groundwater (GW). (**b**) Forage yield was evaluated at 25, 40, 60, and 90 days after sowing (DAS). Grain yield was evaluated 115 DAS. n = 6; p = 0.007 for grain yield and p = 0.006 for forage yield. Different letters indicate significant differences between water sources (Dunnett test; $p < 0.05$) and black bars represent the standard error.

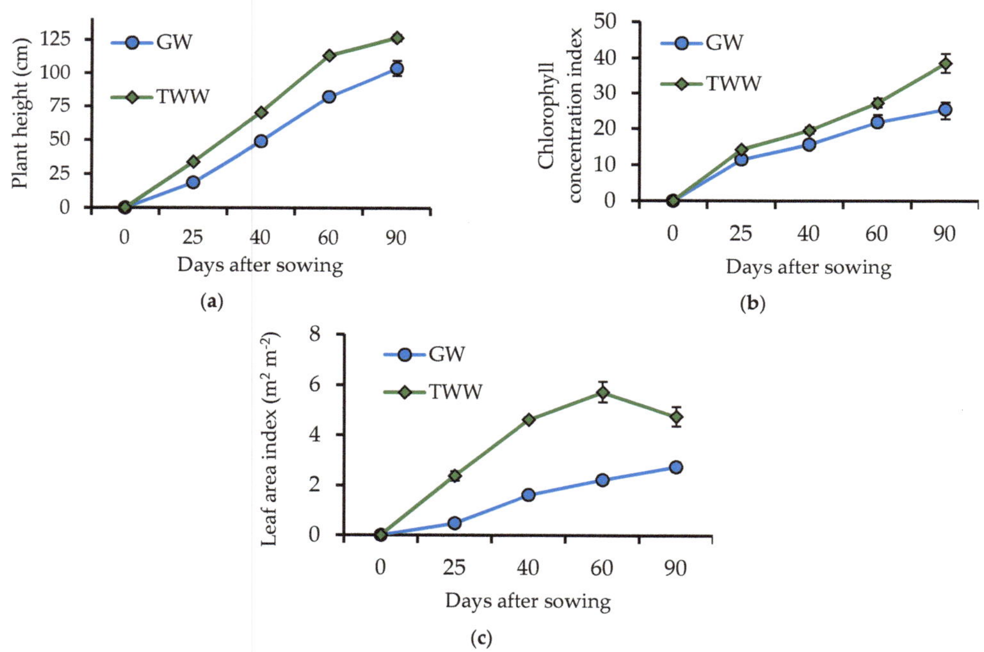

Figure 3. (**a**) Plant height, (**b**) chlorophyll concentration index, and (**c**) leaf area index of barley plants (*Hordeum vulgare*) irrigated with treated wastewater (TWW) and groundwater (GW). Sampling days were 25, 40, 60, and 90 days after sowing. n = 6; p = 0.013 for plant height, p = 0.006 for chlorophyll concentration index, and p = 0.002 for leaf area index. The black bars represent the standard error.

3.2. Treated Wastewater Enhanced the Photosynthetic Efficiency of Barley Plants

The water source for irrigation significantly influenced ($p < 0.05$) all the photosynthetic efficiency parameters evaluated (Figure 4). Plants irrigated with TWW presented higher ($p < 0.001$) maximum photochemical efficiency (Fv/Fm) than those irrigated with GW. This difference was consistent with the highest CCI found in the plants irrigated with TWW. These results agree with those reported by Palliotti et al. [37], who found that high concentrations of chlorophyll benefit the absorption of light and increase Fv/Fm. Chlorophyll is the pigment responsible for capturing solar radiation for photosynthesis; therefore, it is correlated with Fv/Fm [38,39].

The Fv/Fm for plants irrigated with GW was 0.72. This was lower than the 0.80 found in plants irrigated with TWW. These results suggest plants with GW were under stress since values of Fv/Fm lower than 0.80 indicate damage in the photosynthetic apparatus. All the factors causing inhibition of the reaction centers of PSII increase energy dissipation [40–43]. In the same way, the plants irrigated with GW had higher Y(NO) ($p = 0.004$) than the plants irrigated with TWW. The Y(NO) index measures the amount of non-regulated energy dissipated, which is a detrimental form of dissipation [44]. These results are consistent with several studies reporting that plants under stress have lower Fv/Fm and higher Y(NO) compared to healthy plants. For example, Shu et al. [45] found that salinity decreased Fv/Fm and increased Y(NO) in cucumber plants (*Cucumis sativus*). Similarly, Marriboina and Attipalli [46] found the same effect in Indian bean plants (*Pongamia pinnata*) under salinity stress, while Song et al. [47] reported a significant decrease in Fv/Fm in rice plants (*Oryza sativa*) caused by heat stress.

The plants with TWW lost more energy due to heat dissipation, as estimated by Y(NPQ), which is an indicator of regulated energy dissipation and is associated with the xanthophyll cycle and acidification of the thylakoid lumen. Xanthophylls are three carotenoids (violaxanthin, anteroxanthin, and zeaxanthin) involved in heat dissipation [48]. Under radiation stress conditions, violaxanthin is converted to zeaxanthin by the enzyme violaxanthin de-epoxidase. This set of reactions is known as the xanthophyll cycle. The binding of protons and zeaxanthin light-collecting antenna proteins in thylakoids causes conformational changes leading to energy capture and heat dissipation [49]. The violaxanthin de-epoxidase enzyme is located in the lumen of the thylakoids and is activated at acidic pH [50]. For this reason, the acidification of the thylakoid lumen is also involved in the dissipation of excess light energy in the form of heat.

Stress from excess radiation can induce an increase in Y(NPQ) [51]. Plants under stress conditions due to radiation can be exposed to excess energy, which can damage the PSII if the energy is not dissipated in a regulated manner [52]. High temperatures therefore can affect the thylakoid membrane and disrupt the electron donor and acceptor complexes in PSII [53,54]. In addition, excess of undissipated energy can react with molecular oxygen and create free radicals, which damage the photosynthetic apparatus [55]. Then, heat stress commonly causes a significant reduction in Fv/Fm and ΦPSII [47]. Overall, the results suggest that plants irrigated with TWW dissipated excess heat in a better way since they obtained higher Y(NPQ) and lower Y(NO) than plants irrigated with GW.

The plants irrigated with TWW also had higher ΦPSII ($p = 0.011$) and ETR ($p = 0.044$) than plants irrigated with GW, suggesting that those plants transform solar radiation into energy molecules (i.e., ATP) more efficiently. Results of the photosynthetic parameters are consistent with the grain and forage yield since plants irrigated with TWW had the highest values. Tambussi et al. [36] reported that barley cultivars with the highest ΦPSII were the most productive.

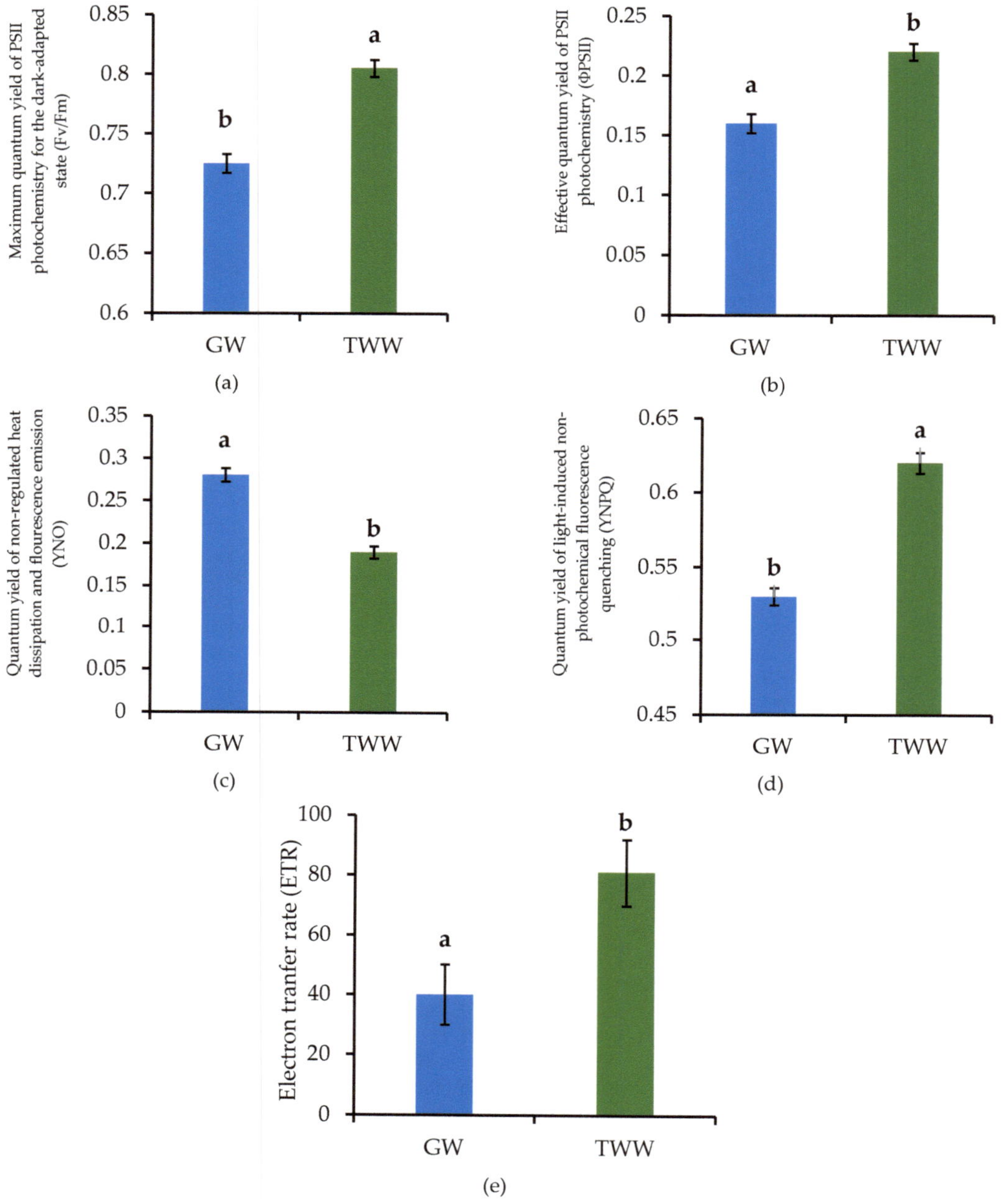

Figure 4. Photosynthetic efficiency parameters of barley plants (*Hordeum vulgare*) irrigated with groundwater (GW) and treated wastewater (TWW). $n = 18$; $p < 0.001$ for Fv/Fm (**a**), $p = 0.011$ for ΦPSII (**b**), $p = 0.004$ for YNO (**c**), $p = 0.004$ for YNPQ (**d**), and $p = 0.044$ for ETR (**e**). Different letters indicate significant differences between water sources (Dunnett test; $p < 0.05$) and black bars represent the standard error.

3.3. Treated Wastewater Produced Changes in Stomatal Density and Area of Barley Plants

The water source for irrigation (GW and TWW) produced different SDs and SAs between the treatments (Figures 5 and 6). The barley plants irrigated with TWW had lower SD than the plants irrigated with GW ($p < 0.001$), both on the adaxial and abaxial leaf surfaces. In contrast, the SA was similar on the abaxial leaf surfaces ($p > 0.05$) but different on the adaxial surface, with the lowest SA in the plants irrigated with TWW ($p < 0.001$).

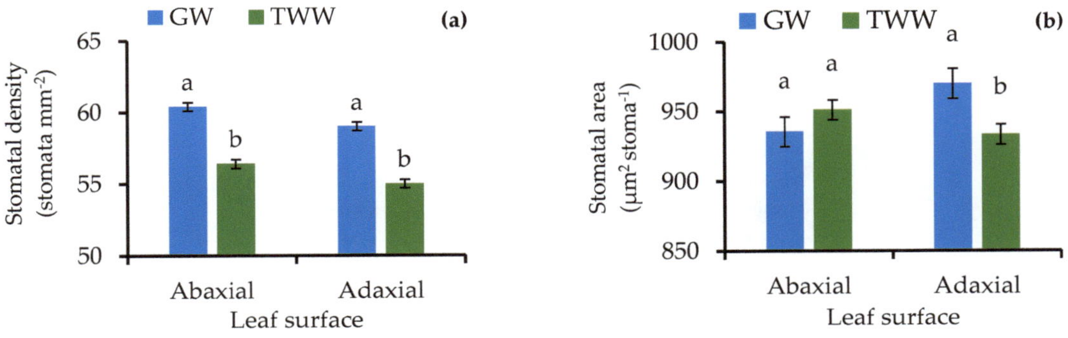

Figure 5. (**a**) Stomatal density and (**b**) stomatal area of barley (*Hordeum vulgare*) irrigated with groundwater (GW) and treated wastewater (TWW). $n = 18$; $p < 0.001$ for abaxial and adaxial stomatal density, $p > 0.05$ for abaxial stomatal area, and $p < 0.001$ for adaxial stomatal area. Different letters indicate significant differences between water sources (Dunnett test; $p < 0.05$) and black bars represent the standard error.

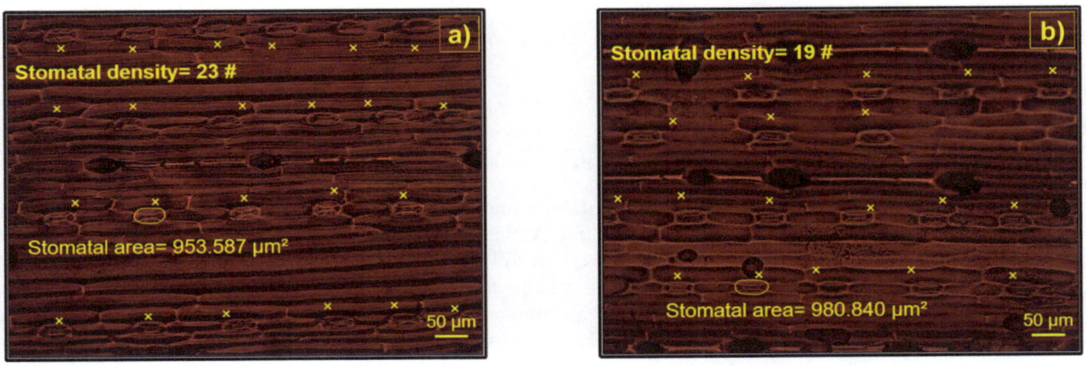

Figure 6. Microstructure of the abaxial leaf surface of barley plants irrigated with (**a**) treated wastewater (TWW) and (**b**) groundwater (GW). "#" = number of stomata per field of view. "×" = stomata accounted.

Stomatal characteristics are important for the plant's physiology since these pores control the loss of water through transpiration and the CO_2 assimilation through photosynthesis [56,57]. The stomatal size and density, therefore, have been important for genotype selection and plant breeding research [58]. For instance, Franks et al. [59] found that the reduction of SD, through overexpression of the EPF2 gene, decreased the stomatal conductance and increased the water-use efficiency in mutant *Arabidopsis* lines. Likewise, Li et al. [60] detected that wheat (*Triticum aestivum*) cultivars with low density and SA showed less transpiration and increased the photosynthetic rate and water-use efficiency. The reason could be that smaller stomata can open and close faster, increasing CO_2 assimilation and decreasing transpiration [61–63]. Hughes et al. [64] also reported that mutant barley lines with low SD showed higher ΦPSII, resulting in better water-use efficiency under stress conditions. The results of the present study then suggest that the high photosynthetic efficiency found in plants irrigated with TWW was related to the low density and SA and have a positive effect on forage and grain yields.

3.4. Treated Wastewater Changed the Nutritional Composition of Barley Plants

Regarding the concentration of nutrients in plants, N had the highest ($p < 0.05$) in leaves and grains of plants irrigated with GW (Table 3). These results are consistent with those from the water analyses, as the concentration of NO_3^- (1.78 mg L^{-1}) was higher in plants irrigated with GW than in plants irrigated with TWW (Table 1). The high content of NO_3^- in GW is an indicator of contamination and may be the result of poor management of agricultural plots, especially related to N fertilizers [65,66], and could be also the effect of irrigation with TWW [8].

Table 3. Nitrogen, phosphorus, potassium, and magnesium content in barley plants (*Hordeum vulgare*) irrigated with groundwater (GW) and treated wastewater (TWW).

Water Source	Nitrogen (N) mg kg^{-1}	Phosphorus (P) mg kg^{-1}	Potassium (K) mg kg^{-1}	Magnesium (Mg) mg kg^{-1}
		Grains		
GW	21,300 a	3508 a	7500 a	1600 a
TWW	11,450 b	2800 a	6300 b	1400 a
		Leaves		
GW	11,320 a	833 a	8300 b	2400 a
TWW	11,020 b	933 a	9200 a	2000 b
		Stems		
GW	7500 a	458 b	8200 a	1500 a
TWW	7100 a	691 a	8700 a	1200 b
		Roots		
GW	5500 a	933 b	8300 a	2600 a
TWW	5200 a	1566 a	8400 a	2000 b

Different letters indicate a significant difference ($p < 0.05$).

Since N is a fundamental part of chlorophyll molecules, it was expected that the high concentration of N in GW plants would lead to a high CCI; however, this index was higher in plants irrigated with TWW. This could be explained by the molybdenum (Mo) concentration as follows: plants absorb N either in the form of ammonium (NH_4^+) or NO_3^-. For N-NO_3^- to be assimilated, it must be reduced to NH_4^+ through the action of the enzymes NO_3^- and NO_2^- reductase [67]. These enzymes require Mo as an enzymatic cofactor, and the absence of this element could cause low assimilation of N-NO_3^-. In the absence of Mo, plants fertilized with NO_3^- present a poor growth and low concentrations of chlorophyll and ascorbic acid and a high content of NO_3^- [68]. The concentration of Mo was 0.01 mg L^{-1} higher in the TWW plants ($p < 0.05$) than in the GW plants (Table 1), and this may explain the high CCI and photosynthetic efficiency of plants irrigated with TWW. Meanwhile, the plants irrigated with GW may have had assimilated less N due to lower concentrations of Mo in the plants' tissue.

Nonetheless, the characteristics of GW plants could be of interest to the beer industry. The N content of the grain influences the quality of malt production because it is related to nitrogenous compounds, such as proteins, amino acids, amines, and purines. For this reason, the N content significantly influences beer production, as it is important for yeast fermentation [69]. In addition, it is related to beer quality parameters, especially color, texture, turbidity, foam formation, CO_2 retention, and microbial nutrition [70]. Then, the grain produced with GW may have a higher malting quality than that generated with TWW since it presented a higher N concentration. Regarding P, the stem and root tissues of the plants irrigated with TWW presented the highest concentration of this nutrient ($p < 0.05$). In contrast, the stem, root, and leaf tissues of the plants irrigated with GW had the highest concentration of Mg ($p < 0.05$). These results are consistent with the nutrient concentrations found in the water since the TWW plants had the lowest concentration of PO_4^{3-} (6.80 mg L^{-1}; Table 2) and the highest concentration of Mg in soil (2.07 meq 100 g^{-1}). The high content of P-PO_4^{3-} in TWW may be a result of the degradation of organic materials contained in these water sources, as it has been reported in previous research [71,72]. Conversely, the high Mg content has been more related to the type of rock in the subsoil [66,73].

Plants irrigated with TWW had a higher concentration of P in stems and roots ($p < 0.05$) than the GW plants, which is consistent with the high concentration of P found in these water sources [74,75]. Previous research has found a strong relationship between chlorophyll concentration and P content in various crops [76–80], which suggests that the biosynthesis of chlorophyll molecules depends partially on the assimilation of P. These results are consistent with the ones from the present study since the CCI was higher in the plants irrigated with TWW. Furthermore, the higher concentration of P in stems of TWW plants could be partially explained by the higher photosynthetic efficiency since the P concentration is related to different photosynthetic parameters, such as Fv/Fm and ΦPSII [77,81]. The concentration of P in leaves was similar between the two treatments ($p > 0.05$). P is a highly mobile element within the plant [82]; thus, it could have been in the leaves at the time of measurement of photosynthetic parameters. The photosynthetic parameters were evaluated during the flowering stage, while the nutritional evaluation was evaluated on mature plants.

The K concentration in leaves was higher in plants irrigated with TWW ($p < 0.05$). The nutrient K controls the entry and exit of water to the cells and therefore the opening and closing of the stomata [83,84]. Hence, the high concentration of K in plants irrigated with TWW could have influenced the differences found in SD and SA. Previous research has reported a relationship between K and stomata. For instance, Shabala et al. [85] found that a mechanism to tolerate salinity is to increase the concentration of K in the leaves and decrease SD in quinoa plants (*Chenopodium quinoa*). Benlloch-González et al. [86] reported that K deficiency inhibited stomatal closure, which induced a state of water stress and affected growth in olive trees (*Olea europaea*) and sunflower plants (*Helianthus annuus*). In contrast, the high concentration of K in plants irrigated with TWW may have also contributed to produce the differences found in grain yield, forage yield, LAI, and CCI. Accordingly, Zhang et al. [87] reported a significant interaction between K, LAI, and CCI in potatoes (*Solanum tuberosum*). According to the authors, a high concentration of K promoted high tuber yield and quality, which agrees with our findings. It is important to point out that the N, P, and K concentrations in leaves were below the ranges indicated by Havlin et al. [88] for barley. In this sense, Jones [82] mentioned that N, P, and K are elements of high mobility in the plant, and as the age of the crop advances, the concentration of these nutrients decreases. The sampling in the present study was carried out during harvest, and this may explain the low concentration of these nutrients. However, the concentration of nutrients in the grains was higher than the values found in previous research. For instance, Dung et al. [89] reported values of 4900, 1000, and 2200 mg Kg^{-1} for K, Mg, and P, respectively. Similar results are reported by the USDA with 2800, 790, and 2210 mg Kg^{-1} of K, Mg, and P in the grain, respectively. The high concentration of nutrients in the grain and the low concentration in the leaf then suggest that the nutrients (K, Mg, and P) were moved from the leaves to the grains at the time of the measurements.

4. Conclusions

TWW is an important source of nutrients since it induces important agronomic, photosynthetic, stomatal, and nutritional changes in barley plants. Plants irrigated with TWW showed higher grain yield, forage yield, PH, CCI, and LAI. The photosynthetic efficiency increased in plants irrigated with TWW. Furthermore, plants irrigated with TWW had lower SD and SA than GW plants. That may be a mechanism of adaptation to the stress generated by the increase in biomass and leaf area. Agronomic, biochemical, and physiological attributes in plants irrigated with TWW appear to be linked to better assimilation of K and P. Overall, the use of TWW represent a sustainable alternative for barley production in arid and semi-arid regions. Nevertheless, the evaluation of concentrations of toxic elements, emergent pollutants, and microplastics in the plant, water, and soil after using TWW is highly recommended.

This study may contribute to the sustainability of water resources and agricultural practices in northern Mexico since it was part of a government project that aimed to develop a new metropolitan irrigation district.

Author Contributions: A.A.-H. participated in the entire research process, including experimental design, sample analysis, data processing, and writing the paper. G.S.-P. contributed to experimental design and writing—review and editing. O.C.P.-G. participated in the data processing and discussion of results. C.R.L.-M. contributed in sample analysis. F.V.-G. discussion of results as well as writing the manuscript. C.G.M.-B. contributed in sample analysis. J.M.O.-R. contributed to conceptualization, methodology, supervision, validation, and acquisition of financial support. All authors have read and agreed to the published version of the manuscript.

Funding: This research was funded by the Chihuahua's Municipal Government in the project "Validación de aguas residuales para siembra de forrajes para factibilidad del proyecto de Distrito de Riego Metropolitano, grant number 121394264", INIFAP.

Institutional Review Board Statement: Not applicable.

Informed Consent Statement: Not applicable.

Data Availability Statement: Data are available upon request, but if used in another manuscript, the authors wish to be included as co-authors.

Acknowledgments: The authors thank the National Research Institute for Forestry, Agriculture, and Animal Production (INIFAP-Mexico) and the Autonomous University of Chihuahua (UACH). We sincerely thank Mauro Parada Muñoz, Humberto Molinar Hernandez, Yazmin Azalea Acosta Rodriguez, Norma Eugenia Delgado Prieto, Jose Eduardo Cabada Estrada, Abelardo Lugo Ochoa, Cristo Omar Ponce Valenzuela, Baltazar Corral-Diaz, and Kimberly Anai Alcala Carmona for the assistance provided while this research was conducted. We deeply thank Romualdo Manuel Nieto Gonzalez, owner of the agricultural area in Aldama, Mexico, for providing the experimental site.

Conflicts of Interest: The authors declare no conflict of interest.

References

1. Qadir, M.; Drechsel, P.; Jiménez Cisneros, B.; Kim, Y.; Pramanik, A.; Mehta, P.; Olaniyan, O. Global and regional potential of wastewater as a water, nutrient and energy source. *Nat. Resour. Forum.* **2020**, *44*, 40–51. [CrossRef]
2. Lado, M.; Ben-Hur, M. Treated domestic sewage irrigation effects on soil hydraulic properties in arid and semiarid zones: A review. *J. Soil. Tillage Research* **2009**, *106*, 152–163. [CrossRef]
3. AQUASTAT. FAO Global Information System on Water and Agriculture. Wastewater Section. Available online: http://www.fao.org/nr/water/aquastat/wastewater/index.stm (accessed on 14 May 2021).
4. Zang, J.; Kumar, M.; Werner, D. Real-world sustainability analysis of an innovative decentralized water system with rainwater harvesting and wastewater reclamation. *J. Environ. Manag.* **2021**, *280*, 111639. [CrossRef] [PubMed]
5. Mahfooz, Y.; Yasar, A.; Guijian, L.; Islam, Q.U.; Akhtar, A.B.T.; Rasheed, R.; Irshad, S.; Naeem, U. Critical risk analysis of metals toxicity in wastewater irrigated soil and crops: A study of a semi-arid developing region. *Sci. Rep.* **2020**, *10*, 12845. [CrossRef] [PubMed]
6. Connor, R.; Renata, A.; Ortigara, C.; Koncagül, E.; Uhlenbrook, S.; Lamizana-Diallo, B.M.; Zadeh, S.M.; Qadir, M.; Kjellén, M.; Sjödin, J. *The United Nations World Water Development Report 2017. Wastewater: The Untapped Resource*; The United Nations Educational, Scientific and Cultural Organization: Paris, France, 2017.
7. Jaramillo, M.F.; Restrepo, I. Wastewater reuse in agriculture: A review about its limitations and benefits. *Sustainability* **2017**, *9*, 1734. [CrossRef]
8. Ofori, S.; Puskacova, A.; Ruzickova, I.; Wanner, J. Treated wastewater reuse for irrigation: Pros and cons. *Sci. Total Environ.* **2021**, *760*, 144026. [CrossRef]

9. Al-Rashed, M.F.; Sherif, M.M. Water resources in the GCC countries: An overview. *Water Resour. Manag.* **2000**, *14*, 59–75. [CrossRef]
10. Mohammad, M.J.; Mazahreh, N. Changes in soil fertility parameters in response to irrigation of forage crops with secondary treated wastewater. *Commun. Soil Sci. Plant Anal.* **2003**, *34*, 1281–1294. [CrossRef]
11. Mendoza-Espinosa, L.G.; Cabello-Pasini, A.; Macias-Carranza, V.; Daessle-Heuser, W.; Orozco-Borbon, M.V.; Quintanilla-Montoya, A.L. The effect of reclaimed wastewater on the quality and growth of grapevines. *Water Sci. Technol.* **2008**, *57*, 1445–1450. [CrossRef]
12. Tahtouh, J.; Mohtar, R.; Assi, A.; Schwab, P.; Jantrania, A.; Deng, Y.; Munster, C. Impact of brackish groundwater and treated wastewater on soil chemical and mineralogical properties. *Sci. Total Environ.* **2019**, *647*, 99–109. [CrossRef]
13. Lahlou, F.-Z.; Mackey, H.R.; Al-Ansari, T. Wastewater reuse for livestock feed irrigation as a sustainable practice: A socio-environmental-economic review. *J. Clean. Prod.* **2021**, *294*, 126331. [CrossRef]
14. Rusan, M.J.M.; Hinnawi, S.; Rousan, L. Long term effect of wastewater irrigation of forage crops on soil and plant quality parameters. *Desalination* **2007**, *215*, 143–152. [CrossRef]
15. Khalid, S.; Shahid, M.; Bibi, I.; Sarwar, T.; Shah, A.H.; Niazi, N.K. A review of environmental contamination and health risk assessment of wastewater use for crop irrigation with a focus on low and high-income countries. *Int. J. Environ. Res. Public Health* **2018**, *15*, 895. [CrossRef] [PubMed]
16. Hettiarachchi, H.; Caucci, S.; Ardakanian, R. The golden example of nexus approach. In *Safe Use of Wastewater in Agriculture*; Springer: Berlin/Heidelberg, Germany, 2018; pp. 1–11.
17. Inyinbor, A.A.; Bello, O.S.; Oluyori, A.P.; Inyinbor, H.E.; Fadiji, A.E. Wastewater conservation and reuse in quality vegetable cultivation: Overview, challenges and future prospects. *Food Control* **2019**, *98*, 489–500. [CrossRef]
18. Elfanssi, S.; Ouazzani, N.; Mandi, L. Soil properties and agro-physiological responses of alfalfa (*Medicago sativa* L.) irrigated by treated domestic wastewater. *Agric. Water Manag.* **2018**, *202*, 231–240. [CrossRef]
19. Shahrivar, A.A.; Rahman, M.M.; Hagare, D.; Maheshwari, B. Variation in kikuyu grass yield in response to irrigation with secondary and advanced treated wastewaters. *Agric. Water Manag.* **2019**, *222*, 375–385. [CrossRef]
20. Tekaya, M.; Mechri, B.; Dabbaghi, O.; Mahjoub, Z.; Laamari, S.; Chihaoui, B.; Boujnah, D.; Hammami, M.; Chehab, H. Changes in key photosynthetic parameters of olive trees following soil tillage and wastewater irrigation, modified olive oil quality. *Agric. Water Manag.* **2016**, *178*, 180–188. [CrossRef]
21. Samarah, N.H.; Bashabsheh, K.Y.; Mazahrih, N.T. Treated wastewater outperformed freshwater for barley irrigation in arid lands. *Ital. J. Agron.* **2020**, *15*, 183–193. [CrossRef]
22. Langridge, P. Economic and Academic Importance of Barley. In *The Barley Genome*; Springer: Cham, Switzerland; New York, NY, USA, 2018; pp. 1–10.
23. INEGI. Instituto Nacional de Estadística, Geografía e Historia. Pronatuario de información Geográfica Municipal de los Estados Unidos Mexicanos. Aldama, Chihuahua. Available online: https://www.inegi.org.mx/contenidos/app/mexicocifras/datos_geograficos/08/08002.pdf (accessed on 13 May 2021).
24. CONAGUA-SMN. Comisión Nacional del Agua-Servicio Meteorológico Nacional. Información Climatológica por Estado. Available online: https://smn.conagua.gob.mx/es/informacion-climatologica-por-estado?estado=chih (accessed on 13 May 2021).
25. Baird, R.B. *Standard Methods for the Examination of Water and Wastewater*, 23rd ed.; Water Environment Federation, American Public Health Association, American Water Works Association, 2017. Available online: https://connect.wef.org/s/store#/store/browse/detail/a2n1S000000KVk1 (accessed on 6 February 2022).
26. CONAGUA. Ley Federal de Derechos. Disposiciones Aplicables en Materia de Aguas Nacionales 2020. Available online: https://www.gob.mx/cms/uploads/attachment/file/555838/CGRF-1-20_LFD.pdf (accessed on 11 May 2021).
27. Ayers, R.; Westcott, D. *Water Quality for Agriculture*; No. 29; Food and Agriculture Organization of the United Nations, 1985. Available online: https://www.fao.org/3/t0234e/t0234e00.htm (accessed on 6 February 2022).
28. Kitajima, M.; Butler, W.L. Quenching of chlorophyll fluorescence and primary photochemistry in chloroplasts by dibromothymoquinone. *Biochim. Biophys. Acta* **1975**, *376*, 105–115. [CrossRef]
29. Genty, B.; Briantais, J.-M.; Baker, N.R. The relationship between the quantum yield of photosynthetic electron transport and quenching of chlorophyll fluorescence. *Biochim. Biophys. Acta-Biomembr.* **1989**, *990*, 87–92. [CrossRef]
30. Miller, J.; Miller, J.C. *Statistics and Chemometrics for Analytical Chemistry*; Pearson Education: Harlow, UK, 2018.
31. Bru, I.P.; Lozano, L.L.; Bru, A.P.; Fernández, M.A.; González-Estecha, M. Procedimiento de validación de un método para cuantificar cromo en suero por espectroscopía de absorción atómica con atomización electrotérmica. *Rev. Lab. Clin.* **2015**, *8*, 52–57. [CrossRef]
32. EPA. Methods for Chemical Analysis of Water and Wastes. Available online: https://www.wbdg.org/FFC/EPA/EPACRIT/epa600_4_79_020.pdf (accessed on 10 May 2021).
33. Alkhamisi, S.A.; Abdelrahman, H.; Ahmed, M.; Goosen, M. Assessment of reclaimed water irrigation on growth, yield, and water-use efficiency of forage crops. *Appl. Water Sci.* **2011**, *1*, 57–65. [CrossRef]
34. Aghtape, A.A.; Ghanbari, A.; Sirousmehr, A.; Siahsar, B.; Asgharipour, M.; Tavssoli, A. Effect of irrigation with wastewater and foliar fertilizer application on some forage characteristics of foxtail millet (*Setaria italica*). *Int. J. Plant Physiol. Biochem.* **2011**, *3*, 34–42. [CrossRef]

35. Ramírez-Novoa, U.; Rodríguez-Guillén, A.; Morán-Vázquez, N.; Cervantes-Ortíz, F.; Mendoza-Elos, M.; Rangel-Lucio, J. Manejo agronómico de cebada maltera. Rendimiento de semilla y componentes. *Cienc y Tecnol. Agrop. México* **2014**, *2*, 24–29.
36. Tambussi, E.; Nogues, S.; Ferrio, P.; Voltas, J.; Araus, J. Does higher yield potential improve barley performance in Mediterranean conditions?: A case study. *Field Crops Res.* **2005**, *91*, 149–160. [CrossRef]
37. Palliotti, A.; Silvestroni, O.; Petoumenou, D.; Viticulture. Photosynthetic and photoinhibition behavior of two field-grown grapevine cultivars under multiple summer stresses. *AJEV* **2009**, *60*, 189–198.
38. Hailemichael, G.; Catalina, A.; González, M.; Martin, P. Relationships between water status, leaf chlorophyll content and photosynthetic performance in Tempranillo vineyards. *SAJEV* **2016**, *37*, 149–156. [CrossRef]
39. Xu, Y.; Yang, M.; Cheng, F.; Liu, S.; Liang, Y. Effects of LED photoperiods and light qualities on in vitro growth and chlorophyll fluorescence of *Cunninghamia lanceolata*. *BMC Plant Biol.* **2020**, *20*, 269. [CrossRef]
40. Osmond, C.B.; Chow, W.S.; Robinson, S.A. Inhibition of non-photochemical quenching increases functional absorption cross-section of photosystem II as excitation from closed reaction centres is transferred to open centres, facilitating earlier light saturation of photosynthetic electron transport. *Funct. Plant Biol.* **2021**. [CrossRef]
41. Lu, T.; Meng, Z.; Zhang, G.; Qi, M.; Sun, Z.; Liu, Y.; Li, T. Sub-high Temperature and High Light Intensity Induced Irreversible Inhibition on Photosynthesis System of Tomato Plant (*Solanum lycopersicum* L.). *Front. Plant Sci.* **2017**, *8*, 365. [CrossRef]
42. Webb, J.A.; Fletcher, R. Paclobutrazol protects wheat seedlings from injury due to waterlogging. *Plant Growth Regul.* **1996**, *18*, 201–206. [CrossRef]
43. Odasz-Albrigtsen, A.M.; Tømmervik, H.; Murphy, P. Decreased photosynthetic efficiency in plant species exposed to multiple airborne pollutants along the Russian-Norwegian border. *Can. J. Bot.* **2000**, *78*, 1021–1033. [CrossRef]
44. Brestic, M.; Zivcak, M. PSII Fluorescence Techniques for Measurement of Drought and High Temperature Stress Signal in Crop Plants: Protocols and Applications. In *Molecular Stress Physiology of Plants*; Springer: Berlin/Heidelberg, Germany, 2013; pp. 87–131.
45. Shu, S.; Yuan, L.-Y.; Guo, S.-R.; Sun, J.; Liu, C.-J. Effects of exogenous spermidine on photosynthesis, xanthophyll cycle and endogenous polyamines in cucumber seedlings exposed to salinity. *Afr. J. Biotechnol.* **2012**, *11*, 6064–6074. [CrossRef]
46. Marriboina, S.; Attipalli, R.R. Hydrophobic cell-wall barriers and vacuolar sequestration of Na+ ions are among the key mechanisms conferring high salinity tolerance in a biofuel tree species, Pongamia pinnata L. pierre. *EEB* **2020**, *171*, 103949. [CrossRef]
47. Song, L.; Yue, L.; Zhao, H.; Hou, M. Protection effect of nitric oxide on photosynthesis in rice under heat stress. *Acta Physiol. Plant* **2013**, *35*, 3323–3333. [CrossRef]
48. Maoka, T. Carotenoids as natural functional pigments. *J. Nat. Med.* **2020**, *74*, 1–16. [CrossRef]
49. Xu, P.; Tian, L.; Kloz, M.; Croce, R. Molecular insights into Zeaxanthin-dependent quenching in higher plants. *Sci. Rep.* **2015**, *5*, 13679. [CrossRef]
50. Dlouhý, O.; Kurasová, I.; Karlický, V.; Javornik, U.; Šket, P.; Petrova, N.Z.; Krumova, S.B.; Plavec, J.; Ughy, B.; Špunda, V. Modulation of non-bilayer lipid phases and the structure and functions of thylakoid membranes: Effects on the water-soluble enzyme violaxanthin de-epoxidase. *Sci. Rep.* **2020**, *10*, 11959. [CrossRef]
51. Samson, G.; Bonin, L.; Maire, V. Dynamics of regulated YNPQ and non-regulated YNO energy dissipation in sunflower leaves exposed to sinusoidal lights. *Photosynth. Res.* **2019**, *141*, 315–330. [CrossRef]
52. Xue, D.W.; Jiang, H.; Hu, J.; Zhang, X.Q.; Guo, L.B.; Zeng, D.L.; Dong, G.J.; Sun, G.C.; Qian, Q. Characterization of physiological response and identification of associated genes under heat stress in rice seedlings. *Plant Physiol. Biochem.* **2012**, *61*, 46–53. [CrossRef]
53. Djanaguiraman, M.; Boyle, D.; Welti, R.; Jagadish, S.; Prasad, P. Decreased photosynthetic rate under high temperature in wheat is due to lipid desaturation, oxidation, acylation, and damage of organelles. *BMC Plant Biol.* **2018**, *18*, 1–17. [CrossRef] [PubMed]
54. Farooq, M.; Bramley, H.; Palta, J.A.; Siddique, K.H. Heat stress in wheat during reproductive and grain-filling phases. *Crit. Rev. Plant Sci.* **2011**, *30*, 491–507. [CrossRef]
55. Foyer, C.H. Reactive oxygen species, oxidative signaling and the regulation of photosynthesis. *Environ. Exp. Bot.* **2018**, *154*, 134–142. [CrossRef] [PubMed]
56. Lawson, T.; Blatt, M.R. Stomatal size, speed, and responsiveness impact on photosynthesis and water use efficiency. *Plant Physiol.* **2014**, *164*, 1556–1570. [CrossRef] [PubMed]
57. Lawson, T.; Simkin, A.J.; Kelly, G.; Granot, D. Mesophyll photosynthesis and guard cell metabolism impacts on stomatal behaviour. *New Phytol.* **2014**, *203*, 1064–1081. [CrossRef]
58. Bertolino, L.T.; Caine, R.S.; Gray, J.E. Impact of Stomatal Density and Morphology on Water-Use Efficiency in a Changing World. *Front. Plant Sci.* **2019**, *10*, 225. [CrossRef]
59. Franks, P.J.W.; Doheny-Adams, T.; Britton-Harper, Z.J.; Gray, J.E. Increasing water-use efficiency directly through genetic manipulation of stomatal density. *New Phytol.* **2015**, *207*, 188–195. [CrossRef]
60. Li, H.; Li, M.; Wei, X.; Zhang, X.; Xue, R.; Zhao, Y.; Zhao, H. Transcriptome analysis of drought-responsive genes regulated by hydrogen sulfide in wheat (*Triticum aestivum* L.) leaves. *Mol. Genet. Genomics* **2017**, *292*, 1091–1110. [CrossRef]
61. Drake, P.L.; Froend, R.H.; Franks, P.J. Smaller, faster stomata: Scaling of stomatal size, rate of response, and stomatal conductance. *J. Exp. Bot.* **2013**, *64*, 495–505. [CrossRef]
62. Raven, J.A. Speedy small stomata? *J. Exp. Bot.* **2014**, *65*, 1415–1424. [CrossRef]

63. Fanourakis, D.; Giday, H.; Milla, R.; Pieruschka, R.; Kjaer, K.H.; Bolger, M.; Vasilevski, A.; Nunes-Nesi, A.; Fiorani, F.; Ottosen, C.-O. Pore size regulates operating stomatal conductance, while stomatal densities drive the partitioning of conductance between leaf sides. *Ann. Bot.* **2015**, *115*, 555–565. [CrossRef] [PubMed]
64. Hughes, J.; Hepworth, C.; Dutton, C.; Dunn, J.A.; Hunt, L.; Stephens, J.; Waugh, R.; Cameron, D.D.; Gray, J.E. Reducing stomatal density in barley improves drought tolerance without impacting on yield. *Plant Physiol.* **2017**, *174*, 776–787. [CrossRef] [PubMed]
65. Andrade, E.P.; Bonmati, A.; Esteller, L.J.; Brunn, S.; Jensen, L.S.; Meers, E.; Anton, A. Selection and application of agri-environmental indicators to assess potential technologies for nutrient recovery in agriculture. *Ecol. Indic.* **2022**, *134*, 108471. [CrossRef]
66. Reyes, V.M.G.; Gutiérrez, M.; Haro, B.N.; López, D.N.; Herrera, M.T.A. Groundwater quality impacted by land use/land cover change in a semiarid region of Mexico. *Groundw. Sustain. Dev.* **2017**, *5*, 160–167. [CrossRef]
67. Pilbeam, D.J. *Handbook of Plant Nutrition*; CRC Press: Boca Raton, FL, USA, 2016.
68. Broadley, M.; Brown, P.; Cakmak, I.; Rengel, Z.; Zhao, F. Function of Nutrients: Micronutrients. In *Marschner's Mineral Nutrition of Higher Plants*; Elsevier: Stuttgart, Germany, 2012; pp. 191–248.
69. Bettenhausen, H.M.; Barr, L.; Broeckling, C.D.; Chaparro, J.M.; Holbrook, C.; Sedin, D.; Heuberger, A.L. Influence of malt source on beer chemistry, flavor, and flavor stability. *Food Res. Int.* **2018**, *113*, 487–504. [CrossRef]
70. Dos Santos, M.T.R.; de Mello, P.P.M.; Sérvulo, E.F.C. Nitrogen compounds in brewing wort and beer: A review. *J. Brew. Distilling* **2014**, *5*, 10–17. [CrossRef]
71. Chan, S.S.; Khoo, K.S.; Chew, K.W.; Ling, T.C.; Show, P.L. Recent advances biodegradation and biosorption of organic compounds from wastewater: Microalgae-bacteria consortium-A review. *Bioresour. Technol.* **2022**, *344*, 126159. [CrossRef]
72. Ravelo, S.G.M.; Reyes, F.G.; Cabriales, J.J.P.; Moreno, J.P.; Chávez, L.T. Evaluación de la recuperación del nitrógeno y fósforo de diferentes fuentes de fertilizantes por el cultivo de trigo irrigado con aguas residuales y de pozo. *Acta Agron.* **2014**, *63*, 25–30. [CrossRef]
73. Márquez, L.V.; González, M.R.; Bayter, Y.O.; Sarabia, A.B.C. Caracterización microbiológica y fisicoquímica de aguas subterráneas de los municipios de La Paz y San Diego, Cesar, Colombia. *RIAA* **2012**, *3*, 27–35. [CrossRef]
74. Egle, L.; Rechberger, H.; Zessner, M. Overview and description of technologies for recovering phosphorus from municipal wastewater. *Resour. Conserv. Recy.* **2015**, *105*, 325–346. [CrossRef]
75. Wilfert, P.; Kumar, P.S.; Korving, L.; Witkamp, G.J.; van Loosdrecht, M.C. The Relevance of Phosphorus and Iron Chemistry to the Recovery of Phosphorus from Wastewater: A Review. *Environ. Sci. Technol.* **2015**, *49*, 9400–9414. [CrossRef] [PubMed]
76. Liang, Z.; Soranno, P.A.; Wagner, T. The role of phosphorus and nitrogen on chlorophyll a: Evidence from hundreds of lakes. *Water Res.* **2020**, *185*, 116236. [CrossRef] [PubMed]
77. Singh, S.; Reddy, V.; Fleisher, D.; Timlin, D. Relationship between photosynthetic pigments and chlorophyll fluorescence in soybean under varying phosphorus nutrition at ambient and elevated CO_2. *Photosynthetica* **2017**, *55*, 421–433. [CrossRef]
78. Dayal, J.; Goswami, C.; Munjal, R. Influence of phosphorus application on water relations, biochemical parameters and gum content in cluster bean under water deficit. *Biol. Plant.* **2004**, *48*, 445–448. [CrossRef]
79. Jiang, H.M.; Yang, J.C.; Zhang, J.F. Effects of external phosphorus on the cell ultrastructure and the chlorophyll content of maize under cadmium and zinc stress. *Environ. Pollut.* **2007**, *147*, 750–756. [CrossRef] [PubMed]
80. Ahmad, Z.; Waraich, E.A.; Ahmad, R.; Shahbaz, M. Modulation in water relations, chlorophyll contents and antioxidants activity of maize by foliar phosphorus application under drought stress. *Pak. J. Bot.* **2017**, *49*, 11–19.
81. Singh, S.K.; Reddy, V.R. Response of carbon assimilation and chlorophyll fluorescence to soybean leaf phosphorus across CO_2: Alternative electron sink, nutrient efficiency and critical concentration. *J. Photochem. Photobiol. B Biol.* **2015**, *151*, 276–284. [CrossRef]
82. Jones, J.J.B. *Plant Nutrition and Soil Fertility manual*; CRC Press: Boca Raton, FL, USA, 2012.
83. Humble, G.D.; Raschke, K. Stomatal opening quantitatively related to potassium transport: Evidence from electron probe analysis. *Plant Physiol.* **1971**, *48*, 447–453. [CrossRef]
84. Johnson, R.; Vishwakarma, K.; Hossen, M.S.; Kumar, V.; Shackira, A.M.; Puthur, J.T.; Abdi, G.; Sarraf, M.; Hasanuzzaman, M. Potassium in plants: Growth regulation, signaling, and environmental stress tolerance. *Plant Physiol. Biochem.* **2022**, *172*, 56–69. [CrossRef]
85. Shabala, S.; Hariadi, Y.; Jacobsen, S.-E. Genotypic difference in salinity tolerance in quinoa is determined by differential control of xylem Na+ loading and stomatal density. *Plant Physiol.* **2013**, *170*, 906–914. [CrossRef]
86. Benlloch-González, M.; Arquero, O.; Fournier, J.M.; Barranco, D.; Benlloch, M. K+ starvation inhibits water-stress-induced stomatal closure. *Plant Physiol.* **2008**, *165*, 623–630. [CrossRef] [PubMed]
87. Zhang, S.; Fan, J.; Zhang, F.; Wang, H.; Yang, L.; Sun, X.; Cheng, M.; Cheng, H.; Li, Z. Optimizing irrigation amount and potassium rate to simultaneously improve tuber yield, water productivity and plant potassium accumulation of drip-fertigated potato in northwest China. *Agric. Water Manag.* **2022**, *264*, 107493. [CrossRef]
88. Havlin, J.L.; Tisdale, S.L.; Nelson, W.L.; Beaton, J.D. *Soil Fertility and Fertilizers*; Prentice Hall: Newark, NJ, USA, 2016.
89. Dung, D.D.; Godwin, I.; Nolan, J.V. Nutrient content and in sacco digestibility of barley grain and sprouted barley. *J. Anim. Vet. Adv.* **2010**, *9*, 2485–2492. [CrossRef]

International Journal of
Environmental Research and Public Health

MDPI

Review

From Conventional Disinfection to Antibiotic Resistance Control—Status of the Use of Chlorine and UV Irradiation during Wastewater Treatment

Muhammad Umar

Norwegian Institute for Water Research (NIVA), Økernveien 94, 0579 Oslo, Norway; muhammad.umar@niva.no

Abstract: Extensive use of antibiotics for humans and livestock has led to an enhanced level of antibiotic resistance in the environment. Municipal wastewater treatment plants are regarded as one of the main sources of antibiotic-resistant bacteria (ARB) and antibiotic resistance genes (ARGs) in the aquatic environment. A significant amount of research has been carried out to understand the microbiological quality of wastewater with respect to its antibiotic resistance potential over the past several years. UV disinfection has primarily been used to achieve disinfection, including damaging DNA, but there has been an increasing use of chlorine and H_2O_2-based AOPs for targeting genes, including ARGs, considering the higher energy demands related to the greater UV fluences needed to achieve efficient DNA damage. This review focuses on some of the most investigated processes, including UV photolysis and chlorine in both individual and combined approaches and UV advanced oxidation processes (AOPs) using H_2O_2. Since these approaches have practical disinfection and wastewater treatment applications globally, the processes are reviewed from the perspective of extending their scope to DNA damage/ARG inactivation in full-scale wastewater treatment. The fate of ARGs during existing wastewater treatment processes and how it changes with existing treatment processes is reviewed with a view to highlighting the research needs in relation to selected processes for addressing future disinfection challenges.

Keywords: wastewater treatment; antibiotic resistance genes; DNA; UV treatment; advanced oxidation processes

Citation: Umar, M. From Conventional Disinfection to Antibiotic Resistance Control—Status of the Use of Chlorine and UV Irradiation during Wastewater Treatment. *Int. J. Environ. Res. Public Health* **2022**, *19*, 1636. https://doi.org/10.3390/ijerph19031636

Academic Editors: Yung-Tse Hung, Hamidi Abdul Aziz and Issam A. Al-Khatib

Received: 13 December 2021
Accepted: 29 January 2022
Published: 31 January 2022

1. Introduction

Wastewater reuse has traditionally been considered safe after treatment by a combination of physicochemical and biological treatments that target the removal of organic matter, chemical contaminants, and microorganisms. Consequently, the benchmarks for assessing treatment performance that are currently incorporated into the regulations generally include total suspended solids, chemical oxygen demand, biochemical oxygen demand, ammonia, nitrate, total phosphorus, and inactivation of microbes such as fecal coliforms or *Escherichia coli* (*E. coli*) [1]. However, microbial inactivation alone is not sufficient to ensure the safety of treated wastewater since genes may still be present even after disinfection [2]. This requires a shift from conventional microbiological inactivation to sufficient DNA damage to minimize the spread and development of antibiotic resistance. Importantly, the current disinfection processes are not designed to damage genes, which makes antibiotic resistance genes (ARGs) a serious concern since they enable microorganisms to become resistant to antibiotics.

Municipal wastewater is regarded as one of the main sources of antibiotic-resistant bacteria (ARB) and ARGs in the aquatic environment. In fact, wastewater treatment plants (WWTPs) are considered a major "hotspot" of ARB and ARGs since these have been frequently found in WWTP effluents [3]. This is primarily because the current WWTPs are not equipped with appropriate technologies for their mitigation/inactivation prior to or upon effluent reuse or discharge. It is well known that the final effluents (particularly biologically treated effluents) release a high number of bacteria and genes that may be resistant to antibiotics [4]. Existing literature suggests that the high growth rates and microbial densities that are fundamental to conventional biological treatments at WWTPs, along with the presence of residual antibiotics, could create a highly suitable environment for promoting ARG transfer and multiantibiotic resistance among bacteria [5]. Biological processes create an environment that is potentially suitable for resistance development and spread due to the continuous interaction of bacteria with antibiotics at sub-inhibitory concentrations. For example, the Class 1 integron gene *intI*1 has been reported to increase in WWTP effluents, indicating the fitness gain of bacteria harboring this mobile element [6]. Moreover, the potential for co-selection of ARGs is increased when the bacterial communities are exposed to a chemical stress (heavy metals, antibiotics, or both) in a continuously changing WWTP environment [7]. Various investigations have shown a correlation between the presence of genes encoding for resistances against different metals and of ARGs in plasmids and integrons associated with contaminated soils and WWTPs [8–10].

Such concerns may impede efforts to provide safe drinking water as well as the discharge and reuse of the treated wastewater. Since there is an increasing interest in and demand for the safe discharge and/or reuse of wastewater for applications such as irrigation, aquaculture, aquifer recharge, indirect potable reuse, and even direct potable reuse, an understanding and mitigation of the risks arising from ARB and associated ARGs are very important. It is particularly relevant considering that wastewater reuse (especially for agricultural purposes) and biosolids that are generated during treatment are considered important means of AMR spread.

UV disinfection as an alternative to chlorine is of particular interest since UV light is directly absorbed by DNA and therefore has the potential to inactivate ARGs and limit their release and transfer to bacteria. Although the research on the efficiency of UV disinfection in damaging ARGs is recent, it is one of the most promising technologies to address this emerging issue [2]. However, the process is inherently expensive due to the high energy requirements. Moreover, UV photolysis alone may not be as efficient for damaging ARGs since the extent of DNA damage is dependent on the UV fluence, which is generally higher for achieving DNA damage than microbiological inactivation, cell structure, and water matrix [11]. For real wastewaters, the process can be enhanced by using oxidants such as H_2O_2 and Cl_2 in what are termed advanced oxidation processes (AOPs). Direct UV photolysis, UV/H_2O_2, and UV/Cl_2 are focused on in this article considering their practical applications and greater future potential compared with other disinfectants and AOPs for ARG inactivation. Since these processes and/or their combinations are widely used in practical water and wastewater treatment for a range of contaminants, they stand out as some of the best potential technologies for inactivation of ARGs on a large scale. Considering that an AOP produces radicals that can break down organic matter while simultaneously inactivating bacteria and ARGs, this process can be useful for both water treatment and wastewater reclamation and reuse [12]. Several recent investigations have shown these processes to be effective but to a varying degree depending on a range of process and operational parameters. This article specifically focuses on the potential of these processes for practical application at full scale for the purpose of inactivating ARGs considering the findings recently reported on the topic. This review specifically focuses on selected technologies to: (1) summarize and review the progress over the past several years; and (2) identify and discuss areas of importance in relation to the state of the art with particular reference to challenges needed to be addressed for full-scale applications. Knowledge gaps related to the significance of target ARGs and the effectiveness of UV and UV-based AOPs in various water matrices with respect to changes in the resistance shift and impact of co-existing contaminants are identified and reviewed in the context of knowledge needs for future applications of these processes.

2. Antibiotic Resistance—Mechanisms of Resistance Spread and Wastewater Treatment

2.1. Mechanism of Spread and Selection of ARGs

Antibiotic resistance is the ability of bacteria to survive and potentially thrive in the presence of antibiotics [12]. Extensive use of antibiotics in human and animal health, and their use for the promotion of growth in animals, has accelerated the process of microbial resistance. Resistance to antibiotics is encoded in ARGs, enabling bacteria to fight antibiotics through various mechanisms. The two mechanisms responsible for transfer of antimicrobial resistance include: vertical gene transmission (VGT), which involves inheriting the genetic information from the parent cells; and horizontal gene transfer (HGT), in which a non-resistant bacterium becomes resistant by gaining the resistance genes other than from its parent cell [2]. Antibiotic resistance spread mechanisms are shown in Figure 1. HGT occurs through conjugation, transduction, and transformation. Notably, VGT and conjugation occur when the gene-carrying bacterium is viable whereas successful transduction occurs when the virus carrying the gene is also infective [2]. Transformation, however, can occur without a viable or infective donor microorganism since bacteria can obtain ARGs from cell-free DNA [13]. Inactivation of microorganisms alone, without effective DNA damage, as a primary objective, is therefore not sufficient because it does not guarantee control over the spread of antimicrobial resistance [2].

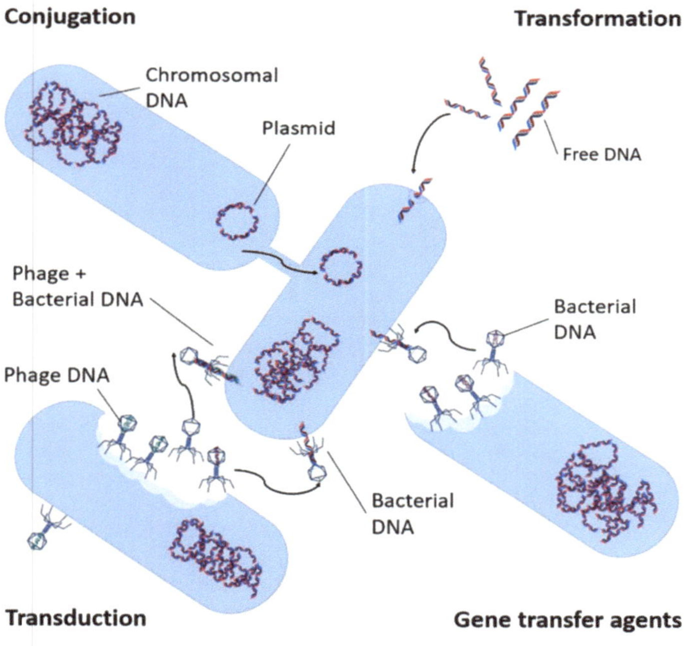

Figure 1. Antibiotic resistance transport mechanisms (adapted from von Wintersdorff et al. [14]).

ARGs are therefore considered a contaminant of emerging concern (CEC) considering that they present "new" environmental and public health concerns. Contrary to chemical CECs, which have been widely focused on over the past several years, perspectives on antibiotic resistance in general and with respect to water reuse in particular are unclear. For example, knowledge on the fate and concentrations of chemical CECs is well established and suitable approaches and technologies have been developed for mitigating their impact. However, very little is known about ARGs in terms of the safe level, the types of genes to prioritize in the broader perspective, and methods for and approaches to mitigating resistance risk. It is attributed predominantly to the lack of knowledge and understanding of the complexity of and the multiple dimensions to the issue of microbial resistance. Although the existing water reuse regulations and guidelines do not address concerns related to AMR, this trend is poised to change with increasing focus on effective treatments for the safe reuse of the treated wastewater. Currently, most wastewater treatment processes lack the ability to sufficiently destroy nucleic acids and thus reduce the downstream spread of resistance. In fact, some treatment processes, including commonly used processes such as disinfection (e.g., chlorination), have been reported to increase the level of resistance through a selective increase in certain ARBs [15]. For example, chlorine has been known to select for certain ARB and ARGs under conditions typically used in water treatment and elevate the average resistance of ARB [15], which may lead to the selection of organisms resistant to higher concentrations of antibiotics.

Therefore, the optimization and incorporation of alternative, and advanced, wastewater processes are important steps in the mitigation of the antibiotic resistance spread. This is even more important considering that the treatments designed for the removal of conventional contaminants could potentially be contributing to an increased risk of ARB and ARG selection and spread. Consequently, inefficient removal of ARB and ARGs could further compound the problem of antibiotic resistance due to a selective increase in more resistant bacteria. It therefore depends on the type of treatment as well as the type of drug family present in the wastewater [16]. For example, the presence of ampicillin, sulfamethoxazole, ciprofloxacin, and tetracycline-resistant *E. coli* during different treatment stages of a WWTP (raw sewage, post-secondary, post-UV, and post-chlorination) was investigated [16]. Resistance to three or more antibiotics (multidrug resistance) was reported in 21% of the *E. coli* isolates. The minimum inhibitory concentration (MIC) values suggested that the resistance to ampicillin was most common among the multidrug-resistant *E. coli* with four of the isolates showing an ampicillin MIC > 256 μg/mL. Three of these high MIC values were after UV and chlorination treatment of the finished water. Two *E. coli* isolates were also found to be resistant to tetracycline with a MIC > 256 μg/mL [16]. Ampicillin, ciprofloxacin, and trimethoprim–sulfamethoxazole multidrug resistance was observed in 15 *E. coli* isolates (75%), where one isolate was resistant to all four antibiotics. Overall, the authors concluded that tetracycline resistance was the least among the isolates whereas ampicillin-resistant *E. coli* was the most predominant (63% of the total *E. coli* population) after UV and chlorine disinfection. It was further noted that the ABR *E. coli* concentrations in the effluent were higher after the rainfall event.

These findings are supported by another investigation [17] that performed a metagenomic study on sewage-derived microorganisms from WWTP influent, showing that the diversity in the microbial community increased after the rainfall. It was attributed to low disinfection efficiency due to increased flow and a reduced retention time. It shows that the role of environmental conditions is important in addition to the type of treatment applied. Not much work, however, has been done to determine the changes in the concentration of ARB and ARGs due to rainfall, requiring more investigations to understand the impact of rainfall events. Nonetheless, the effect of different treatments on the fate of ARGs has been increasingly evaluated and is the focus of the following section.

2.2. Fate of ARGs in Wastewater Treatment

A number of investigations have been performed identifying the fate of ARGs in conventional wastewater treatment scenarios. For example, quantitative and qualitative changes in the level of various ARGs (*tet*X, *tet*M, *tet*A, *sul*1, *sul*2, *erm*B, *qnr*D, and bla_{TEM}) were evaluated in two municipal WWTPs receiving influent mixed with pretreated livestock water (WWTP1) or industrial wastewater (WWTP2) [18]. The level of ARGs shifted markedly during different treatment stages with significant differences between the two WWTPs as well as ARGs. For example, the total number of ARGs in final effluent showed an increase of 10% for WWTP1 whereas a decrease of 75% was observed for WWTP2. The differences at the individual gene level were also reported to be significant. For example, *sul*, *qnr*D, and bla_{TEM} increased markedly during the treatment processes only in

WWTP1 (receiving pretreated livestock water), whereas no such trend was observed for WWTP2. Most of the ARGs showed an increase ranging between 113% and 564%, except for *tet* and *erm* for WWTP1. This trend was quite different to WWTP2, for which most of the ARGs showed a decrease of 22–92% except for *tet*, which increased by 29%. These differences could be attributed to the different qualities of the influent wastewaters for both the WWTPs as well as the level of antibiotics. Importantly, the greatest shift in the ARG abundance was noted during coagulation, secondary sedimentation, and biological treatments, emphasizing the need to closely monitor these processes for changes in the composition of ARGs.

In another study, total bacterial abundance (estimated from copy numbers of the bacterial 16S rRNA gene) at five different municipal WWTPs did not reduce after wastewater treatment [19]. While the relative abundance of ARGs was generally similar before and after treatment, the $bla_{CTX\text{-}M}$, bla_{TEM}, and qnr_S genes were higher in the effluent of one of the WWTPs. Notably, this particular WWTP received untreated hospital wastewater in addition to domestic and industrial wastewater. Overall, their results agreed with some previous findings that the ARGs and the *intI*1 gene are inefficiently removed during conventional wastewater treatment [20,21].

Some studies have looked at the interaction of various classes of ARGs from a co-selection perspective. Correlation between six ARGs (*tet*A, *sul1*I, bla_{TEM}, bla_{CTXM}, *erm*B, and *qnr*S), two heavy metal resistance genes (HMRGs; *czc*A, *ars*B), and the mobile genetic element class I integron was investigated for three WWTPs during different treatment steps [22]. Class 1 integrons are closely correlated to co-selection mechanisms and are often associated with gene cassettes having both HMRGs and ARGs [10,22]. It has been shown that the bacterial strains having class I integrons have a selective advantage compared with the rest of the bacterial community [23]. An increase in the level of class I integron genes in the effluent therefore indicates the fitness gain of bacteria carrying this mobile genetic element. The authors noted two well-defined groups, which included (1) *tet*A, *erm*B, and *qnr*S and (2) *sul1*I, *czc*A, *ars*B, and *int*1 [22]. Overall, a strong correlation between *sul1*I, HMRGs, and *int*1 was noted. Furthermore, the authors concluded that *czc*A and *sul1*I could be used as model genes for investigating co-selection in WWTPs. Both *tet*A and *qnr*S are harbored by Gram-negative bacteria, whereas *erm*B is associated with Gram-positive bacteria. Therefore, it was unlikely that the co-presence and similar distribution of these genes indicate co-occurrence in the same cells. It is plausible to hypothesize that these clustered genes were present within the microbial community of WWTPs comprised of both Gram-negative and Gram-positive bacteria. Another study found *qnr*S-like genes in Gram-positive bacteria, i.e., *Enterococcus faecalis* and *Enterococcus faecium*, which could lead to the potential co-presence of *qnr*S-like genes and *erm*B in the same cells [24]. For the second cluster, a very strong correlation (Pearson's; $p < 0.0001$) between *czc*A and *sul1*I and *ars*B and *int*1 was found, whereas the correlation between *sul1*I and *ars*B and *czc* and *int*I was weak yet significant ($p < 0.01$).

These findings indicate the significance of determining the mechanisms of co-selection to better understand the correlations between different genes and mobile genetic elements during different treatment stages in WWTPs. Future investigations should however be carried out since the above-described study is one of the very few to have reported this correlation. However, it can be hypothesized that the chemical stress associated with heavy metals, antibiotics, or both could lead to enhanced resistance of the bacterial community during wastewater treatment. It is worth noting that the heavy metals are not biodegradable and their concentration in wastewater is generally 2–4 fold higher than that of antibiotics [10], which could result in persistent selection for heavy metal resistance [25]. A greater understanding and knowledge of stress factors are crucial for determining potential model genes for determining co-selection scenarios.

The presence of organic matter is also important from the perspective of its interaction with ARGs. Recently, the role of organic carbon in shifting the relative abundance of ARGs was investigated in a sand filter biofilm [26]. A decrease in the concentration of total organic carbon (TOC) was found to increase the diversity and relative abundance of ARGs, suggesting that lower levels of TOC were more favorable for enhanced antibiotic resistance. Although this study was carried out during sand filtration of drinking water, the changes in richness, absolute abundance, and relative abundance of ARGs associated with changing concentrations in TOC could be similar during other wastewater treatment processes. In fact, a higher concentration of organic matter in wastewater compared with drinking water could potentially imply a greater richness of the bacterial community, which might lead to a greater increase in the diversity and relative abundance of ARGs upon the gradual decrease in organic carbon during different wastewater treatment processes. The authors also analyzed the changes in the antibiotic resistome with the depth of the sand filter and found that the relative abundance of ARGs increased with the depth and richness of ARGs, correlating positively with the respective TOC levels [26]. In addition to a reduced concentration of organic carbon, the oligotrophic environment was also found to be favorable for the growth and survival of ARBs.

There is a whole range of factors that could change the antibiotic resistome during various stages in a WWTP. The type of wastewater, concentration of organic carbon, and treatment train applied affect the level of ARB and ARGs and resultantly correlations between different genes and mobile genetic elements at a WWTP. It is therefore important to assess the role of each factor in contributing to the final antibiotic resistance scenario considering the impacts related to the discharge or reuse of wastewater.

3. UV Radiation for Controlling Antibiotic Resistance

3.1. UV Photolysis for DNA Damage

UV photolysis is among the most investigated processes for ARG inactivation at lab-scale, which is largely due to its effective application for conventional disinfection. Several studies have recently investigated inactivation of various classes of ARGs (Table 1). As shown in Table 1, the extent of damage to ARGs varies significantly between different studies. The range of UV doses applied also shows significant variations as discussed later in this section.

The impact of UV irradiation has been investigated for a range of ARGs. For example, changes in the *tet* and *sul* ARGs located on chromosomes were investigated after exposing ARBs to two different UV fluences (5 and 10 mJ/cm^2) [27]. The impact of UV irradiation was studied for both chromosomal and plasmid DNA in terms of average harboring frequency. The trends were found to be different for both types of DNA. The average harboring frequency of ARGs located on chromosomal DNA was 2–3% for *tet* whereas for *sul* genes it was 14% ($p < 0.05$). These changes were primarily attributed to the changes in the microbial community post-UV irradiation. In addition to changes in the bacterial community, it was postulated that there might be interactions between ARGs and genes related to UV defense (i.e., co-selection), resulting in an increase in chromosomal ARGs. A similar finding was reported in an earlier investigation in which genes related to oxidative stress and protective mechanisms, including cellular repair, were found to be upregulated in multiantibiotic-resistant *E. coli* after solar irradiation [28]. Mechanisms of UV disinfection affecting ARG conjugation and transfer are shown in Figure 2. It shows that UV alone has little effect on the cell membrane but results in damage to the plasmid containing ARGs, resulting in the death of the donor or the recipient [29].

The authors noted that the trends for plasmid DNA were different to that for chromosomal DNA. For example, six ARGs (*tet*A, *tet*C, *tet*M, *tet*W, *tet*X, and *sul*1) showed a decrease upon UV irradiation with an average reduction of 15% and 6% for the *tet* and *sul* ARGs, respectively [27]. A 30% reduction in the concentration of bacteria harboring three to five *tet* genes was reported whereas the ratio of bacteria simultaneously carrying both *sul*1 and *sul*2 genes also reduced, although no data were provided on the level of reduction. The authors hypothesized that the reduction in the plasmid ARGs was predominantly caused by the loss of plasmids. Moreover, chromosomal DNA was concluded to be more stable to UV exposure and thus require a greater UV fluence compared with the plasmid DNA. Furthermore, the authors calculated the MAR index according to the following formula [27].

$$\text{MAR index} = a/(b \times c)$$

where a is the aggregate resistance score of all isolates from one sample (if one isolate was observed to resist one antibiotic, it will obtain one point; if not, it will obtain zero points), b is the number of antibiotics used, and c is the number of isolates.

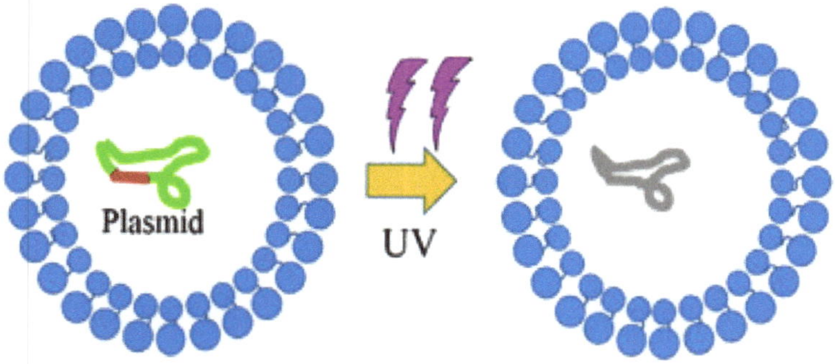

Figure 2. Mechanisms of UV disinfection affecting ARG conjugation and transfer (adapted from Guo et al. [29]).

The MAR index was determined for phenotypes of cultivable isolates that were resistant to ten antibiotics (tetracycline, sulfadiazine, cephalexin, penicillin, erythromycin, vancomycin, gentamicin, chloramphenicol, ofloxacin, and ciprofloxacin [27]). The authors found that the MAR index increased markedly at a lower UV fluence (5 mJ/cm^2), whereas increasing the UV fluence to 10 mJ/cm^2 did not lead to any changes in the MAR index. It was related to the significant changes in the bacterial community at a higher UV fluence with genera having a lower MAR index dominating at the higher UV fluence.

Acknowledging the significance of UV fluence for different ARGs, the effectiveness of UV irradiation in inactivating ARGs (bla_{TEM1}, *tet*A, *sul*, and *mph*A) was investigated at various UV fluences by Destiani and Templeton [30]. The lowest level of inactivation was noted for the *mph*(A) gene with 0.05-log inactivation at 20 mJ/cm^2 increasing to 0.42-log at the highest UV fluence (200 mJ/cm^2) used in their study. The *tet*(A) gene showed a slightly greater level of inactivation than that for the *mph*(A) gene with 0.05-, 0.36-, 0.38-, and 0.74-log inactivation at UV fluences of 20, 50, 100, and 200 mJ/cm^2, respectively. The $bla_{\text{-}TEM1}$ gene was the least resistant to UV irradiation with 1.18-log inactivation at a UV fluence of 200 mJ/cm^2. Hence, the order of susceptibility of ARGs was $bla_{TEM1} \geq$ *tet*A $\geq$ *sul*1 $\geq$ *mph*A, essentially following the order of potential dimers, specifically TT dimers. The number of thymine dimer sites has previously been correlated with UV susceptibility, i.e., the greater the number of sites the greater the UV damage [30]. For example, *amp*C, which has a lower number of thymine dimer sites, was reported to be much more resistant to UV damage compared with *mec*A with a higher number of thymine sites. Furthermore, these findings agree with another investigation in which a comparable level of inactivation of both ampicillin and kanamycin ARGs was correlated to the comparable number of thymine sites [31]. Additionally, it must be noted that the actual damage to ARGs could be higher depending on the amplicon length (which the authors did not report) used in the qPCR assay. It has recently been shown that qPCR could underestimate the damage to ARGs [32], particularly for short amplicons.

The impact of UV irradiation on ampicillin-resistant *E. coli* (CGMCC 1.1595) as well as the plasmid-encoding ampicillin resistance gene $bla_{TEM\text{-}1}$ was assessed after exposure to different UV fluences [33]. A range of UV fluences was tested and the damage to ARG was found to be 0.5-log at a UV fluence of 20 mJ/cm^2, whereas the inactivation ratio of viable *E. coli* was >2.0-log at a similar UV fluence. The ARG damage started to become more prominent with increasing UV fluence such that it increased to 1.2-log at 80 mJ/cm^2. The inactivation ratio of viable *E. coli* was much higher (6-log) at half the UV fluence (40 mJ/cm^2), demonstrating that the extent of *E. coli* inactivation was significantly higher than the ARG damage.

A relatively higher UV fluence value of 600 mJ/cm^2 was used in another study [34]. The authors used drinking water and various ARGs in *E. faecium* and *E. coli*. Despite a 3-fold increase in the UV fluence compared with Destiani and Templeton [30], the authors reported a lower reduction (0.33-log) in the gene copy number of various ARGs (*erm*B, *van*A) using *E. faecium*. When using *E. coli*, they reported no reduction in the tetracycline resistance gene (*tet*A), whereas the reduction in the β-lactam resistance gene (*amp*C) was 1-log. Furthermore, the authors looked at the removal of selected ARGs at two different amplicon lengths, i.e., *tet*A (160 and 1054 bp) and *van*A (377 and 1030 bp). In agreement with others [32,35], a greater log reduction (up to 1-log) in *tet*A and *van*A was reported when using long-amplicon qPCR compared with lower amplicon lengths, for which the reduction was negligible.

A much higher UV fluence of 12,477 mJ/cm^2 was reported to obtain ~2.5-log damage of four ARGs (*sul*1, *tet*G, *int*I1, and 16S rDNA) in wastewater containing COD of 13–29 mg/L [36]. The authors reported that the damage to 16S rDNA and *intl1* flattened with increasing UV fluence from 1248 to 3743 mJ/cm^2, whereas other ARGs showed greater damage with increasing UV fluence. No explanation was provided as to why increasing the UV fluence did not result in increased damage to 16S rDNA and *intl1*. The authors also noted an increase in the relative abundance (the gene copy numbers of ARGs and *int*I1 normalized to that of 16S rDNA) under a UV fluence of up to 1248 mJ/cm^2. It could be attributed to the corresponding changes in the 16S rDNA upon UV exposure. Contrary to the damage to ARGs reported by Zhuang et al. [34], no reduction in *tetG* and *tetQ* was reported even after subjecting the wastewater to a UV fluence of 30,100 mJ/cm^2 [37].

Most of the studies thus far have been conducted at lab-scale, with a very few at full-scale [18,36]. One of the full-scale studies investigated different UV irradiation times from 4 to 18 s (UV fluence not provided) [36]. As expected, increased damage to DNA and to 16S rRNA and ARGs was reported. However, the overall damage as determined by qPCR was <1-log [33]. These results corroborate previous studies including full-scale applications regarding the limited efficacy of UV photolysis due to quenching by organic materials in wastewater [17]. At a UV fluence of 27 mJ/cm^2, no significant change in the concentration of ARGs (*tet*X, *tet*M, *tet*A, *sul*1, *sul*2, *erm*B, *qnr*D, and bla_{TEM}) was found as determined by their relative abundance (ARG copies/16S rRNA gene copies) [18]. Although both these investigations reported a negligible effect on the level of ARGs upon UV irradiation, it is difficult to directly compare the effectiveness of the UV process since it is not possible to compare the UV fluence values between these studies. It highlights the need to report the UV fluence in a standardized unit (mJ/cm^2) to enable comparative assessment of the process performance. Although UV fluence data were not provided by Rodríguez-Chueca et al. [38], the time of UV irradiation is quite short and hence the results are not surprising. Similarly, the UV fluence used by Lee et al. [18] is also not very high (27 mJ/cm^2). In another investigation, UV disinfection post-activated sludge treatment at a WWTP in Tunis was reported to be ineffective in reducing the abundance of ARGs [19]. This was not unexpected, since the current UV fluences applied in practice rarely exceed 40 mJ/cm^2, which is not high enough to damage ARGs.

Although a direct comparison with regard to the ARG damage cf. UV fluence is not possible, it is plausible to argue that the differences in the level of ARG inactivation in different studies could be due to the type of strain (single vs. mixed) and the differences in the characteristics of wastewater. For example, McKinney and Pruden [35] showed an insignificant difference in the UV inactivation of ARGs in two different matrices, i.e., phosphate buffer and wastewater effluent containing TOC of 4.61 mg/L. However, the UV fluence needed to achieve a 3–4-log inactivation of ARGs (*mec*A, *van*A, *tet*A, *amp*C) was much higher (200–400 mJ/cm^2) compared with the <150 mJ/cm^2 reported by Yoon et al. [31]. Furthermore, it must be noted that the impact of the matrix could not be truly determined from the findings of McKinney and Pruden [35] since the samples were pre-filtered for turbidity removal. These findings demonstrate that while UV radiation can potentially be useful for damaging ARGs in real-water matrices containing organics and inorganics, the energy and cost associated with these processes are major factors that need to be considered. None of the studies have so far looked at the electrical energy needed for various inactivation/reduction levels of ARGs.

3.2. UV Radiation and Chlorine for DNA Damage

The use of chlorine both individually and in combination with UV radiation has been investigated for inactivation of ARGs. Chlorine alone has been reported to be ineffective in damaging most ARGs, except for some, as discussed later in this section. UV radiation can be combined with chlorine to achieve greater damage to microorganisms. UV photolysis of chlorine produces a wide range of highly reactive species, such as $HO^•$ and $Cl^•$ (a redox potential of 2.4 V) [39,40]. However, it is known that $HO^•$ is >5-fold the concentration of $Cl^•$ and thus contributes more to the disinfection efficiency [41]. The photodecay rates during the UV/chlorine AOP are related to the wavelength-dependent molar absorption coefficient [42]. The photodecay rate of chlorine has also been shown to increase with increasing pH at any wavelength. Therefore, the pH has to be controlled when using HOCl in the UV/chlorine AOP because it significantly affects the molar absorption coefficient [43]. Mechanisms of chlorination disinfection affecting ARG conjugation and transfer are shown in Figure 3. Contrary to UV radiation alone, which results in damage to the plasmid containing ARGs, HClO reacts first with NH^{4+}, leading to the generation of chloramine (mainly NH_2Cl) that results in cell permeability and ARG transfer. Consequently, chlorination could amplify the risk of ARG transfer, particularly in wastewater with a high concentration of ammonia nitrogen [29].

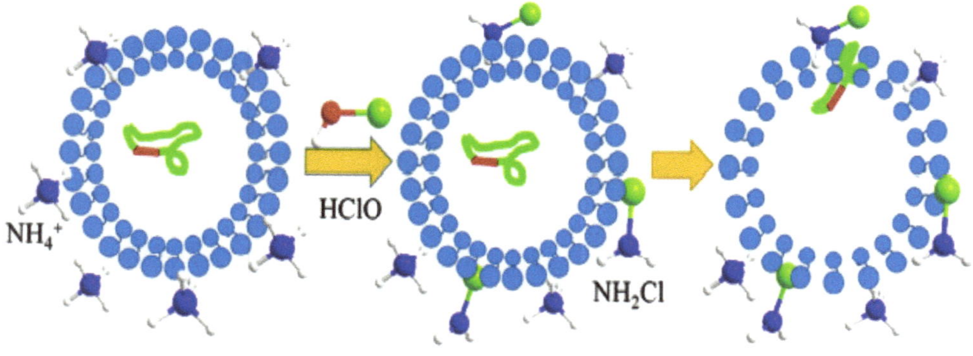

Figure 3. Mechanisms of chlorination disinfection affecting ARG conjugation and transfer (adapted from Guo et al. [29]).

A sequential UV/chlorine process was investigated for *sul*1, *tet*X, *tet*G, *intI*1, and 16S rRNA genes in municipal wastewater effluent [36]. Compared with UV radiation alone, the sequential UV/chlorine process was found to achieve synergy that ranged between 0.006 and 0.031-log removals for the investigated genes. The greatest energy was achieved for 16S rRNA, with *tet*X showing the least. The amplicon length used in the study by Zhuang et al. [36] was short (163–280 bp), which makes it hard to evaluate the overall gene damage. Furthermore, the concentration of chlorine used was 25 mg/L, which is much higher than the concentration used in practice, which rarely exceeds 2 mg/L [44]. A summary of the literature studies focusing on chlorine alone and the UV/chlorine process is given in Table 2.

A range of ARGs have been investigated under different UV radiation and chlorine dose conditions with a fairly high log removal value (LRV) as shown in Table 2.

Most studies have looked at the combined UV and chlorine process with some evaluating the efficiency of chlorine alone for comparison with the combined process. A recent study investigated the reduction of *sul*1 and *intI*1 within Pseudomonas HLS-6, a multiple-antibiotic-resistant bacterium [45]. While the UV/chlorine process showed a greater amount of gene damage in the first 20 min of treatment when compared with chlorination alone, the final gene damage efficiency of *sul*1 and *intI*1 was comparable for both processes. Moreover, both chlorination alone and UV/chlorine treatment gave higher removal efficiencies for both genes (*sul*1, >3.50 log; *intI*1, >4.00 log). The efficiency of damage to the genes was also analyzed by gel electrophoresis, which confirmed the benefit of UV/chlorine treatment on DNA damage. For example, the band intensity of *sul*1 for the UV/chlorine sample treated for 60 min was much lower than that of the samples treated by UV or chlorine alone under comparable conditions. It was further shown that $\sim 10^2$ copies/mL of *intI*1 remained after UV/chlorine treatment for the shorter amplicon (146 bp) but, for the larger amplicon (484 bp), the gene was found to be completely damaged since no band appeared after 30 min of disinfection [45]. It is therefore plausible to conclude that the disinfection process would be much more effective if the complete sequence size were to be examined for damage detection. Under the condition of a low chlorine dosage, *sul*1 was easier to damage than *intI*1 by the UV/chlorine process. The log damage to both *sul*1 and *intI*1 decreased with increasing pH during the UV/chlorine treatment.

Furthermore, the authors investigated the damage to *sul*1 and *intI*1 under different chlorine concentrations (0–40 mg/L) [45]. A clear increase in the damage to *sul*1 was observed when the chlorine concentration increased to 5 mg/L, but no improvement was seen with a further increase in the dose of chlorine. For *intI*1, however, an increase in the chlorine dose resulted in increasing damage up to 20 mg/L with no further damage observed when the dose was doubled (i.e., 40 mg/L). These results demonstrate that *sul*1 has lower chlorine dose requirements compared with *intI*1, which could be associated with the larger size of *sul*1. Notably, the final log inactivation for *sul*1 and *intI*1 was fairly similar (~4 log) when 5 and 20 mg/L chlorine were used in the UV/chlorine process. It is worth emphasizing that optimizing the chlorine dose is not only important with respect to minimizing the use of chemicals to avoid the formation of disinfection byproducts (DBPs) but also to minimize the self-scavenging of radicals ($HO^\bullet$, $Cl^\bullet$) under high chlorine dose conditions [46]. Another investigation looked at the damage to the plasmid-encoding ampicillin resistance gene $bla_{TEM\text{-}1}$ after chlorination alone [33]. The authors reported no damage to $bla_{TEM\text{-}1}$ at a chlorine dose of up to 10 mg Cl_2/L. However, they did not look at the higher chlorine doses that many other authors have investigated to investigate the impact of higher chlorine concentrations on ARG damage.

The impact of pH on the UV/chlorine process was also investigated by Zhang et al. [45]. The damage to both genes decreased with an increase in pH from 5 to 9. Since the quantum yields of HOCl and OCl^- were same the during UV/chlorine process, the pH variation was not deemed to affect the formation of $HO^\bullet$ and $Cl^\bullet$. However, since the consumption of $HO^\bullet$ and $Cl^\bullet$ by OCl^- was several folds faster compared with HOCl, the damage to both genes reduced at a high pH [45]. Moreover, the concentration of HOCl, which has a higher oxidizing capability compared with OCl^-, was greater at a low pH, leading to higher gene damage. The authors further investigated the role of free radicals ($HO^\bullet$) and reactive free chlorine species ($Cl^\bullet$, $ClO^\bullet$, and $Cl_2^{\bullet-}$) generated during the UV/chlorine treatment [45]. Nitrobenzene (NB) was used a scavenger of $HO^\bullet$ ($k_{HO^\bullet\text{-}NB} = 3.9 \times 10^9$ M^{-1} s^{-1}). In agreement with an earlier study [43], the authors found no influence of $HO^\bullet$ on *sul*1 and *intI*1 damage since the level of inactivation was similar before and after the addition of NB in the UV/chlorine process. The reduction in both genes, however, was greater for the UV/chlorine process than for the UV + chlorine process ($k_{UV/Cl2} \approx k_{UV/Cl2+NB} > k_{UV+Cl2}$), indicating the role of RCS radials in damaging genes. However, these results differ from the findings of Rattanakul and Oguma [47], who investigated the damage to the viral genome using the UV/chlorine process and found that $HO^{\bullet}$ did result in damage to the genome. The difference in the findings could be attributed to the level of $HO^{\bullet}$ generated during the process and other operational conditions.

The use of different target genes (intracellular bacterial genes vs. the viral genome) in the above-mentioned studies could also be contributing to the different findings. For example, bacteria could consume $HO^{\bullet}$ quicker than bacteriophages due to their complex cellular matrix, minimizing their impact on damaging the genes within bacteria [45]. It can therefore be concluded that the $HO^{\bullet}$ does damage the genes but its impact is dependent on the process's efficiency and the experimental conditions. In fact, in a recent investigation, it was found that $HO^{\bullet}$ exhibited a very high nonselective reactivity (k $\sim 10^9$–10^{10} $M^{-1}s^{-1}$) towards all nucleobases [48]. The damage to DNA by $HO^{\bullet}$ predominantly occurs by strand fragmentation (via phosphate backbone cleavage) that leads to a reduction in the length of DNA [49]. This ultimately results in a weakened attachment and/or donor–acceptor complexation if exposure to $HO^{\bullet}$ is extended.

Destiani and Templeton [30] found a synergistic effect for inactivating ARGs using sequential UV and chlorine compared with individual treatments. As expected, and in agreement with others [32], the log inactivation of ARGs increased with increasing UV fluence. Sequential UV and chlorine resulted in a synergistic inactivation of the target ARGs [30]. Synergy in the inactivation of the *sul*1 gene was noted for all the ARGs, although some differences were observed. For example, synergy in the inactivation of the $bla_{TEM\text{-}1}$ gene was observed at UV fluences of 50 and 200 mJ/cm^2, whereas the synergy for *tet*A was only observed at a UV fluence of 100 mJ/cm^2 and at UV fluences of 100 and 200 mJ/cm^2 for *mph*A. The extent of synergy was also dependent on the chlorine dose, with the maximum synergy of 0.25-log occurring for *mph*A at a chlorine dose of 1 mg/L. Increasing the chlorine dose to 2 mg/L (a UV fluence of 200 mJ/cm^2) resulted in an increased synergy with a 0.6-log greater inactivation of *mph*A compared with the sum of the individual treatments with a final inactivation of 2.8-log. Since free chlorine predominantly reacts with amino acids and membrane-bound proteins and UV radiation with nucleic acids, the synergy in the ARG inactivation during the sequential UV and chlorination process could be attributed to the decrease in bioactivity due to UV irradiation resulting in enhanced reaction of chlorine with cells [11]. The authors also investigated the inactivation of ARBs and found that the required UV fluence was much higher for ARGs than for ARB. Using a chlorine concentration of 30 mg/L, the inactivation of *tet*A, $bla_{TEM\text{-}1}$, *sul*1, and *mph*A was 1.7-log, which was higher than using UV irradiation alone at a fluence of 200 mJ/cm^2 (1.2-log).

It could be concluded that a pre-disinfection step (e.g., ozone or chlorine dioxide) leading to substantial cell envelope damage followed by a downstream post-disinfection step using an oxidant for DNA damage might lead to a synergistic effect [50]. Hence, it is plausible to hypothesize that the synergy could be attributed to multiple mechanisms related to two different disinfections. For example, dual damage mechanisms in which chlorine inflicts damage to cell walls and UV irradiation to purines, pyrimidines, and nucleic acid could result in synergistic ARG inactivation [11].

Table 1. Inactivation of ARGs by UV irradiation.

Mode	Peak λ	Volume (mL)/Flow Rate	Target	Log Inactivation	UV Dose (mJ/cm^2)	k (cm^2/mJ)	Matrix	Reference
Full-scale	NG	130 mgd	*tet*R	0	30,100	NG	WW effluent	[35]
Bench-scale collimated beam	254	10	*mec*A, *van*A, *tet*A, and *amp*C	3–4	200–400	0.4–0.25, 0.015–0.01	PB and WW effluent	[32]
Batch	254	1500	*sul*1, *tet*G, and *intl*1	2.5–2.7	12,477	0.0002	WW effluent	[34]
Batch	254	1800	*sul*1, *tet*X, *tet*G, *intI*1, and 16S rRNA	* <1	62.4, 124.8, 249.5	0.016, 0.008, 0.004	WW effluent	[41]
Full-scale	254	-	*tet*X, *tet*M, *tet*A, *sul*1, *sul*2, *erm*B, *qnr*D, and *bla*TEM	0	27	-	WW effluent	[17]
Lab-scale	254	15	*bla*TEM-1	1.2	80	0.015	PBS	[32]
Collimated beam	254	NG	$bla_{\text{TEM-1}}$, *tet*A	** 1	20–25	0.05–0.04	Plasmid suspension in DNase-free water	[13]
Bench-scale quasi-collimated beam	254	120	amp^R, Kan^R	4	60–140	0.11–0.07,0.15–0.09	PB	[29]
Collimated beam system	254	10	*tet*A, *tet*B, *str*B, *sul*2, and *aac*C2	1.6	320	0.005	Hospital WW	[49]
Bench-scalecollimated beam	254	100	*tet*(A), $bla\text{-}_{\text{TEM1}}$, *sul*1, and *mph*(A)	0.42–1.18	200	0.0021–0.0059	PB	[28]
Batch	254		*tet*A, *van*A, and *erm*B	1	600	0.001	DW	[31]

NG, not given; mgd, million gallons per day; * value considered 1; ** 1-log reduction per UV fluence of 20–25 mJ/cm^2; WW, wastewater; PB, phosphate buffer; DW, drinking water.

Table 2. Inactivation of ARGs by chlorine and UV/chlorine.

Target	Wavelength (nm)	UV Fluence (mJ/cm2)	LRV	Output Power (W)	Cl_2 Dose (mg/L)	Volume (mL)	Reference
*sul*1, *tet*X, *tet*G, *intI*1, and *16S rRNA*	254	62.4, 124.8, 249.5	2	16	30	1800	[41]
*sul*1, *intI*1	254	120	~3.5–4	NG	20	50	[43]
tet(A), *bla*-$_{TEM1}$, *sul*1, and *mph*(A)	254	200	2.2–2.8	NG	30	100	[28]
*tet*A, *tet*B, *str*B, *sul*2, *aac*C2	254	320	2.7–3.1	NG	2	10	[49]
sulI, and *intI1*	-	-	* 1.5–2.4	-	20	50	[44]
bla$_{TEM-1}$	-	-	* 0	-	10	15	[32]
*tet*A, *bla*-$_{TEM1}$, *sul*1, and *mph*A	-	-	* 3.4–3.6	-	5	200	[29]

* Chlorine alone without UV irradiation.

Another recent investigation looked at UV/chlorine synergy for bacterial inactivation and ARG damage [51]. Using chlorine alone (2 mg/L), the copy number of 16S rRNA genes in *Morganella morganii* and *Enterococcus faecalis* decreased by 4.19-log and 3.99-log, respectively. When UV radiation (320 mJ/cm^2) was combined with chlorine (2 mg/L), the reduction in the copy number of ARGs increased up to 1.5-log. The authors further noted that the inactivation of the *tet*B gene in *Enterococcus faecalis* was greater than that in *Morganella morganii* under similar conditions ($p < 0.05$). In agreement with Destiani and Templeton [30], ARG inactivation needed higher UV fluences compared with ARB. Overall, it was concluded that the inactivation of ARGs was more pronounced at a higher chlorine concentration and was of greater significance than a higher UV fluence. The effect was further enhanced when UV and chlorine were combined. For example, at a UV fluence of 40 mJ/cm^2, the reduction in ARGs ranged between 0.15 and 0.38-log, whereas adding 2 mg/L chlorine increased it to 0.41–0.94-log.

Combining UV radiation with chlorine is an emerging AOP that potentially could be retrofitted in most water and wastewater facilities. Compared with stand-alone UV and UV/H_2O_2 treatments, UV/chlorine treatment could lead to minimizing the selection of resistant genes by reducing chlorine requirements, which would also be beneficial in lowering the formation of DBPs. Simultaneously, the UV fluence could also be reduced when chlorine is used in the UV/chlorine AOP, which would reduce the energy requirements and hence improve the economic efficiency of the combined treatment. However, research in this area is quite limited and further studies looking at the optimization of the process are needed. Further work to determine the role of different conditions (such as the radical generating agent, its concentration in a UV/chlorine AOP, the impact of gene-carrying organisms (bacteriophages or bacteria), and the nature of genes (intracellular or extracellular)) is important in determining the impact of oxidizing species, particularly $HO^\bullet$. Additionally, more research optimizing the chlorine dose in relation to UV fluence would be useful to understand and enhance the synergy for ARG inactivation. Nonetheless, the UV/chlorine AOP could be a promising practical alternative to not only improve the inactivation of ARGs and ARB, but to also simultaneously reduce the possibility of bacterial regrowth in water distribution systems as well as minimize microbial selection and the formation of DBPs [2].

3.3. *UV/H_2O_2 AOP for Damaging DNA*

The UV/H_2O_2 process relies on photolysis of H_2O_2 to generate $HO^\bullet$ and is among the most widely investigated and applied AOPs in water and wastewater treatment at laboratory and/or small scale with huge potential for full-scale applications. The process has been increasingly used for damaging ARGs in recent years with a few full-scale investigations (Table 3). A recent full-scale study employed the UV/H_2O_2 process for damaging different ARGs in tertiary treated wastewater, i.e., wastewater post-coagulation/flocculation/decantation followed by filtration by a rotofilter [36]. The UV/H_2O_2 process was the most efficient in damaging 16S rRNA and other investigated ARGs when compared with other AOPs including UV/peroxymonosulfate (PMS) with or without Fe (II) [36]. Although the generation of radicals was not investigated by the authors, it was hypothesized to be higher during UV/PMS-based processes compared with the UV/H_2O_2 process. It must, however, be noted that the species of radicals produced are different during these processes, with $HO^{\bullet}$ being non-selective in nature compared with the selective $SO_4^{\bullet-}$ generated during UV irradiation of PMS. It is also worth noting that the redox potential of PMS (+1.82 V) is higher than that of H_2O_2 (+1.76 V) [52]. Some lab-scale investigations have demonstrated UV/PMS to be superior to the UV/H_2O_2 process [53,54]. UV/PMS is outside the scope of this review but it is apparent that the efficiency of UV/PMS is more dependent on the type of water matrix when compared with UV/H_2O_2 [55]. Further work is therefore needed to fully understand the UV/PMS process and its robustness with respect to damaging ARGs, particularly in representative water matrices.

Table 3. Inactivation of ARGs by UV/H_2O_2.

Target	Wavelength (nm)	UV Fluence	LRV	Output Power (W)	H_2O_2 Dose (mg/L)	COD (mg/L)	Volume (mL)	Reference
16S rRNA, *sul*1, *sul*2, *qnr*S, bla_{TEM}, bla_{OXA-A}, and *intI*1	254	40–170 J/L	<1	330 W	17	27 ± 3	140,000	[33]
*sul*1, *tet*X, *tet*G, *intI*1, and 16s rRNA	254	NG	2.8–3.5	16	340	13–39	1800 mL	[55]
bla_{TEM}	320–450	25 mJ/cm^2	0	250	20	NG	500 mL	[53,54]
*amp*R, *kan*R	254	44–140 mJ/cm^2	4	NG	10	(DOC, 5.2)	120 mL	[30]

The impact of $HO^{\bullet}$ on the inactivation of ARGs is one of the most focused-on aspects of most studies investigating the UV/H_2O_2 process, with varying findings. For example, Yoon et al. [31] reported a negligible contribution of $HO^{\bullet}$ to the inactivation of e-ARGs (*amp* and *kan*) during UV/H_2O_2 treatment of wastewater effluent. It must be noted that the wastewater effluent samples were collected from the conventional activated sludge process and therefore scavenging of radicals was expected. According to the authors, the average UV fluence delivered in the case of wastewater was 1.4-fold lower than the UV-transparent water, such as phosphate buffer solution. The rates of e-ARG damage were fairly similar for the UV-only and UV/H_2O_2 treatments (p = 0.56 and 0.75, respectively) due to the reduced $HO^{\bullet}$ oxidation efficiency in the wastewater effluent matrix due to radical scavenging by organic matter (DOC = 5.2 mg/L). Although the authors did not support their results with an actual measurement of radicals, the fact that the e-ARG damage in the phosphate buffer was greater during the UV/H_2O_2 treatment compared with the UV treatment corroborates their findings of radical scavenging. Therefore, despite the high $HO^{\bullet}$ reactivity to e-ARGs (k = ~10^{10} M^{-1} s^{-1}), the resulting degradation of ARGs could be insignificant during UV/H_2O_2 treatment of complex water matrices [31].

Another study investigated the impact of a 250 W lamp equipped with a UV filter (emission range: 320–450 nm) in UV/H_2O_2 (20 mg/L) for its potential to reduce resistance transfer [56]. The antibiotic-resistant *E. coli* strain was isolated from the effluent of an activated sludge process of a WWTP and cultivated on selective culture medium. Qualitative PCR was performed on total DNA and on DNA extracted from cell cultures to investigate the bla_{TEM}, *qnr*S, and *tet*W genes in *E. coli*-spiked samples. While the bla_{TEM} gene was detected in both samples, *qnr*S and *tet*W were not detected in the PCR assay;

they were either absent or present at very low concentrations. The results showed no change in the DNA extracted from cell cultures after UV/H_2O_2 treatment for up to 90 min. Similar results were reported for total DNA with the bla_{TEM} gene copy number remaining unchanged during treatment for up to 300 min despite total inactivation of *E. coli* after 240 min. Although the treatment time was very high, the UV fluence delivered at this time was quite low (25 mJ/cm^2) compared with other investigations reporting several hundreds of UV fluence needed for a 2–4 log ARG inactivation. Moreover, the wavelength range used by the authors is not commonly used in real applications. Additionally, the efficiency of the UV/H_2O_2 process is highly dependent on the concentration of H_2O_2 and other water quality parameters that need to be optimized for achieving effective inactivation.

The impact of operating parameters such as H_2O_2 dose and pH has been well reported during the UV/H_2O_2 process for conventional applications, i.e., degradation of organics in water and wastewater. Some recent studies have also been carried out to determine their impact on the ARG damage in wastewater. For example, an optimized UV/H_2O_2 process taking pH, H_2O_2 dose, and time of UV irradiation into account was investigated for various ARGs (*sul*1, *tet*X, *tet*G, *intI*1, and 16S rRNA) in secondary effluent using a 254 nm UV lamp [57]. At a much higher concentration of H_2O_2 (340 mg/L) and a pH of 3.5 compared with the study of Ferro et al. [57], a much higher reduction in the investigated genes in the range of 2.64–3.48 after 30 min of irradiation was reported by Zhang et al. [58]. Increasing the pH to 7 resulted in a reduced log reduction of ARGs to 1.55–2.32. Therefore, a pH of 3 and a H_2O_2 concentration of 340 mg/L were considered best for damaging ARGs. Such a lower pH is not practically feasible. Therefore, optimization of process parameters for ARG inactivation needs to be carried out specifically considering practical application of the UV/H_2O_2 process. As the concentration of H_2O_2 is one of the most important factors determining the efficiency of the UV/H_2O_2 process, the impact of H_2O_2 concentration on the inactivation of two ARGs (*mec*A and *amp*C) has also been investigated by others [59]. For a UV fluence of 120 mJ/cm^2, approximately 2.3–2.9- and 1.4–2.7-log inactivation of *amp*C and *mec*A, respectively, was achieved with different concentrations (340, 1700, and 3400 mg/L) of H_2O_2. With the addition of thin TiO_2 film, the inactivation of *mec*A and *amp*C improved to 2.7–3.4- and 2.7–3.2-log, respectively.

It is clear from the studies reported thus far that the UV/H_2O_2 process is an effective measure to damage ARGs. However, the UV fluence and H_2O_2 dose required could be very high for a real-water matrix with organics. The focus therefore needs to be placed on appropriate pre-treatment technologies to reduce the organic load to minimize costs and enhance process efficiency. Considerable research has been carried out to determine the efficacy of different pre-treatments prior to UV AOPs for the removal of contaminants to improve the subsequent process performance, which could make an excellent starting point for investigating the ARG damage in real matrices by combining UV/H_2O_2 with suitable pre-treatments. A biological treatment as a post-UV/H_2O_2 treatment is generally very effective in improving the energy efficiency and overall treatment performance. It is, however, uncertain how the combined UV/H_2O_2 and biological process performs with respect to changes in ARGs, which needs to be investigated in future studies. UV-LEDs as emerging sources of UV irradiation could prove more effective considering the possibility of combining different wavelengths in novel reactor designs. Only one study has thus far looked at the application of UVC-LEDs for damaging ARGs [60].

A shared concern with most disinfection processes as well as UV-based processes is the selection of ARGs that could result in transfer of antibiotic resistance even after treatment. The focus of future studies irrespective of the UV source needs to be placed on determining the optimum conditions to minimize or avoid selection pressure. Within this context, benchmarking UV fluences and the role of $HO^\bullet$ in combination with other operational conditions should be considered in future investigations with a view to reducing the overall ability of ARGs to transfer resistance.

4. Current and Future Perspectives

Considering that antimicrobial resistance is a recent concern when compared with traditional contaminants in water and wastewater, there are several unknowns and uncertainties that need to be understood to enable the control or mitigation of antibiotic resistance. One of the greatest challenges is to identify practical and relevant target types in addition to levels of antibiotic resistance benchmarks specifically for reuse applications [1], which are important to providing or formulating a basis for potential future regulations. Moreover, standardized methods for both the identification and quantification of ARB and ARGs are needed. Since identifying all the ARB and ARGs is not practical, there is a need to identify targets both in terms of the types and final level of inactivation considering the deleterious health impacts—an approach analogous to the one adopted for pathogens as microbiological water quality indicators. Although it is uncertain if such a strategy could be applied to DNA/ARGs, it could be a good starting point in identifying potential approaches to answer some of the key questions related to tackling the challenge of antibiotic resistance.

There is a considerable knowledge gap in relation to the fate of ARB and ARGs during conventional wastewater treatment and the future role of advanced treatment technologies in addressing the important issue of antibiotic resistance. For example, little is known about the impact of potential factors on the selection of and changes in ARB and ARGs during different wastewater treatment stages. Moreover, knowledge on how the overall bacterial community, antibiotics, metals, and other selective agents relate to the abundance or removal of clinically relevant ARB and ARGs under selected treatment schemes is lacking. This information is important to understand the conditions that favor the growth of ARBs and to understand the associated risks. Furthermore, because of DNA repair mechanisms in the cells, it is uncertain how much DNA damage is needed to make ARGs useless to bacteria, i.e., to inactivate ARGs permanently or make them unsuitable for transfer. Therefore, it is important to establish quantitative DNA assays and relate them to culture-based assays after UV irradiation. Advanced treatment technologies, including UV and membrane filtration, have been successfully implemented for wide-ranging applications in the water industry and these have been shown to be effective for controlling antibiotic resistance. In regard to sources of UV irradiation, UV-LEDs as robust and energy-efficient sources of UV irradiation are of particular interest and need to be investigated in the future.

Wastewater disinfection processes operated under typical treatment conditions are markedly effective in minimizing the overall ARB levels; however, they could lead to ARB selection (i.e., increased relative proportions of ARB amongst the surviving bacterial cells) [61,62]. This could lead to an increase in the potential for antibiotic resistance dissemination. Therefore, processes aimed at achieving disinfection need to shift the focus from microbial inactivation to achieving ARG inactivation. It is important to note that the selection pressure exerted on environmental bacteria depends on several factors, including the type of selective agent, their concentrations and chemical speciation, co-exposure to other selective agents, exposure time, and environmental conditions for bacterial growth [63]. It is therefore critical to understand the complexities and uncertainties involved in these factors to completely recognize the role that selective agents play in promoting the antibacterial resistance [64].

Conventional UV lamp systems are a well-established technology for water disinfection, but most studies conducted thus far have investigated "log inactivation of ARGs". A higher or a specific log inactivation could be a useful measure for bacterial inactivation, but it does not necessarily reflect the safety of the treated water in terms of the remaining resistance potential. It is therefore important to consider how the log ARG inactivation correlates with the ARG transformation potential after treatment, which requires simultaneous application of both DNA quantification and plate count methods for determining transformation potential. Considering that AOPs can achieve the dual objective of organics degradation, including several antibiotics, and microbial disinfection, including DNA damage, these processes need to be thoroughly investigated under different conditions. This is important to understand the role of different radicals ($Cl^{\bullet}$, $Cl_2^{\bullet-}$, $HO^{\bullet}$) in simultaneously oxidizing organics and damaging DNA to minimize opportunities for antibiotic resistance spread. The effectiveness of AOPs has been proven for the inactivation of ARB and damage to ARGs at lab-scale. However, their application on real wastewater matrices at full-scale is challenging and requires research in a broad range of areas, including the optimization of the processes in the presence of contaminants, the impact of pre- and post-treatments on the process efficiency, and the overall operational cost and energy requirements under different scenarios (oxidant dose, type and quality of wastewater, UV fluence). UV-LEDs could become practically applicable in the future, opening up more opportunities to combine different wavelengths for the purpose of effective microbial and DNA damage.

The challenges that remain to be addressed are multidisciplinary in nature. From initial clinical control of antibiotic administration to their spread in the aquatic environment, questions related to the types of antibiotics and associated ARGs that are of most concern include: What is their removal and bacterial uptake potential? Which ARGs pose the maximum level of risk? How is the maximum level to be defined? How does abundance relate to risk in a wider environmental perspective? All these aspects are important to be considered for an accurate risk assessment and devising comprehensive and economically viable treatment strategies. Since treated wastewater could be reused multiple times and for diverse applications, such as indirect potable and industrial use, irrigation, and aquifer recharge, assessing the relevant risks is even more complex when compared with drinking water. Therefore, the need to adequately address the relationship between anthropogenic and environmental factors in assessing health risks cannot be overemphasized.

5. Conclusions

Antimicrobial-resistant bacteria and antimicrobial resistance genes present an emerging challenge to treated wastewater reuse applications considering that these contaminants are unregulated. These concerns are further exacerbated by the fact that: (1) wastewater treatment plants are considered a hotspot of microbial resistance; (2) the fate of antimicrobial resistance genes is not understood; and (3) the current treatment approaches are inefficient in inactivating genes. UV-based processes are some of the most investigated advanced treatments that have been tested and validated at large-scale with many full-scale applications globally. UV and chlorine as stand-alone processes could be effective, but modifications (combined use, use of oxidants, pre-treatments) need to be made for their efficient application. Nonetheless, the efficiency of these processes needs to be investigated with an approach different to what has been prevalent in the water industry over the past few decades, i.e., from inactivating indicator microorganisms to DNA damage. Hence, there is a need for a paradigm shift from conventional disinfection, with the primary aim of inactivating pathogens as an indicator of the microbiological safety of water, to effective damage to the DNA and resistance genes that could still be present after microbial inactivation. This approach requires an understanding of the types of genes and their fate during wastewater treatment, potential health and ecological impacts, the transformation potential, and the impact of conventional treatments cf. advanced treatments both on the inactivation and relative abundance of different types of genes. Considering the limited knowledge on the real impact (on human health, on ecosystems) of the presence of environmental ARB and ARGs present in source water, treatment plants, and distribution systems and upon wastewater reuse, future research needs to place emphasis on understanding and quantifying the risks. Furthermore, it is critical to understand how the technological interventions would impact the overall treatment and economic outlook of wastewater reuse applications.

Funding: This work was funded by NIVA's publication funding initiative.

Institutional Review Board Statement: Not applicable.

Informed Consent Statement: Not applicable.

Data Availability Statement: Not applicable.

Acknowledgments: The author would like to thank NIVA for providing funding for this work.

Conflicts of Interest: The author declares no conflict of interest.

References

1. Hong, P.-Y.; Julian, T.R.; Pype, M.-L.; Jiang, S.C.; Nelson, K.L.; Graham, D.; Pruden, A.; Manaia, C.M. Reusing Treated Wastewater: Consideration of the Safety Aspects Associated with Antibiotic-Resistant Bacteria and Antibiotic Resistance Genes. *Water* **2018**, *10*, 244. [CrossRef]
2. Umar, M.; Roddick, F.; Fan, L. Moving from the traditional paradigm of pathogen inactivation to controlling antibiotic resistance in water—Role of ultraviolet irradiation. *Sci. Total Environ.* **2019**, *662*, 923–939. [CrossRef] [PubMed]
3. Rizzo, L.; Manaia, C.; Merlin, C.; Schwartz, T.; Dagot, C.; Ploy, M.C.; Michael, I.; Fatta-Kassinos, D. Urban wastewater treatment plants as hotspots for antibiotic resistant bacteria and genes spread into the environment: A review. *Sci. Total Environ.* **2013**, *447*, 345–360. [CrossRef] [PubMed]
4. Pruden, A.; Larsson, D.G.J.; Amézquita, A.; Collignon, P.; Brandt, K.K.; Graham, D.W.; Lazorchak, J.M.; Suzuki, S.; Silley, P.; Snape, J.R.; et al. Management options for reducing the release of antibiotics and antibiotic resistance genes to the environment. *Environ. Health Perspect.* **2013**, *121*, 878–885. [CrossRef]
5. Moura, A.; Oliveira, C.; Henriques, I.; Smalla, K.; Correia, A. Broad diversity of conjugative plasmids in integron-carrying bacteria from wastewater environments. *FEMS Microbiol. Lett.* **2012**, *330*, 157–164. [CrossRef]
6. Gillings, M.R.; Gaze, W.H.; Pruden, A.; Smalla, K.; Tiedje, J.M.; Zhu, Y.-G. Using the class 1 integron-integrase gene as a proxy for anthropogenic pollution. *ISME J.* **2015**, *9*, 1269–1279. [CrossRef]
7. Di Cesare, A.; Eckert, E.; D'Urso, S.; Bertoni, R.; Gillan, D.C.; Wattiez, R.; Corno, G. Co-occurrence of integrase 1, antibiotic and heavy metal resistance genes in municipal wastewater treatment plants. *Water Res.* **2016**, *94*, 208–214. [CrossRef]
8. Graham, D.W.; Olivares-Rieumont, S.; Knapp, C.W.; Lima, L.; Werner, D.; Bowen, E. Antibiotic resistance gene abundances associated with waste discharges to the Almendares River near Havana, Cuba. *Environ. Sci. Technol.* **2011**, *45*, 418–424. [CrossRef]
9. Knapp, C.W.; Mccluskey, S.; Singh, B.; Campbell, C.; Hudson, G.; Graham, D.W. Antibiotic Resistance Gene Abundances Correlate with Metal and Geochemical Conditions in Archived Scottish Soils. *PLoS ONE* **2011**, *6*, e27300. [CrossRef]
10. Seiler, C.; Berendonk, T.U. Heavy metal driven co-selection of antibiotic resistance in soil and water bodies impacted by agriculture and aquaculture. *Front. Microbiol.* **2012**, *3*, 399. [CrossRef]
11. Dodd, M.C. Potential impacts of disinfection processes on elimination and deactivation of antibiotic resistance genes during water and wastewater treatment. *J. Environ. Monit.* **2012**, *14*, 1754–1771. [CrossRef] [PubMed]
12. Pruden, A. Balancing Water Sustainability and Public Health Goals in the Face of Growing Concerns about Antibiotic Resistance. *Environ. Sci. Technol.* **2014**, *48*, 5–14. [CrossRef] [PubMed]
13. Chang, P.H.; Juhrend, B.; Olson, T.M.; Marrs, C.F.; Wigginton, K.R. Degradation of Extracellular Antibiotic Resistance Genes with UV254 Treatment. *Environ. Sci. Technol.* **2017**, *51*, 6185–6192. [CrossRef] [PubMed]
14. von Wintersdorff, C.J.; Penders, J.; Van Niekerk, J.M.; Mills, N.D.; Majumder, S.; van Alphen, L.B.; Savelkoul, P.H.M.; Wolffs, P.F.G. Dissemination of Antimicrobial Resistance in Microbial Ecosystems through Horizontal Gene Transfer. *Front. Microbiol.* **2016**, *7*, 173. [CrossRef] [PubMed]
15. Huang, J.J.; Hu, H.Y.; Wu, Y.H.; Wei, B.; Lu, Y. Effect of chlorination and ultraviolet disinfection on tetA-mediated tetracycline resistance of Escherichia coli. *Chemosphere* **2013**, *90*, 2247–2253. [CrossRef]
16. Aslan, A.; Cole, Z.; Bhattacharya, A.; Oyibo, O. Presence of Antibiotic-Resistant Escherichia coli in Wastewater Treatment Plant Effluents Utilized as Water Reuse for Irrigation. *Water* **2018**, *10*, 805. [CrossRef]
17. McLellan, S.L.; Huse, S.M.; Mueller-Spitz, S.R.; Andreishcheva, E.N.; Sogin, M.L. Diversity and population structure of sewage-derived microorganisms in wastewater treatment plant influent. *Environ. Microbiol.* **2010**, *12*, 378–392. [CrossRef]
18. Lee, J.; Jeon, J.H.; Shin, J.; Jang, H.M.; Kim, S.; Song, M.S.; Kim, Y.M. Quantitative and qualitative changes in antibiotic resistance genes after passing through treatment processes in municipal wastewater treatment plants. *Sci. Total Environ.* **2017**, *605–606*, 906–914. [CrossRef]
19. Rafraf, I.D.; Lekunberri, I.; Sànchez-Melsió, A.; Aouni, M.; Borrego, C.M.; Balcázar, J.L. Abundance of antibiotic resistance genes in five municipal wastewater treatment plants in the Monastir Governorate, Tunisia. *Environ. Pollut.* **2016**, *219*, 353–358. [CrossRef]
20. Czekalski, N.; Berthold, T.; Caucci, S.; Egli, A.; Buergmann, H. Increased levels of multiresistant bacteria and resistance genes after wastewater treatment and their dissemination into lake geneva, Switzerland. *Front. Microbiol.* **2012**, *3*, 106. [CrossRef]
21. Mao, D.; Yu, S.; Rysz, M.; Luo, Y.; Yang, F.; Li, F.; Hou, J.; Mu, Q.; Alvarez, P.J.J. Prevalence and proliferation of antibiotic resistance genes in two municipal wastewater treatment plants. *Water Res.* **2015**, *85*, 458–466. [CrossRef] [PubMed]
22. Di Cesare, A.; Eckert, E.M.; Teruggi, A.; Fontaneto, D.; Bertoni, R.; Callieri, C.; Corno, G. Constitutive presence of antibiotic resistance genes within the bacterial community of a large subalpine lake. *Mol. Ecol.* **2015**, *24*, 3888–3900. [CrossRef] [PubMed]
23. Rosewarne, C.P.; Pettigrove, V.; Stokes, H.W.H.; Parsons, Y.M. Class 1 integrons in benthic bacterial communities: Abundance, association with Tn402-like transposition modules and evidence for coselection with heavy-metal resistance. *FEMS Microbiol. Ecol.* **2010**, *72*, 35–46. [CrossRef] [PubMed]

24. Rodríguez-Martínez, J.M.; Velasco, C.; Briales, A.; García, I.; Conejo, M.C.; Pascual, A. Qnr-like pentapeptide repeat proteins in gram-positive bacteria. *J. Antimicrob. Chemother.* **2008**, *61*, 1240–1243. [CrossRef] [PubMed]
25. Stepanauskas, R.; Glenn, T.C.; Jagoe, C.H.; Tuckfield, R.C.; Lindell, A.H.; McArthur, J.V. Elevated microbial tolerance to metals and antibiotics in metal-contaminated industrial environments. *Environ. Sci. Technol.* **2005**, *39*, 3671–3678. [CrossRef]
26. Wan, K.; Zhang, M.; Ye, C.; Lin, W.; Guo, L.; Chen, S.; Yu, X. Organic carbon: An overlooked factor that determines the antibiotic resistome in drinking water sand filter biofilm. *Environ. Int.* **2019**, *125*, 117–124. [CrossRef]
27. Zhang, S.; Han, B.; Gu, J.; Wang, C.; Wang, P.; Ma, Y.; Cao, J.; He, Z. Fate of antibiotic resistant cultivable heterotrophic bacteria and antibiotic resistance genes in wastewater treatment processes. *Chemosphere* **2015**, *135*, 138–145. [CrossRef]
28. Al-Jassim, N.; Mantilla-Calderon, D.; Wang, T.; Hong, P.-Y. Inactivation and Gene Expression of a Virulent Wastewater Escherichia coli Strain and the Nonvirulent Commensal Escherichia coli DSM1103 Strain upon Solar Irradiation. *Environ. Sci. Technol.* **2017**, *51*, 3649–3659. [CrossRef]
29. Guo, M.T.; Yuan, Q.B.; Yang, J. Distinguishing effects of ultraviolet exposure and chlorination on the horizontal transfer of antibiotic resistance genes in municipal wastewater. *Environ. Sci. Technol.* **2015**, *49*, 5771–5778. [CrossRef]
30. Destiani, R.; Templeton, M.R. Chlorination and ultraviolet disinfection of antibiotic-resistant bacteria and antibiotic resistance genes in drinking water. *AIMS Environ. Sci.* **2019**, *6*, 222–241. [CrossRef]
31. Yoon, Y.; Chung, H.J.; Wen Di, D.Y.; Dodd, M.C.; Hur, H.G.; Lee, Y. Inactivation efficiency of plasmid-encoded antibiotic resistance genes during water treatment with chlorine, UV.; UV/H_2O_2. *Water Res.* **2017**, *123*, 783–793. [CrossRef] [PubMed]
32. Umar, M.; d'Auriac, M.A.; Wennberg, A.C. Application of UV-LEDs for antibiotic resistance genes inactivation—Efficiency monitoring with qPCR and transformation. *J. Environ. Chem. Eng.* **2021**, *9*, 105260. [CrossRef]
33. Pang, Y.; Huang, J.; Xi, J.; Hu, H.; Zhu, Y. Effect of ultraviolet irradiation and chlorination on ampicillin-resistant Escherichia coli and its ampicillin resistance gene. *Front. Environ. Sci. Eng.* **2016**, *10*, 522–530. [CrossRef]
34. Stange, C.; Sidhu, J.P.S.; Toze, S.; Tiehm, A. Comparative removal of antibiotic resistance genes during chlorination, ozonation; UV treatment. *Int. J. Hyg. Environ. Health* **2019**, *222*, 541–548. [CrossRef]
35. McKinney, C.W.; Pruden, A. Ultraviolet disinfection of antibiotic resistant bacteria and their antibiotic resistance genes in water and wastewater. *Environ. Sci. Technol.* **2012**, *46*, 13393–13400. [CrossRef] [PubMed]
36. Zhuang, Y.; Ren, H.; Geng, J.; Zhang, Y.; Zhang, Y.; Ding, L.; Xu, K. Inactivation of antibiotic resistance genes in municipal wastewater by chlorination, ultraviolet; ozonation disinfection. *Environ. Sci. Pollut. Res. Int.* **2015**, *22*, 7037–7044. [CrossRef]
37. Auerbach, E.A.; Seyfried, E.E.; McMahon, K.D. Tetracycline resistance genes in activated sludge wastewater treatment plants. *Water Res.* **2007**, *41*, 1143–1151. [CrossRef]
38. Rodríguez-Chueca, J.; Della Giustina, S.V.; Rocha, J.; Fernandes, T.; Pablos, C.; Encinas, Á.; Barceló, D.; Rodríguez-Mozaz, S.; Manaia, C.M.; Marugán, J. Assessment of full-scale tertiary wastewater treatment by UV-C based-AOPs: Removal or persistence of antibiotics and antibiotic resistance genes? *Sci. Total Environ.* **2019**, *652*, 1051–1061. [CrossRef]
39. Beitz, T.; Bechmann, W.; Mitzner, R. Investigations of Reactions of Selected Azaarenes with Radicals in Water. 2. Chlorine and Bromine Radicals. *J. Phys. Chem. A* **1998**, *102*, 6766–6771. [CrossRef]
40. Watts, M.J.; Linden, K.G. Chlorine photolysis and subsequent OH radical production during UV treatment of chlorinated water. *Water Res.* **2007**, *41*, 2871–2878. [CrossRef]
41. Chuang, Y.H.; Chen, S.; Chinn, C.J.; Mitch, W.A. Comparing the UV/Monochloramine and UV/Free Chlorine Advanced Oxidation Processes (AOPs) to the UV/Hydrogen Peroxide AOP Under Scenarios Relevant to Potable Reuse. *Environ. Sci. Technol.* **2017**, *51*, 13859–13868. [CrossRef] [PubMed]
42. Yin, R.; Ling, L.; Shang, C. Wavelength-dependent chlorine photolysis and subsequent radical production using UV-LEDs as light sources. *Water Res.* **2018**, *142*, 452–458. [CrossRef] [PubMed]
43. Miklos, D.B.; Remy, C.; Jekel, M.; Linden, K.G.; Drewes, J.E.; Hübner, U. Evaluation of advanced oxidation processes for water and wastewater treatment—A critical review. *Water Res.* **2018**, *139*, 118–131. [CrossRef] [PubMed]
44. Health Canada. *Guidelines for Canadian Drinking Water Quality: Guideline Technical Document—Chlorine*; Health Canada: Ottawa, ON, Canada, 2016.
45. Zhang, T.; Hu, Y.; Jiang, L.; Yao, S.; Lin, K.; Zhou, Y.; Cui, C. Removal of antibiotic resistance genes and control of horizontal transfer risk by UV, chlorination and UV/chlorination treatments of drinking water. *Chem. Eng. J.* **2019**, *358*, 589–597. [CrossRef]
46. Dong, H.; Qiang, Z.; Hu, J.; Qu, J. Degradation of chloramphenicol by UV/chlorine treatment: Kinetics, mechanism and enhanced formation of halonitromethanes. *Water Res.* **2017**, *121*, 178–185. [CrossRef] [PubMed]
47. Rattanakul, S.; Oguma, K. Analysis of Hydroxyl Radicals and Inactivation Mechanisms of Bacteriophage MS2 in Response to a Simultaneous Application of UV and Chlorine. *Environ. Sci. Technol.* **2017**, *51*, 455–462. [CrossRef]
48. He, H.; Zhou, P.; Shimabuku, K.K.; Fang, X.; Li, S.; Lee, Y.; Dodd, M.C. Degradation and Deactivation of Bacterial Antibiotic Resistance Genes during Exposure to Free Chlorine, Monochloramine, Chlorine Dioxide, Ozone, Ultraviolet Light; Hydroxyl Radical. *Environ. Sci. Technol.* **2019**, *53*, 2013–2026. [CrossRef] [PubMed]
49. Balasubramanian, B.; Pogozelski, W.K.; Tullius, T.D. DNA strand breaking by the hydroxyl radical is governed by the accessible surface areas of the hydrogen atoms of the DNA backbone. *Proc. Natl. Acad. Sci. USA* **1998**, *95*, 9738–9743. [CrossRef] [PubMed]
50. Cho, M.; Kim, J.H.; Yoon, J. Investigating synergism during sequential inactivation of Bacillus subtilis spores with several disinfectants. *Water Res.* **2006**, *40*, 2011–2920. [CrossRef]

51. Wang, H.; Wang, J.; Li, S.; Ding, G.; Wang, K.; Zhuang, T.; Huang, X.; Wang, X. Synergistic effect of UV/chlorine in bacterial inactivation, resistance gene removal; gene conjugative transfer blocking. *Water Res.* **2020**, *185*, 116290. [CrossRef]
52. Betterton, E.A.; Hoffmann, M.R. Kinetics and mechanism of the oxidation of aqueous hydrogen sulfide by peroxymonosulfate. *Environ. Sci. Technol.* **1990**, *24*, 1819–1824. [CrossRef]
53. Khan, J.A.; He, X.; Shah, N.S.; Khan, H.M.; Hapeshi, E.; Fatta-Kassinos, D.; Dionysiou, D.D. Kinetic and mechanism investigation on the photochemical degradation of atrazine with activated H_2O_2, $S_2O_8^{2-}$ and HSO_5^{-}. *Chem. Eng. J.* **2014**, *252*, 393–403. [CrossRef]
54. Xiao, Y.; Zhang, L.; Zhang, W.; Lim, K.Y.; Webster, R.D.; Lim, T.T. Comparative evaluation of iodoacids removal by UV/persulfate and UV/H_2O_2 processes. *Water Res.* **2016**, *102*, 629–639. [CrossRef] [PubMed]
55. Ahn, Y.; Lee, D.; Kwon, M.; Choi, I.-H.; Nam, S.-N.; Kang, J.-W. Characteristics and fate of natural organic matter during UV oxidation processes. *Chemosphere* **2017**, *184*, 960–968. [CrossRef] [PubMed]
56. Ferro, G.; Guarino, F.; Cicatelli, A.; Rizzo, L. β-lactams resistance gene quantification in an antibiotic resistant Escherichia coli water suspension treated by advanced oxidation with UV/H_2O_2. *J. Hazard. Mater.* **2017**, *323 (Pt A)*, 426–433. [CrossRef]
57. Ferro, G.; Guarino, F.; Castiglione, S.; Rizzo, L. Antibiotic resistance spread potential in urban wastewater effluents disinfected by UV/H2O2 process. *Sci. Total Environ.* **2016**, *560–561*, 29–35. [CrossRef]
58. Zhang, Y.; Zhuang, Y.; Geng, J.; Ren, H.; Xu, K.; Ding, L.-L. Reduction of antibiotic resistance genes in municipal wastewater effluent by advanced oxidation processes. *Sci. Total Environ.* **2016**, *550*, 184–191. [CrossRef] [PubMed]
59. Guo, C.; Wang, K.; Hou, S.; Wan, L.; Lv, J.; Zhang, Y.; Qu, X.; Chen, S.; Xu, J. H_2O_2 and/or TiO_2 photocatalysis under UV irradiation for the removal of antibiotic resistant bacteria and their antibiotic resistance genes. *J. Hazard. Mater.* **2017**, *323 (Pt B)*, 710–718. [CrossRef]
60. Umar, M. Reductive and Oxidative UV Degradation of PFAS—Status, Needs and Future Perspectives. *Water* **2021**, *13*, 3185. [CrossRef]
61. Alexander, J.; Knopp, G.; Dötsch, A.; Wieland, A.; Schwartz, T. Ozone treatment of conditioned wastewater selects antibiotic resistance genes, opportunistic bacteria; induce strong population shifts. *Sci. Total Environ.* **2016**, *559*, 103–112. [CrossRef]
62. Lüddeke, F.; Heß, S.; Gallert, C.; Winter, J.; Güde, H.; Löffler, H. Removal of total and antibiotic resistant bacteria in advanced wastewater treatment by ozonation in combination with different filtering techniques. *Water Res.* **2015**, *69*, 243–251. [CrossRef] [PubMed]
63. Baquero, F.; Alvarez-Ortega, C.; Martinez, J.L. Ecology and evolution of antibiotic resistance. *Environ. Microbiol. Rep.* **2009**, *1*, 469–476. [CrossRef] [PubMed]
64. Larsson, D.G.J.; Andremont, A.; Bengtsson-Palme, J.; Brandt, K.K.; de Roda Husman, A.M.; Fagerstedt, P.; Fick, J.; Flach, C.-F.; Gaze, W.H.; Kuroda, M.; et al. Critical knowledge gaps and research needs related to the environmental dimensions of antibiotic resistance. *Environ. Int.* **2018**, *117*, 132–138. [CrossRef] [PubMed]

International Journal of
Environmental Research and Public Health

MDPI

Article

Mild Hydrothermal Synthesis of 11Å-TA from Alumina Extracted Coal Fly Ash and Its Application in Water Adsorption of Heavy Metal Ions (Cu(II) and Pb(II))

Jingjie Yang [1,2,3], Hongjuan Sun [1,2,*], Tongjiang Peng [1,2], Li Zeng [1,2] and Xin Zhou [1,2]

1 Key Laboratory of Ministry of Education for Solid Waste Treatment and Resource Recycle, Southwest University of Science and Technology, Mianyang 621010, China; yangjingjie007@mails.swust.edu.cn (J.Y.); pengtongjiang@swust.edu.cn (T.P.); zengli6041@mails.swust.edu.cn (L.Z.); 18355412280@sina.cn (X.Z.)

2 Institute of Mineral Materials & Application, Southwest University of Science and Technology, Mianyang 621010, China

3 Mian Yang City College, Southwest University of Science and Technology, Mianyang 621010, China

* Correspondence: sunhongjuan@swust.edu.cn

Abstract: Non-biodegradable copper (Cu) and lead (Pb) contaminants in water are highly toxic and have series adverse effects. Therefore, it is very important to extract heavy metals from wastewater before it is discharged into the environment. Adsorption is a cost-effective alternative method for wastewater treatment. Choosing a low-cost adsorbent can help reduce the cost of adsorption. In this study, conversion of reside after extracting aluminum (REA) produced by sub-molten salt method transform high-alumina coal fly ash (CFA) into 11Å-tobermorite to adsorb Cu(II) and Pb(II) from aqueous solutions at room temperature. The synthesis of the adsorbent was confirmed using scanning electron microscope (SEM), X-ray diffractometer (XRD) and Brunauer–Emmett–Teller (BET) method surface analysis. To study the adsorption characteristics, factors such as initial Cu(II) and Pb(II) concentration, pH, contact time, adsorption characteristics and temperature were investigated in batch mode. The maximum adsorption capacity of Cu(II) and Pb(II) was 177.1 $mg \cdot g^{-1}$ and 176.2 $mg \cdot g^{-1}$, respectively. The Langmuir adsorption model was employed to better describe the isothermal adsorption behavior and confirm the monolayer adsorption phenomenon. The pseudo-second-order kinetic model was used to highlight Cu(II) and Pb(II) adsorption kinetics. Thermodynamic analysis indicated the removal Cu(II) and Pb(II) by TA-adsorbent was a nonspontaneous and exothermic reaction. The obtained results are of great significance to the conversion of industrial waste to low-cost adsorbent for Cu(II) and Pb(II) removal from water.

Keywords: residue after extracting aluminum from CFA (REA); synthetic tobermorite; heavy metals; copper; lead; adsorption; kinetics study

Citation: Yang, J.; Sun, H.; Peng, T.; Zeng, L.; Zhou, X. Mild Hydrothermal Synthesis of 11Å-TA from Alumina Extracted Coal Fly Ash and Its Application in Water Adsorption of Heavy Metal Ions (Cu(II) and Pb(II)). *Int. J. Environ. Res. Public Health* **2022**, *19*, 616. https://doi.org/10.3390/ijerph19020616

Academic Editors: Yung-Tse Hung, Hamidi Abdul Aziz and Issam A. Al-Khatib

Received: 14 November 2021
Accepted: 24 December 2021
Published: 6 January 2022

1. Introduction

Water pollution is a major problem humankind faces today [1], large-scale wastewater is inevitable generated by industrial and economic development and places a lot of strain on ecosystem [2,3]. Heavy metals have been recognized as highly toxic contaminants owing to their highly pathogenic and bioaccumulation throughout the food chain [4]. Among the heavy metals, copper (Cu(II)) and lead (Pb(II)) are of particular interest as they are extensively produced in human activities and have been released into the aquatic environment [5,6]. According to the values announced by the World Health organization (WTO) and Environmental Protection Agency (EPA), the maximum permissible limits of Cu(II) and Pb(II) in drinking water are 1 $mg \cdot L^{-1}$ [7] and 0.05 $mg \cdot L^{-1}$ [8], respectively. Once these metals are ingested beyond the maximum permissible limits, they may result in mutagenic and carcinogenic effects could cause further damage to multiple systems and organs of the human body, or even death [9]. Nowadays, remediate Cu(II) and

Pb(II) contaminated water has been developed by various techniques, such as chemical precipitation, flocculation, ion exchange, membrane technology and permeable reactive barriers (PRBs) [10], among them, adsorption has low initial capital and maintenance costs and applicable to many technologies [11].

Over the last decade, many studies have focused on utilization of unique Coal fly ash (CFA) distribution in northern Shanxi and western part of Inner Mongolia China as the potential substituted resource of bauxite for the alumina industry, because it contains 40-50% alumina [12–15]. Among these efforts, the hydro-chemical process has emerged as a more promising alternative [16]. the process was first proposed by researchers in the former Soviet Union, which is believed to be an effective method for "Al-rich" waste product, such as red mud (RM) [17], high-alumina coal fly ash (HAFA) [12] and low-grade bauxite. During the process, the silicon-bearing components react with the slaked lime to generate a new crystal phase ($NaCaHSiO_4$) and enter the solid phase, where alumina is extracted from HAFA under high temperature and pressure. Sun proposed a mild hydro-chemical process for the extraction of alumina from HAFA [18]. The alumina extraction efficiency reached above 90% under optimal conditions. However, it is known that $NaCaHSiO_4$ exhibits strong alkalinity and can potentially damage agricultural land and groundwater if not appropriately treatment [18]. Previous studies have indicated that $NaCaHSiO_4$ is easily digested and converted to tobermorite adsorbent (TA) in diluted alkaline solution [19,20]. Additionally, the physiochemical properties of TA make it a potential adsorbent.

Tobermorite is a layer-lattice calcium silicate hydrate mineral, which consists of infinite layers of Ca-O polyhedra, with wollastonite-type silicate chains condensed on both sides of Ca-O polyhedra [21]. Three different individual tobermorite (14 Å, 11 Å and 9 Å tobermorite) can be defined owing to their c-axis length [22], among these, 11Å-tobermorite is structurally stable at ambient temperature and the most important mineral of the tobermorite family. Apart from their widespread application in the building materials and refractory materials, tobermorite is nowadays extensively applied in organic or inorganic effluent treatment due to its high specific surface area and ion-exchange capacity [23]. In previous studies, 11Å-tobermorite has been synthesized from pure chemical sources. If industrial wastes serve as the sources of calcium or silicon, the synthesis cost of TA can be further reduced. Herein, alumina-extracted residue (AER) from the mild hydro-chemical process, whose principal crystalline constituents were NaCaHSiO4, was used to synthesize 11Å-TA via hydrothermal method. Different Ca/Si molar ratio, hydrothermal temperature and time were designed as the modification parameters. Furthermore, prepared 11Å-TA was applied to Cu(II) and Pb(II) removal from water solutions, where the removal mechanism and kinetics of Cu(II) and Pb(II) were studied in detail. This work not only reduces the cost of Cu(II) and Pb(II) removal but also ameliorates issues related to solid waste disposal.

2. Materials and Methods

2.1. Raw Materials

AER obtained using hydrothermal process from high-alumina coal fly ash (HCFA) (collected from the Inner Mongolia Da Tang Thermal power plant, Hohhot, China). The operation conditions of AER synthesis were previously reported [24]. X-ray fluorescence (XRF, PANalytical, Eindhoven, Netherlands) was used to determine the chemical composition of AER, (Table 1), where SiO_2 content was 35.57% and Na_2O content was 17.91%. Therefore, the high alkalinity of AER without treatment generates environmental pollution. In contrast, exhibits AER potential recycling value for the recovery of alkali, and is a source of silicon and calcium. AER also contains trace quantities of various metal oxides such as Fe_2O_3, Al_2O_3, TiO_2 and MgO. Scanning electron microscope (SEM) analysis of AER is shown in Figure 1a. X-ray diffractometer (XRD) pattern of AER (Figure 1b) shows that the main phases correspond to $NaCaHSiO_4$ and $Ca(OH)_2$. The remaining chemical reagents (analytical grade) were supplied by KULONG Chemical Reagent Co. Ltd. (Chengdu, China). Deionized water was used in all experiments.

Table 1. Chemical composition of AER (ωt%).

Component	CaO	SiO_2	Al_2O_3	Na_2O	Fe_2O_3	TiO_2	MgO	Others
Mass (%)	37.39	35.57	2.10	17.91	2.03	2.76	1.86	0.39

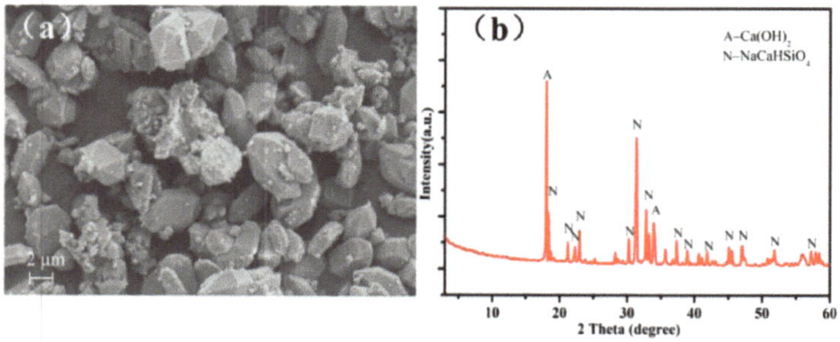

Figure 1. Scanning electron microscope (SEM)image (**a**) and X-ray diffractometer (XRD) pattern (**b**) of alumina-extracted residue (AER).

2.2. Preparation of Adsorbents

11Å-TA from AER was synthesized by hydrothermal synthesis, where AER was sufficiently alkaline to effect the decomposition of itself without addition of a strongly basic reaction medium. AER molar ratio Ca/Si was approx 1.1 (Table 1). However, the optimal molar ratio of Ca/Si was 0.83, which was calculated from the ideal composition of 11Å-TA ($5CaO{\cdot}6SiO_2{\cdot}5H_2O$). Hence, the molar ratio of Ca/Si was adjusted by mix ratio of fumed silica and AER, which also allowed to study the effect of different mix mass ratio of fumed silica and AER on the hydrothermal product, with the designed molar ratio of Ca/Si was being 0.8–1.1. In all batches, the liquid-solid ratio was 25 $mL{\cdot}g^{-1}$. The slurry was transferred to a 100 mL polytetrafluoroethylene (PTFE)-lined Ni steel autoclave after mixing, the autoclave was sealed and finally placed in an oven (WGZ-9040, ever briGht medical treatment instrument Co. Ltd., Beijing, China) to conduct hydrothermal experiments, which were performed over specific times at different temperatures. The overall reaction conditions are summarized in Table 2. When the hydrothermal process was finished, the products were filtered after cooling to room temperature and washing to neutral pH. The synthesized samples were collected and dried at 103 °C to constant weight for characterization.

2.3. Adsorption Batch Tests

In order to explore the adsorption and potential of synthetic 11Å-(TA) to remove Cu(II) or Pb(II), gradient solutions of metal ions between 200 $mg{\cdot}L^{-1}$ and 2000 $mg{\cdot}L^{-1}$ were prepared by subsequent dilution of the stock solutions using deionized water. The stock solution (4000 $mg{\cdot}L^{-1}$) of Cu(II) and Pb(II) ions were prepared by dissolving $CuSO_4{\cdot}5H_2O$ and $Pb(NO_3)_2$ in deionized water, respectively. According to the solubility product principle, in view of the difference between $K_{sp}[Pb(OH)_2]$ (25 °C) and $K_{sp}[Cu(OH)_2]$ (25 °C), Cu(II) and Pb(II) began to precipitate at approx pH 5.3 and 6.6, respectively [25]. Therefore, the initial pH value of Cu(II) and Pb(II) solution were adjusted to 5.0 in the subsequent experiment using dilute HNO_3 (0.1 M) and NaOH (0.1 M) solutions.

A batch equilibrium method was applied to the removal experiments. Typically, 0.1 g of TA was added into a 100 mL centrifuge tube. Then, 25 mL of simulated contaminated water (Cu(II) or Pb(II)) was added and the tube was subjected to a thermostatic oscillator (SHA-B, LiChen Co. Ltd., Shanghai, China) at 200 rpm for 12 h at room temperature (25 ± 2 °C). After adsorption, the slurry was centrifuged for 180 s at 6000 rpm. Liquid supernatant was withdrawn using an injection syringe (25 mL, Kelun Co. Ltd., Chengdu, China) and then filtered through a 0.45 μm polyethersulfone membrane filter

(JinTeng Co. Ltd., Tianjin, China). Then filtrate was immediately diluted 10 times with distilled water. The amount of metal cations was determined using an inductively coupled plasma mass spectrometer (ICP, ICAP6500, Thermo Fisher Scientific, Waltham, MA, USA). To determine the amount of Cu(II) and Pb(II) absorbed on TA, the following equations were used:

$$q_e = \frac{(C_0 - C_e) \times V}{M} \quad (1)$$

$$\%\text{Removal} = \frac{(C_0 - C_e)}{C_0} \times 100 \quad (2)$$

where q_e is the adsorption capacity of absorbent for heavy metal ions ($mg \cdot L^{-1}$); C_0 is the initial concentration of the heavy metal ions ($mg \cdot L^{-1}$) and C_e is the equilibrium concentration of heavy metal ions ($mg \cdot L^{-1}$); V and m are the volume of solution (L) and the mass of the dry adsorbent (g), respectively. All glass and plastic containers in the experiment were soaked in 0.5 M nitric acid solution for 24 h and then washed repeatedly with distilled water and dried under 70 °C.

Table 2. Ranges and values of the experimental parameters for 11Å-TA synthesis using AER.

Samples Label	Experiment Conditions			Liquid-Solid Ratio ($mL \cdot g^{-1}$)
	Ca/Si (Ratio)	Temperature (°C)	Time (h)	
To-1	0.8	160	12	25
To-2	0.9			
To-3	1.0			
To-4	1.1			
To-5	0.8	180		
To-6	0.9			
To-7	1.0			
To-8	1.1			
To-9	0.8	200		
To-10	0.9			
To-11	1.0			
To-12	1.1			
To-13	1.0 [a]	200	3	
To-14			6	
To-15			9	
To-16			12	

[a] refers to suitable parameter values selected from previous set to be applied for the next sets of experiments.

In order to observe the effect of contact time on the adsorption of heavy metals on the obtained TA, all experiments were performed with 0.1g of TA suspended in 25 mL of Cu(II) (1000 $mg \cdot L^{-1}$) or Pb(II) (1000 $mg \cdot L^{-1}$) solutions at initial pH 5.0 ± 0.1. The effect of contact time on the adsorption rate were carried out by performing the adsorption experiments at pre-determined time intervals (5, 20, 40, 60, 90, 120, 180, 240, 300 and 360 min). The samples were collected at various intervals of time in order to monitor the reaction in kinetics.

To test the effects of solution pH on the adsorption of heavy metals on TA, 0.1 g TA was added to 25 mL of 200 $mg \cdot L^{-1}$ Cu(II) or Pb(II) solution at pH 1.0–6.0. pH values of the solutions were measured using a pH meter (PHS-3C, INESA scientific instrument Co. Ltd., Shanghai, China). The residual heavy metal concentration in the upper clear liquid was detected by ICP.

2.4. Characterization Techniques

The crystal structures were determined using X-ray diffractometer (XRD, Ultima IV, Rigaku Co. Ltd., Tokyo, Japan) with a Cu K_α radiation (λ = 0.154 187 nm) operation at 40 kV and 40 mA. 2θ scans were performed from 3° to 80° at a rate of $15° \cdot min^{-1}$. The chemical compositions were analyzed using X-ray fluorescence analyzer spectrometer (XRF, PANalytical, Eindhoven, The Netherlands). The specific surface area of the obtained material was measured using surface area analyzer (Autosorb-iQ, Quantachrome, Boynton Beach, FL, USA) and evaluated by measuring the isothermal adsorption of N_2 based on the single-point Brunauer–Emmett–Teller (BET) method. The microstructural development and chemical characteristics of samples were studied using a scanning electron microscope (SEM, Zeiss Instruments, Oberkochen, Germany) equipped with energy dispersive spectroscopy (EDS, Oxford, UK). The zeta potentials of the obtained TA adsorbent were conducted by Nano zeta sizer (Zs90, Malvern Instruments, Malvern, UK) in electrophoretic light scattering mode. The concentration of the metal ions in solution was measured via an inductively coupled plasma mass spectrometer (ICP, Thermo Fisher Scientific Co. Ltd., Waltham, MA, USA).

3. Results and Discussion

3.1. Characterization of TA from AER

Powder XRD studies were conducted on the materials obtained with increasing Ca/Si molar ratio at 160 °C (Figure 2). XRD patterns of To-1, To-2, To-3, To-4 and AER were the same except for the peaks height ratio, hence, $NaCaHSiO_4$ from AER failed to effectively dissolve and form the calcium silicate hydrate gel (C-S-H) and crystallize to 11Å-TA. The effect of varying reaction temperature from 160 °C to 200 °C was investigated with Ca/Si molar ratio of 1 and treatment time of 12 h (Figure 3). XRD patterns of the samples after hydrothermal reaction in different temperature show a clear decrease in peak intensity corresponding to $NaCaHSiO_4$ and $Ca(OH)_2$ phases, which almost disappear at higher temperature (200 °C), while 11Å-TA had greater intensity in the main reflections of 7.61°, 5.46° and 28.91°(2θ) and $Ca_3Al_2O_6$ was more clear in the main reflections of 33.13° and 47.72° (2θ). Furthermore, attempted synthesize of 11Å-TA at lower hydrothermal temperature was also carried out for comparison. XRD pattern of To-3 was comparable to that of To-7, and the peak intensities of $NaCaHSiO_4$ remained almost unchanged, indicating that $NaCaHSiO_4$ was stable under hydrothermal temperature of 180 °C. It was also observed that hydrothermal temperature played a major role in the dissolution of Si^{4+} species present in $NaCaHSiO_4$ matrix of AER.

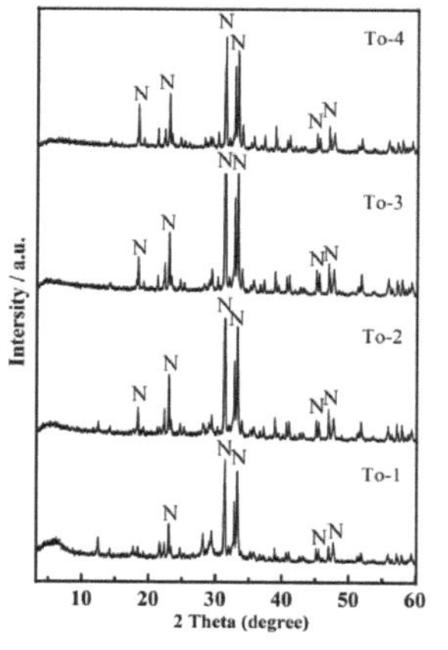

Figure 2. XRD pattern of products synthesized at different Ca/Si molar ratio at 160 °C for 12 h. N—$NaCaHSiO_4$.

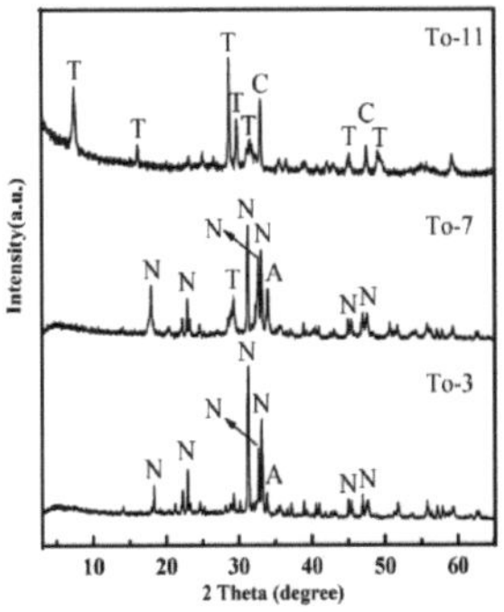

Figure 3. XRD pattern of products synthesized at different temperatures for 12 h. N—$NaCaHSiO_4$; A—$Ca(OH)_2$; T—11Å-TA; C—$Ca_3Al_2O_6$.

XRD patterns of products obtained from the hydrothermal reaction with various Ca/Si molar ratio at 200 °C are shown in Figure 4. The hydrothermal time was fixed at 12 h. The observed peaks of main phases in ARE fully disappeared. Si and Ca were all dissolved and completely used for new mineral formation. All samples show characteristics peaks of 11Å-TA. This highlights the possibility of crystallization of the material over a wider Ca/Si molar ratio range. When the ratio was 0.8, the crystal phases in the product were mainly 11Å-TA, with quartz and analcime detectable only in trace amounts. According to a primitive inference from Hsiao [26], the presence of Na^+ and a basic solution environment are required for the synthesis of analcime. $NaCaHSiO_4$ was sufficiently alkaline to effect the decomposition of itself and produce sufficient $SiO_3{}^{2-}$, Na^+ and alkalinity without addition of any more strongly basic reaction medium during hydrothermal reaction. In addition, some SiO_2 did not transform into TA but into quartz indicating that Ca source was insufficient. As Ca/Si molar ratio increased, the quartz and analcime characteristic peaks disappeared, while relatively strong reflections of $Ca_3Al_2O_6$ were detected. According to the above analyses, hydrothermal treatment of AER without addition of extra silica and calcium source formed 11Å-TA. The reactions were expressed as follows:

$$NaCaHSiO_4 + OH^- \rightarrow SiO_3{}^{2-} + Na + Ca(OH)_2 \tag{3}$$

$$SiO_3{}^{2-} + Ca(OH)_2 + H_2O \rightarrow \text{C-S-H} + OH^- \tag{4}$$

$$\text{C-S-H} \rightarrow \text{11Å-Tobermorite } (Ca_5 \cdot Si_6O_{16}(OH)_2 \cdot nH_2O) \tag{5}$$

The hydrothermal time plays an important role in the synthesis of 11Å-TA. The effect of varying hydrothermal time between 3 and 12 h was investigated with hydrothermal temperature of 200 °C and Ca/Si molar ratio of 1. Figure 5 presents XRD diffraction pattern of the samples obtained after different hydrothermal treatment times. To-13 signals corresponded to the newly generated sodium calcium aluminum oxide silicate hydrate phase in addition to peaks corresponding to $NaCaHSiO_4$ present in the unreacted AER, where most of $NaCaHSiO_4$ remained unreactive after 3 h. After which only the characteristic peaks of 11Å-TA were observed at 6 h. However, after 12 h, the main final phase in the product was 11Å-TA; $Ca_3Al_2O_6$ according to XRD analysis. The patterns remained unchanged, apart from peaks related to 11Å-TA that narrowed as the crystalline phase became more ordered. Hence, the optimal hydrothermal condition was 200 °C for 12 h. The 11Å-TA could be further functionalized to prepare low-cost adsorbing material for water purification.

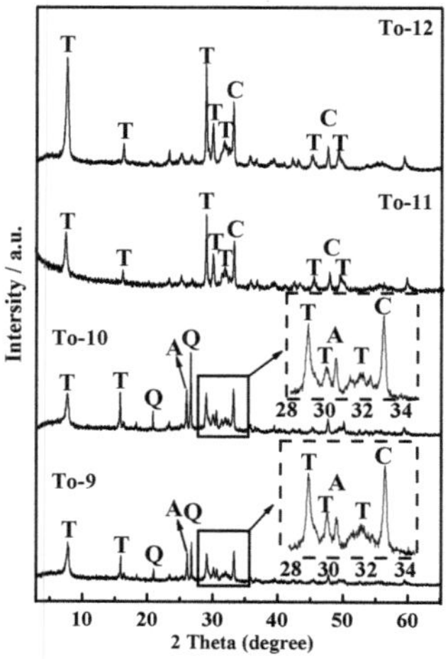

Figure 4. XRD patterns of products synthesized at different Ca/Si molar ratio at 200 °C for 12 h. T—11Å-TA; C—$Ca_3Al_2O_6$; A—Analcime; Q—Quartz.

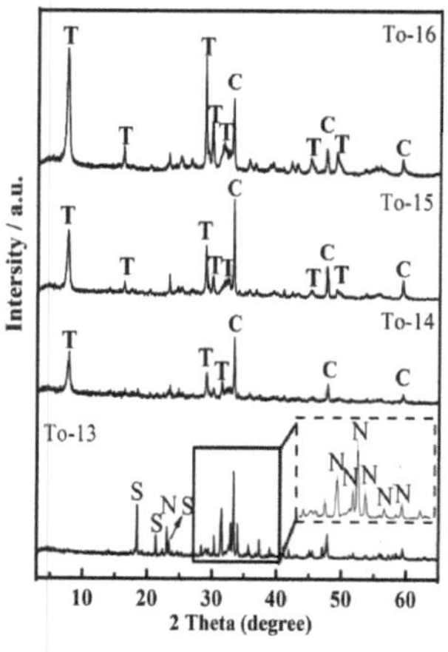

Figure 5. XRD patterns of products synthesized at 200 °C for different time. N—$NaCaHSiO_4$; S—Sodium Calcium Aluminum Oxide Silicate Hydrate; T—11Å-TA; C—$Ca_3Al_2O_6$.

Figure 6 shows the morphological changes of the products at different reaction temperatures. As shown in Figure 6a,b, a clear outline of the angular-like crystals microstructure, and particles as irregular globular coexist, which indicate an incomplete reaction and in agreement with XRD results. With the increase in reaction time. Figure 6c shows that as the reaction time increases to 6 h, $NaCaHSiO_4$ was no longer present, and To-16 sample displayed a shorter and thinner band-like morphology, with a length of approx 1 µm. Additionally, traces of C-H-S phases were observed around 11Å-TA. As shown in Figure 6d, more and more band-like crystallites formed, some of which were coiled together when the reaction proceeded for 9 h. Figure 7 shows the schematic of the TA preparation. The surface area is directly associated with the adsorption capacity for materials using adsorbent. The surface area of TA-adsorbent was determined as 16.77 $m^2 \cdot g^{-1}$ by single-point BET analysis.

Figure 6. SEM images examining the effect of reaction time on the morphology of the product for (**a**) and (**b**) To-13, (**c**) To-14, (**d**) To-15.

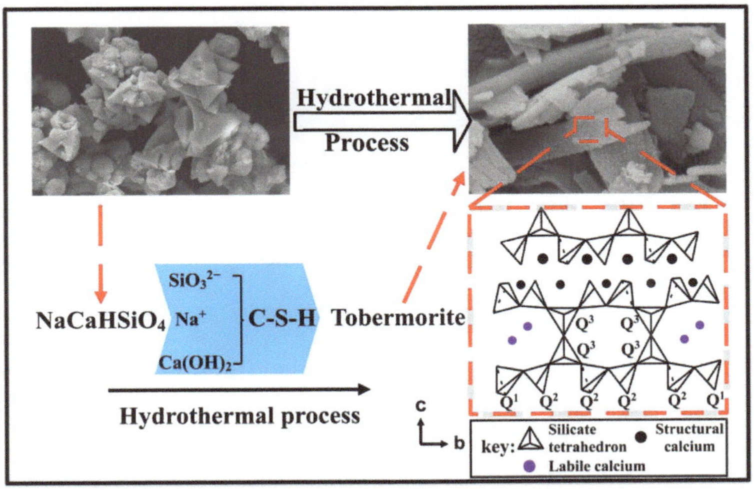

Figure 7. The schematic of the Tobermorite adsorbent (TA) preparation; Q^3 represents the bridging silicate tetrahedra; Q^2 represents mid-chain silicate tetrahedra; Q^1 represents the end group of a chain structure.

3.2. Heavy Metal Removal Studies

3.2.1. Effect of Initial pH

pH of the aqueous system has a dramatic effect on adsorption capacity. Excess H^+ causes the protonation of TA as well as competition with the toxic metal ions to be adsorbed on the active sites. However, pH of the solution also affects the state of toxic metal ions. To prevent the formation of metal hydroxide precipitates in the aqueous system, the effect of the synthesized TA on the uptake capacity of Pb(II) and Cu(II) was studied in pH range of 1.0–6.0 at 25 °C for 6 h.

The results presented in Figure 8 show that the adsorption capacity of Cu(II) increases with pH from 1 to 3, hence, the adsorption was relatively low under strong acidic conditions. This was because TA surface was positively charged as a result of H^+ would have the advantage for the competition compare with Cu(II), electrostatic repulsion forces restricted existence of the Cu(II) (dominant species), together with trace amounts of $Cu_2(OH)_2^{2+}$ and $Cu(OH)^+$ in the proximity of the adsorption interface. Adsorbent surface became deprotonated with increasing pH. Consequently, at pH between 3 and 6, the concentration of OH^- increased in order of magnitude, the amount of negative charges on TA surfaces increased via deprotonation of hydroxyl groups, which promoted strong electrostatic attraction forces resulting in removal of Cu(II) ions (>95%). Furthermore, removal rate of Pb(II) adsorbed by TA at different pH values are shown in Figure 8. The same trends were observed for Pb(II) removal efficiencies with increasing pH, which increased the adsorptive efficiency from pH 1.0 to 3.0. At pH < 3, the removal was low owing to the active sites on TA surface being occupied by H^+. However, the maximum removal reached 98.3% and remained above 97% at pH 3.0–6.0. According to the surface complexation theory, competition between H^+ and Pb(II) decreased with increasing pH.

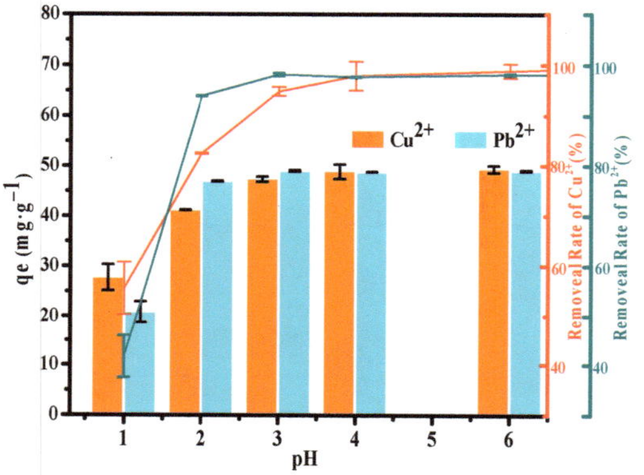

Figure 8. Effect of pH and recycle adsorption performance of TA for Cu(II) and Pb(II). Cation initial concentration was 200 mg·L^{-1}, adsorbent dosage was 0.1 g, 25 °C.

The zero point potential (pH_{zpc}) of TA adsorbent is 7.14 at which the net surface charge of the TA adsorbent is equal to zero. When pH below 7.14, it indicating that the adsorbent were negatively charged, which favored Cu(II) and Pb(II) removal from the solution. In addition, the negative surface charge of TA adsorbent accelerates the adsorption rate at the initial stage. Therefore, the obtained results indicated that the alkaline solution medium was favorable for the removal of toxic metal ions.

3.2.2. Effect of Initial Concentration

As shown in Figure 9, the initial metal ion concentration ranging from 200 mg·L^{-1} to 2000 mg·L^{-1} was examined. Figure 9a depicts that the adsorption capacity of TA increased with Cu(II) concentration. At equilibrium, qe for Cu(II) was achieved by TA at 177.1 mg·g^{-1}. This may be attributed to the number of Cu(II) will be less compared to the available sites on TA, resulting in diminished adsorption at low Cu(II) concentration. At higher initial Cu(II) concentration, Cu(II) transfer from the solution to TA was favored. However, as shown in Figure 9a, the removal rate remained over 99% in initial concentration from 200 mg·L^{-1} to 600 mg·L^{-1}, but decreased with increasing the initial concentration of Cu(II). As shown in Figure 9b, similar trends were observed for Pb(II), where q_e increased with increasing initial concentration and reached a plateau (177.1 mg·g^{-1}) at above 1000 mg·L^{-1}. The increase of the initial concentration of metal ion enhanced the driving force that overcome mass transfer resistance between the liquid and solid phases. q_e reached a plateau due to overlapping of available active sites that adhered the metal ions required for the high concentration of metal ion [27]. Compared with previous reports, the adsorption capacities are significantly higher [28,29], the adsorption capacities are significantly higher. Therefore, TA could effectively treat Cu and Pb-containing wastewater.

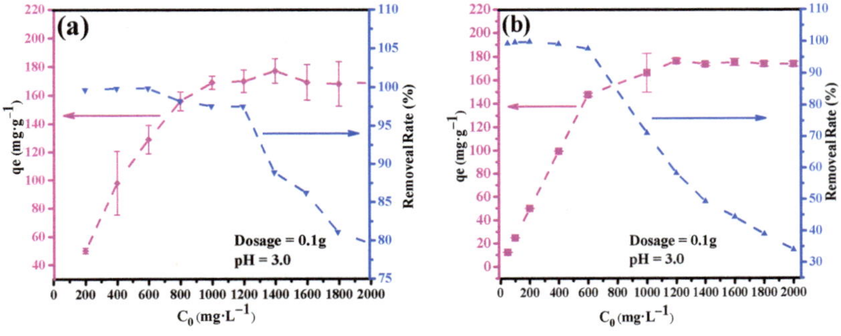

Figure 9. Effect of initial metal ion concentration on adsorption of Cu(II) (**a**) and Pb(II) (**b**) on TA.

3.2.3. Adsorption Isotherm

The effect of initial concentration on the equilibrium adsorption capacity of metal ion adsorbed on TA was evaluated at 25 °C. The linear forms of the Langmuir (Equation (6)) and Freundlich (Equation (7)) isotherms described the relationship of Cu(II) and Pb(II) in the solid and liquid phase at equilibrium.

$$\text{Langmuir model}: \quad \frac{C_e}{q_e} = \frac{1}{q_m b_L} + \frac{C_e}{q_m} \tag{6}$$

$$\text{Freundlich model}: \quad \text{Log}q_e = \text{Log}K_F + \frac{1}{n}\text{Log}C_e \tag{7}$$

where C_e and q_e is consistent with it mentioned above; q_m is the maximum adsorption capacity of metal ion adsorbed per unit weight of TA; K_F and $1/n$ are the Freundlich constants; b_L is the Langmuir constant. The linear plots of the Langmuir and Freundlich isotherms are shown in Figure 10, and the corresponding parameters are calculated and listed in Table 3. Figure 10a,b shows that, the Langmuir model (R_2 = 0.993) fitted better to the Cu(II) adsorption data compared to the Freundlich model (R_2 = 0.921). This indicated that the adsorption of Cu(II) ions on TA adsorbent involved a monolayer coverage process. Additionally, the maximum value of q_m of TA adsorbent for Cu(II) ions was 171.2 mg·g^{-1} which was comparable to the experiment value (177.1 mg·g^{-1}). The separation factor constant (R_L) was used to evaluate the applicability of the adsorbent to the studied adsorbate [30,31], which was calculated from Langmuir isotherm described in Equation (8).

$$\text{Separation factor}: \quad R_L = \frac{1}{1+K_L C_0} \tag{8}$$

where C_o denotes the initial concentration. K_L values were obtained from the Langmuir isotherm model. As shown in Table 3, R_L was between 0.525 and 0.155, and the range from 0 to 1 means that the adsorption process of Cu(II) was theoretically favorable [32].

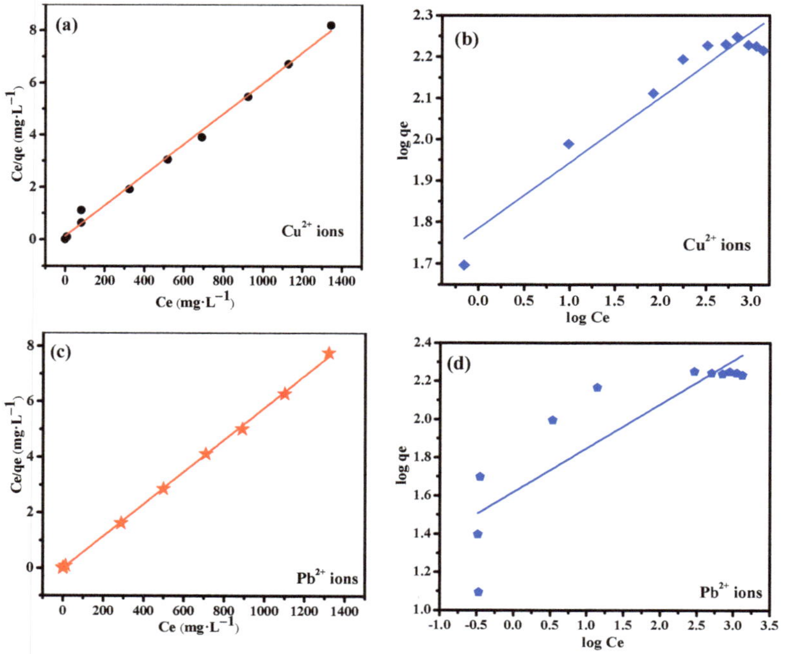

Figure 10. (**a**) Langmuir isotherm; (**b**) Freundlich isotherm (adsorption conditions: Ce (Cu(II)) = 200-2000 mg·L^{-1}, V = 10 mL, T = 25 °C, t = 360 min, pH = 5.0, and adsorbent amount = 0.1 g). (**c**) Langmuir isotherm; (**d**) Freundlich isotherm (adsorption conditions: Ce (Pb(II)) = 200-2000mg·L^{-1}, V = 10 mL, T = 25 °C, t = 360 min, pH = 5.0, and adsorbent amount = 0.1 g).

Table 3. Comparison between Langmuir and Freundlich constants by linear fitting for TA at 25 °C.

Kinetic Model	Parameter	Metal Ions	
		Cu(II)	Pb(II)
Langmuir	q_m (mg·g^{-1})	171.2	173.01
	b_L(min^{-1})	0.00452	1.4522
	R_2	0.993	0.999
	R_L	0.525~0.155	0.0135~0.0003
Freundlich	1/n	0.158	0.229
	K_F (L·mg^{-1})	60.94	41.591
	R_2	0.921	0.732

q_m is the maximum adsorption capacity of metal ion; b_L is the langmuir constant; R_L is the separation factor constant; K_F and 1/n are the freundlich constants.

Figure 10c,d shows the experimental adsorption data of Pb(II) for TA that fitted better using the Langmuir isotherm model (R_2 = 0.999) compared to the Freundlich model (R_2 = 0.732). Hence, the active sites of TA were homogeneously distributed on the surface, and Pb(II) exhibited a more preferred monolayer adsorption. The observed better fitted data also indicated weak interaction between active sites of the adsorbents. The maximum adsorption capacity of Pb(II) for TA obtained from the Langmuir isotherm model was 173.01 mg·g^{-1}, which was similar to the maximum adsorption capacity. Table 3 displays the separation factor (R_L), where adsorption was favorable (0 < RL = (0.0135~0.0003) < 1). Therefore, cost-effectiveness and excellent adsorption performance shows that TA is a promising candidate for the removal of Cu(II) and Pb(II) from wastewater.

3.2.4. Effect of Time and Adsorption Kinetics Analysis

The rate-controlling steps are essential in order to further elucidate the removal mechanism of Cu(II) and Pb(II) by TA. The kinetic data were fitted to either the pseudo-fist-order (Equation (9)) or pseudo-second-order (Equation (10)) kinetic models.

$$\text{Pseudo-first-order model}: \ \mathrm{Ln}(q^e - q^t) = \mathrm{Ln}q^e - K_1 t \quad (9)$$

$$\text{Pseudo-sec ond-model}: \ \frac{t}{q_t} = \frac{t}{q_e} + \frac{1}{q_e^2 k_2} \quad (10)$$

where q_e and q_t are the adsorption capacities of metal ions at equilibrium and different time, respectively; k_1 and k_2 are the rate constants for the pseudo-fist-order model and pseudo-second-order model, respectively. The effect of contact time on the adsorption of Cu(II) by TA is exhibited in Figure 11a. The adsorption process was considerably faster in the first few min and reached equilibrium after 180 min. As shown in Figure 11b,c, the adsorption kinetics of Cu(II) ions on TA was well fitted by both the pseudo-fist-order model (R_2 = 0.9209) and pseudo-second-order (R_2 = 0.9606). The simulation results showed that the calculated q_{max} of the pseudo-second-order model was in agreement with the experimental data. Therefore, the adsorption of Cu(II) ions on TA was mainly controlled by chemisorption. In addition, the effects of the contact time on the adsorption of Pb(II) were also investigated (Figure 11d). The parameters of the models were better fitted to the pseudo-second-order (R_2 = 0.9521) than the pseudo-fist-order model (R_2 = 0.9438), which indicated the chemisorption was the dominant control step for the adsorption process of Pb(II). k_2 for Cu(II) was higher than that of Pb(II) and indicated that Cu(II) was adsorbed at a faster adsorption rate, which was consistent with the time required for reaching equilibrium.

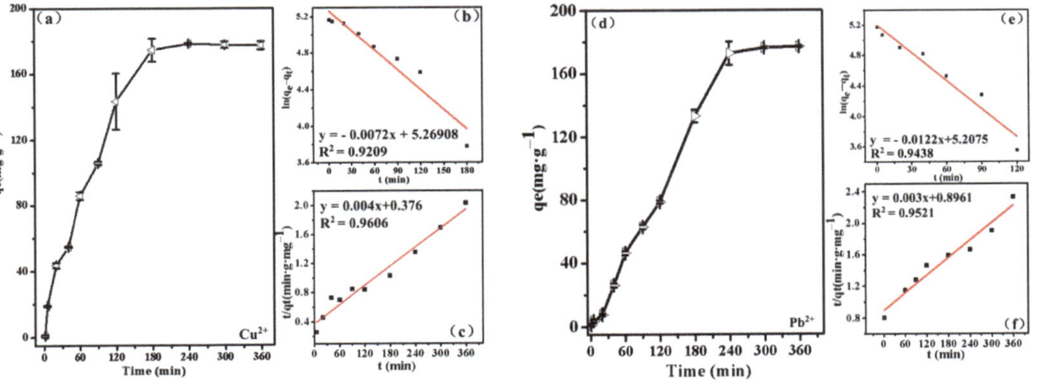

Figure 11. The effect of contact time on the adsorption of Cu(II) ions by TA (**a**); pseudo-fist-order model (**b**) and pseudo-second-order model (**c**); the effect of contact time on the adsorption of Pb(II) ions by TA (**d**); pseudo-fist-order model (**e**) and pseudo-second-order model (**f**).

Compared with other adsorbents, the biggest advantage of TA as adsorbent is lower costs, as shown in Table 4, several solid waste based low cost adsorbents have been used for the removal of Cu(II) and Pb(II) in previous studies. In general, the removal capacity of TA adsorbent for Cu(II) and Pb(II) compares favorably with those of other solid waste based adsorbents under single metal ions batch conditions.

Table 4. Comparison of adsorption capacity of Cu(II) and Pb(II) by TA and other adsorbents.

Adsorbents	q_m (mg·g^{-1})		Reference
	Cu(II) Sorbed	Pb(II) Sorbed	
Magnetic attapulgite composites	189.0	142.8	[33]
Novel geogolymers based on coal gangue and red mud	90.0	137.7	[29]
Tetrazole-bonded bagasse	253.5	89.3	[34]
Zeolites prepared from CFA	57.8	109.9	[35]
Linde F(K) zeolite from CFA	18.5	46.5	[36]
Activated bentonite	9.7	21.3	[37]
TA-adsorbent	177.1	176.2	This study

3.2.5. Adsorption Thermodynamic

Thermodynamic analysis is a scientific theory to study the effect of temperature on the adsorption process. In this study, the temperature range was 25 °C, 45 °C and 65 °C to calculate thermodynamic parameters, actual indicators including Gibbs free energy (ΔG_0) are deduced from Equations (11) and (12), similarly, the enthalpy (ΔH_0) and entropy (ΔS_0) are calculated from the following Equation (13) (Van Hoff's relation) and are summarized in Table 5.

$$K_C = q_e/c_e \quad (11)$$

$$\Delta G^0 = -RTlnK_C \quad (12)$$

$$lnK_C = \Delta S^0/R - \Delta H^0/RT \quad (13)$$

where q_e represents the adsorption capacity of the adsorbent at equilibrium state; ce represents the adsorption equilibrium concentration of metal ions in the solution. R is the gas constant, and T is the temperature (K).

Table 5. Thermodynamic parameter of Cu(II) and Pb(II) adsorption by the obtained TA adsorbents.

Metal	T (K)	K_c	ΔG_0 (kJ·mol^{-1})	ΔH_0 (KJ·mol^{-1})	ΔS_0 (J·mol^{-1}·K^{-1})	R (kJ·mol^{-1}·K^{-1})
Cu(II)	298.15	0.171	4.368	−1.090	−9.047	8.3×10^{-3}
	318.15	0.156	4.906			
	338.15	0.144	5.439			
Pb(II)	298.15	0.173	4.342	−2.760	−5.277	
	318.15	0.155	4.923			
	338.15	0.132	5.683			

Table 5 indicate that the values of ΔG_0 for Cu(II) and Pb(II) at lower temperature are slightly positive. It expresses that the adsorption process was simple but nonspontaneous at lower temperature. The negative values of ΔH_0 indicate that Cu(II) and Pb(II) adsorption processes follow the exothermic reaction. Besides, the negative values of ΔS_0 elucidate a decrease in the degree of randomness of Cu(II) and Pb(II) at the solid adsorbate interface of the TA adsorbent.

4. Conclusions

Using reside after extracting aluminum as raw material, we developed a low-cost TA via hydrothermal process. The experiments were performed under various conditions to determine optimal operation parameters and obtain well crystallized TA within 12 h at 200 °C without addition of extra silica and calcium source. Moreover, synthesized TA was employed to remove toxic substances (Cu(II) and Pb(II)) from wastewater. The adsorption data was in good agreement with the Langmuir model, which indicated that adsorption occurred in a homogeneous adsorption site via a monolayer reaction and maximum adsorption obtained at capacity of 177.1 and 176.2 mg·g^{-1} for Cu(II) and Pb(II), respectively. The kinetic study showed that adsorption followed the pseudo-second-order model, indicating a chemisorption process. Thermodynamic analysis showed the adsorption of Cu(II) and Pb(II) by synthesized TA adsorbent is a nonspontaneous and exothermic process. Therefore, the prepared TA adsorbent has potential application in treatment of heavy metal pollutants. Furthermore, Excellent adsorption properties of adsorbent are not only attribute required, its regeneration is a critical factor as well. For this reason, the regeneration studies of TA adsorbent is a scientific problem worthy of further attention.

Author Contributions: Conceptualization, T.P. and J.Y.; methodology, J.Y.; software, L.Z.; validation, T.P., J.Y. and L.Z.; formal analysis, X.Z.; investigation, J.Y.; resources, X.Z.; data curation, J.Y.; writing—original draft preparation, L.Z.; writing—review and editing, J.Y.; visualization, J.Y.; supervision, H.S.; project administration, H.S.; funding acquisition, H.S. All authors have read and agreed to the published version of the manuscript.

Funding: This work was financially supported by the Longshan academic talent research and Innovation Team Project of SWUST (17LZXT11). We thank all the interviewees for sharing their ideas and experiences with us during our field studies and all the anonymous reviewers who help us to improve the paper.

Institutional Review Board Statement: Not applicable.

Informed Consent Statement: Not applicable.

Data Availability Statement: Not applicable.

Acknowledgments: We thank all the interviewees for sharing their ideas and experiences with us during our field studies and all the anonymous reviewers who help us to improve the paper.

Conflicts of Interest: The authors declare no conflict of interest.

References

1. Bilar, M.; Ihsanullah, I.; Shah, M.U.H.; Younas, M. Enhanced removal of cadmium from water using bio-sorbents synthesized from branches and leaves of Capparis decidua and Ziziphus mauritiana. *Environ. Technol. Innov.* **2021**, *24*, 101922. [CrossRef]
2. Al-Khaldi, F.A.; Abusharkh, B.; Khaled, M.; Atieh, M.A.; Nasser, M.S.; laoui, T.; Saleh, T.A.; Agarwal, S.; Tyagi, I.; Gupta, V.K. Adsorptive removal of cadmium(II) ions from liquid phase using acid modified carbon-based adsorbents. *J. Mol. Liq.* **2015**, *204*, 255–263. [CrossRef]
3. Aigbe, U.O.; Ukhurebor, K.E.; Onyancha, R.B.; Osibote, O.A.; Darmokoesoemo, H.; Kusuma, H.S. Fly ash-based adsorbent for adsorption of heavy metals and dyes from aqueous solution: A review. *J. Mater. Res. Technol.* **2021**, *14*, 2751–2774. [CrossRef]
4. Kim, S.A.; Kamala-Kannan, S.; Lee, K.J.; Park, Y.J.; Shea, P.J.; Lee, W.H.; Kim, H.; Oh, B.T. Removal of Pb(II) from aqueous solution by a zeolite–nanoscale zero-valent iron composite. *Chem. Eng. J.* **2013**, *217*, 54–60. [CrossRef]
5. Hamid, S.A.; Azha, S.F.; Sellaoui, L.; Bonilla-Petriciolet, A.; Ismail, S. Adsorption of copper(II) cation on polysulfone/zeolite blend sheet membrane: Synthesis, characterization, experiments and adsorption modelling. *Colloids. Surf. A.* **2020**, *601*, 124980. [CrossRef]
6. Shang, Z.; Zhang, L.; Zhao, X.; Liu, S.; Li, D. Removal of Pb(II), Cd(II) and Hg(II) from aqueous solution by mercapto-modified coal gangue. *J. Environ. Manag.* **2019**, *231*, 391–396. [CrossRef]
7. Kaçakgil, E.C.; Çetintas, S. Preparation and characterization of a novel functionalized agricultural waste-based adsorbent for Cu^{2+} removal: Evaluation of adsorption performance using response surface methodology. *Sustain. Chem. Pharm.* **2021**, *22*, 100468. [CrossRef]
8. Wang, Y.; Li, M.W.; Hu, J.; Feng, W.P.; Li, J.J.; You, Z.X. Highly efficient and selective removal of Pb^{2+} by ultrafast synthesis of HKUST-1: Kinetic, isotherms and mechanism analysis. *Colloids Surf. A* **2022**, *633*, 127852. [CrossRef]
9. Makhanya, B.N.; Nyandeni, N.; Ndulini, S.F.; Mthembu, M.S. Application of green microalgae biofilms for heavy metals removal from mine effluent. *Phys. Chem. Earth.* **2021**, *124*, 103079. [CrossRef]
10. Faisal, A.A.H.; Al-walel, S.F.A.; Assi, H.A.; Naji, L.A.; Naushad, M. Waterworks sludge-filter sand permeable reactive barrier for removal of toxic lead ions from contaminated groundwater. *J. Water Process. Eng.* **2020**, *33*, 101112. [CrossRef]
11. Dhiman, V.; Kondal, N. ZnO Nanoadsorbents: A potent material for removal of heavy metal ions from wastewater. *Colloid Interface Sci. Commun.* **2021**, *41*, 100380. [CrossRef]
12. Jiang, Z.; Yang, J.; Ma, H.; Wang, L.; Xi, M.A. Reaction behaviour of Al_2O_3 and SiO_2 in high alumina coal fly ash during alkali hydrothermal process. *Trans. Nonferrous Met. Soc. China* **2015**, *25*, 2065–2072. [CrossRef]
13. Qi, L.; Yuan, Y. Characteristics and the behavior in electrostatic precipitators of high-alumina coal fly ash from the Jungar power plant, Inner Mongolia, China. *J. Hazard. Mater.* **2011**, *192*, 222–225. [CrossRef] [PubMed]

14. Yang, C.; Zhang, J.; Li, S.; Li, H.; Hou, X.; Zhu, G. Mechanisms of mechanochemical activation during comprehensive utilization of high-alumina coal fly ash. *Waste. Manag.* **2020**, *116*, 190–195. [CrossRef]
15. Zhang, J.; Li, S.; Li, H.; Wu, Q.; Xi, X.; Li, Z. Preparation of Al-Si composite from high-alumina coal fly ash by mechanical-chemical synergistic activation. *Ceram. Int.* **2017**, *43*, 6532–6541. [CrossRef]
16. Zhong, L.; Zhang, Y.; Zhang, Y. Extraction of alumina and sodium oxide from red mud by a mild hydro-chemical process. *J. Hazard. Mater.* **2009**, *172*, 1629–1634. [CrossRef]
17. Zhang, R.; Zheng, S.; Ma, S.; Zhang, Y. Recovery of alumina and alkali in Bayer red mud by the formation of andradite-grossular hydrogarnet in hydrothermal process. *J. Hazard. Mater.* **2011**, *189*, 827–835. [CrossRef]
18. Sun, Z.; Li, H.; Bao, W.; Wang, C. Mineral phase transition of desilicated high alumina fly ash with alumina extraction in mixed alkali solution. *Int. J. Miner. Process.* **2016**, *153*, 109–117. [CrossRef]
19. Li, H.; Hui, J.; Wang, C.; Bao, W.; Sun, Z. Removal of sodium (Na_2O) from alumina extracted coal fly ash by a mild hydrothermal process. *Hydrometallurgy* **2015**, *153*, 1–5. [CrossRef]
20. Wang, Z.; Ma, S.; Tang, Z.; Wang, X.; Zheng, S. Effects of particle size and coating on decomposition of alumina-extracted residue from high-alumina fly ash. *J. Hazard. Mater.* **2016**, *308*, 253–263. [CrossRef]
21. Mandaliev, P.; Wieland, E.; Dähn, R.; Tits, J.; Churakov, S.V.; Zaharko, O. Mechanisms of Nd(III) uptake by 11Å tobermorite and xonotlite. *Appl. Geochem.* **2010**, *25*, 763–777. [CrossRef]
22. Merlino, S.; Bonaccorsi, E.; Armbruster, T. Tobermorites; their real structure and order-disorder (OD) character. *Am. Miner.* **1999**, *84*, 1613–1621. [CrossRef]
23. Cao, P.X.; Li, G.H.; Luo, J.; Rao, M.J.; Jiang, H.; Peng, Z.W.; Jiang, T. Alkali-reinforced hydrothermal synthesis of lathy tobermorite fibers using mixture of coal fly ash and lime. *Constr. Build. Mater.* **2020**, *238*, 117655. [CrossRef]
24. Ding, J.; Tang, Z.H.; Ma, S.H.; Wang, Y.J.; Zheng, S.L.; Zhang, Y.; Shen, S.; Xie, Z.L. A novel process for synthesis of tobermorite fiber from high-alumina fly ash. *Cem. Concr. Compos.* **2016**, *65*, 11–18. [CrossRef]
25. Li, Y.; Bai, P.; Yan, Y.; Yan, W.; Shi, W.; Xu, R. Removal of Zn^{2+}, Pb^{2+}, Cd^{2+}, and Cu^{2+} from aqueous solution by synthetic clinoptilolite. *Microporous Mesoporous Mater.* **2019**, *273*, 203–211. [CrossRef]
26. Hsiao, Y.; Ho, T.; Shen, Y.; Ray, D. Synthesis of analcime from sericite and pyrophyllite by microwave-assisted hydrothermal processes. *Appl. Clay Sci.* **2017**, *143*, 378–386. [CrossRef]
27. Mzinyane, N.N.; Ofomaja, A.E.; Naidoo, E.B. Synthesis of poly (hydroxamic acid) ligand for removal of Cu(II) and Fe(II) ions in a single component aqueous solution. *S. Afr. J. Chem. Eng.* **2021**, *35*, 137–152. [CrossRef]
28. Gossuin, Y.; Hantson, A.L.; Vuong, Q.L. Low resolution benchtop nuclear magnetic resonance for the follow-up of the removal of Cu^{2+} and Cr^{3+} from water by amberlite IR120 ion exchange resin. *J. Water Process. Eng.* **2020**, *33*, 101024. [CrossRef]
29. Wang, C.; Yang, Z.; Song, W.; Zhong, Y.; Sun, M.; Gan, T.; Bao, B. Quantifying gel properties of industrial waste-based geopolymers and their application in Pb^{2+} and Cu^{2+} removal. *J. Clean. Prod.* **2021**, *315*, 128203. [CrossRef]
30. Anush, S.M.; Chandan, H.R.; Gayathri, B.H.; Asma Manju, N.; Vishalakshi, B.; Kalluraya, B. Graphene oxide functionalized chitosan-magnetite nanocomposite for removal of Cu(II) and Cr(VI) from waste water. *Int. J. Biol. Macromol.* **2020**, *164*, 4391–4402. [CrossRef]
31. Zhang, W.; Sun, P.; Liu, D.; Zhao, Q.; Zou, B.; Zhou, L.; Ye, Z. Method to fabricate porous multifunction β-cyclodextrin modified resin for ultrafast and efficient removal of Cu(II) and bisphenol A. *J. Taiwan Inst. Chem. Eng.* **2021**, *119*, 286–297. [CrossRef]
32. Liu, J.; Yang, X.; Liu, H.; Cheng, W.; Bao, Y. Modification of calcium-rich biochar by loading Si/Mn binary oxide after NaOH activation and its adsorption mechanisms for removal of Cu(II) from aqueous solution. *Colloids. Surf. A* **2020**, *601*, 124960. [CrossRef]
33. Sun, P.; Zhang, W.; Zou, B.Z.; Wang, X.Y.; Zhou, L.C.; Ye, Z.F.; Zhao, Q.L. Efficient adsorption of Cu(II), Pb(II) and Ni(II) from waste water by PANI@APTS-magnetic attapulgite composites. *Appl. Clay Sci.* **2021**, *209*, 106151. [CrossRef]
34. Chen, Y.N.; Zhao, W.; Zhao, H.; Dang, J.Q.; Jin, R.; Chen, Q. Efficient removal of Pb(II), Cd(II), Cu(II) and Ni(II) from aqueous solutions by tetrazole-bonded bagasse. *Chem. Phys.* **2020**, *529*, 110550. [CrossRef]
35. Joseph, I.V.; Tosheva, L.; Doyle, A.M. Simultaneous removal of Cd(II), Co(II), Cu(II), Pb(II), and Zn(II) ions from aqueous solutions via adsorption on FAU-type zeolites prepared from coal fly ash. *J. Environ. Chem. Eng.* **2020**, *8*, 103895. [CrossRef]
36. Cheng, T.; Chen, C.; Tang, R.; Han, C.H.; Tian, Y. Competitive Adsorption of Cu, Ni, Pb, and Cd from Aqueous Solution Onto Fly Ash-Based Linde F(K) Zeolite. *Iran. J. Chem. Chem. Eng.* **2018**, *37*, 61–72.
37. Pawar, R.R.; Lalhmunsiama Bajaj, H.C.; Lee, S.M. Activated bentonite as a low-cost adsorbent for the removal of Cu(II) and Pb(II) from aqueous solutions: Batch and column studies. *J. Ind. Eng. Chem.* **2016**, *34*, 213–223. [CrossRef]

International Journal of
Environmental Research and Public Health

MDPI

Article

The Potential Use of *Nephelium lappaceum* Seed as Coagulant–Coagulant Aid in the Treatment of Semi-Aerobic Landfill Leachate

Hamidi Abdul Aziz [1,2,*], Nur Syahirah Rahmat [1] and Motasem Y. D. Alazaiza [3]

1 School of Civil Engineering, Engineering Campus, Universiti Sains Malaysia, Nibong Tebal 14300, Malaysia; nursyahirahrahmat92@gmail.com
2 Solid Waste Management Cluster, Science and Technology Research Centre, Engineering Campus, Universiti Sains Malaysia, Nibong Tebal 14300, Malaysia
3 Department of Civil and Environmental Engineering, College of Engineering (COE), A'Sharqiyah University (ASU), Ibra 400, Oman; my.azaiza@gmail.com
* Correspondence: cehamidi@usm.my

Abstract: Chemical-based coagulants and flocculants are commonly used in the coagulation–flocculation process. However, the drawbacks of using these chemical materials have triggered researchers to find natural materials to substitute or reduce the number of chemical-based coagulants and flocculants. This study examines the potential application of *Nephelium lappaceum* seeds as a natural coagulant–coagulant aid with Tin (IV) chloride ($SnCl_4$) in eliminating suspended solids (SS), colour, and chemical oxygen demand (COD) from landfill leachate. Results showed that the efficiency of *Nephelium lappaceum* was low when used as the main coagulant in the standard jar test. When $SnCl_4$ was applied as a single coagulant, as much as 98.4% of SS, 96.8% of colour and 82.0% of COD was eliminated at an optimal dose of 10.5 g/L and pH 7. The higher removal efficiency of colour (88.8%) was obtained when 8.40 g/L of $SnCl_4$ was applied with a support of 3 g/L of *Nephelium lappaceum*. When $SnCl_4$ was utilised as a coagulant, and *Nephelium lappaceum* seed was used as a flocculant, the removal of pollutants generally improved. Overall, this research showed that *Nephelium lappaceum* seed is a viable natural alternative for treating landfill leachate as a coagulant aid.

Keywords: landfill leachate; coagulation–flocculation; *Nephelium lappaceum*; COD; solid waste

Citation: Aziz, H.A.; Rahmat, N.S.; Alazaiza, M.Y.D. The Potential Use of *Nephelium lappaceum* Seed as Coagulant–Coagulant Aid in the Treatment of Semi-Aerobic Landfill Leachate. *Int. J. Environ. Res. Public Health* **2022**, *19*, 420. https://doi.org/10.3390/ijerph19010420

Academic Editor: Paul B. Tchounwou

Received: 18 November 2021
Accepted: 23 December 2021
Published: 31 December 2021

1. Introduction

Landfilling is a very famous, recognised, and effective method for municipal solid waste management globally, owing to its low cost and simple operational mechanisms [1–3]. Landfill leachate is one of the major issues of landfilling method [3]. Leachate is usually generated by water infiltration through the waste; hence, it often includes suspended or dissolved solids from the disposed of materials. Generally, leachate is characterised in terms of pH, total suspended solids (TSS), dissolved oxygen (DO), chemical oxygen demand (COD), biological oxygen demand (BOD), total Kjeldahl nitrogen (TKN), ammonium nitrogen (NH_3-N), heavy metals, and others [4]. The composition of the leachate and the standard effluent quality vary by landfill location and legislation. The composition of leachate is affected by a number of parameters, including waste content, local meteorological conditions, landfill physicochemical conditions, and landfill age [5]. The structure and characteristics of the leachate are the most important elements influencing treatment method selection [6,7].

Biological treatment has proved its efficiency in treating different leachate types [8]. However, for old leachate with high COD and ammonium concentrations, numerous heavy metals, and low biodegradability (high COD/ BOD ratio), the efficacy of this approach did not show good removal results [9]. On the other hand, old leachate was found to be removed better using chemical and physical methods as compared to young leachate [10,11].

In general, it is easy to predict the quantity of long-term leachate; however, since the precipitation affects the short-term leachate, it is difficult to predict its quantity [12].

Generally, at the early stages of leachate production from waste decomposition, leachate has a large content of BOD_5, biodegradable and non-biodegradable compounds like volatile fatty acids [11]. However, the leachate from old landfills typically contains a significant percentage of non-biodegradable compounds such as fulvic and humic-like chemicals [13]. Furthermore, inorganic molecules such as NH_3-N are produced by the hydrolysis and fermentation of nitrogen-containing fractions of biodegradable waste substrates in stabilised leachate formed from older landfills. [14]. When the waste is stabilised, and leachate is collected and transported for treatment, washout influences the accumulating concentration of NH_3 [15].

Physicochemical methods are commonly employed to pre-treat or polish the leachate. Coagulation–flocculation is a well-known chemical technique for leachate pre-treatment [16]. This method is normally used to remove suspended particles and recalcitrant compounds like humic acids, heavy metals, polychlorinated biphenyls (PCBs), and absorbable organic halides from the leachate (AOX). The mechanism of the pre-treatment method primarily entails charge neutralisation between the negative charge colloids and cationic hydrolysis products, followed by impurity amalgamation by flocculation [17]. The main parameters removed by this method are total suspended solids (TSS) and colloidal particles [18].

Several materials are used as coagulants and flocculants in the coagulation–flocculation process, where inorganic metal salts are extensively used for this purpose. Excessive use of these inorganic metal salts, on the other hand, may have negative impacts on the environment and represent a health concern [19]. To lessen and remove the negative impacts on living beings and the environment, natural coagulants/flocculants are critical to be developed and applied for landfill leachate treatment.

Malaysia is a tropical country that has a wide variety of fruits. As a result, numerous food companies make canned fruits to take advantage of the availability. In Malaysia and Thailand, the *Nephelium lappaceum* canning industry is well-established, and it involves the manufacture of *Nephelium lappaceum* fruits in syrup [20]. The *Nephelium lappaceums* are deseeded during the canning process, and the seeds are generally discarded as waste by-products. It is, therefore, beneficial to value add the application of the *Nephelium lappaceum* seed from being just disposed of. Zurina et al. [20] used *Nephelium lappaceum* seed polysaccharide as a natural coagulant to remove water turbidity. However, to the author's knowledge, no published research has specifically focused on the use of *Nephelium lappaceum* seed as a natural coagulant or flocculant in the treatment of landfill leachate.

The main objective of this study was to scrutinise the effectiveness of using *Nephelium lappaceum* seed as a natural coagulant in leachate treatment. Specifically, *Nephelium lappaceum* seed was used in two states; where the first state involved the *Nephelium lappaceum* seed as a sole coagulant, whereas the second state used the *Nephelium lappaceum* seed as a flocculant associated with Tin (IV) chloride ($SnCl_4$). The application of *Nephelium lappaceum* seed combined with $SnCl_4$ has not been used in leachate treatment to date. The target parameters were suspended solids (SS), colour, and COD, which generally present at high concentrations in many old landfill sites.

2. Materials and Methods

2.1. Sampling and Characterisation of Leachate

The case study leachate samples were taken from one of the old landfills in Malaysia, i.e., the Alor Pongsu Landfill Site (APLS). APLS is one of the existing landfills in Malaysia located in Bagan Serai, Perak, with coordinates of 5°05′00.6″ N 100°35′53.1″ E. APLS was constructed in 2000 as an open dumping site and categorised as a stabilised and matured landfill as it has been operating for more than 10 years [21]. The total area of APLS is about 10 acres where it consists of a large dumping site and three leachate collection ponds. These ponds are in an anaerobic condition which acts as raw collection ponds. Leachate samples were collected six times within four consecutive months, starting from January

2020 to April 2020. All procedures of sampling (grab method), storage, and preservation of landfill leachate followed the Standard Methods for the Examination of Water and Wastewater [22]. The pH, temperature, dissolved oxygen (DO), and total dissolved solids (TDS) were analysed on-site using YSI 556-Probe System. All samples were transported to the laboratory and maintained in a cool chamber at 4 °C for further use. This procedure is to minimise any reaction, including chemical and biological reactions that may affect leachate properties [2]. The COD, BOD_5, NH_3-N, colour, and SS were conducted in the laboratory immediately upon arrival.

2.2. Extraction and Characterisation of *Nephelium lappaceum* Seed

Fresh *Nephelium lappaceum* fruits were bought from a roadside store at Kulim, Kedah, Malaysia. The fruits were deseeded and washed using tap water before drying them in an oven at a temperature of 105 °C for 10 to 15 min. This is to ensure easy peeling off of the seed coat of *Nephelium lappaceum* seeds, thus preventing the increase of moisture content in the seeds that may damage its natural content. The seeds were then crushed into powder by using a domestic blender and kept in airtight containers until further use.

A 50 g package of *Nephelium lappaceum* seed powder was mixed with one litre of distilled water to produce 50 g/L of *Nephelium lappaceum* seed stock solution. After that, the mix solution was blended again for two minutes. The suspension was then filtered using muslin cloth in a beaker to become the stock solution before being characterised in terms of pH, zeta potential, molecular weight, particle size, surface morphology, and functional groups. The molecular weight of *Nephelium lappaceum* seed was determined by using Malvern Zetasizer Nano ZS. Surface morphology was conducted using ZEISS SUPRA 35VP Field Emission Scanning Electron. Meanwhile, determination of the functional group of *Nephelium lappaceum* seed was conducted using Perkin-Elmer System 2000 FTIR spectrometer. $SnCl_4$ as the main coagulant was the analytical grade of pentahydrate, $SnCl_4.5H_2O$ with 98% purity grade, supplied by Telaga Madu Sdn. Bhd. Malaysia.

2.3. Coagulation–Flocculation Experiment

A jar test apparatus was used to conduct the coagulation–flocculation procedures. The jar test device has six agitators together with 2.5 cm × 7.5 cm rectangular blades for stirring and mixing. The velocity of the instrument and mixing duration were controlled manually. Before the coagulation–flocculation process, samples of landfill leachate were taken from the storage chamber and placed outside until they were conditioned to room temperature. After that, the sample was agitated thoroughly to ensure uniform mixing. Six 1000 mL beakers were filled with 500 mL of agitated sample. The pH of the samples was adjusted with 3 M HCl or 3 M NaOH until the desired pH was obtained. The desired pH was determined according to the optimum dosage, as shown in the next section.

The coagulation–flocculation process involves three important stages: rapid mixing, slow mixing, and settlement. A combination of fast mixing (100 rpm in 8 min), slow mixing (30 rpm in 20 min), and a settlement for 30 min was used for the experiments using *Nephelium lappaceum* seed as the main coagulant [23]. For the studies employing $SnCl_4$ as coagulant without and with *Nephelium lappaceum* seed as flocculant, the rapid mixing of 200 rpm in 1.5 min, slow mixing of 40 rpm in 20 min, and settlement for 20 min were used [24]. The removal efficiencies of key parameters such as SS, colour, and COD in landfill leachate samples were used to estimate the efficacy of the coagulation–flocculation process. Equation (1) [25] was used to compute the removal efficiencies of each pollutant:

$$\text{Removal (\%)} = [(C_i - C_f)/C_i] \times 100 \quad (1)$$

where C_i: is the initial concentration of the samples, C_f: is the final concentration of samples.

2.4. Determination of Optimum pH and Coagulant Dosage

Preliminary tests were carried out to obtain the pre-determined dosage and pH conditions. A preliminary test for *Nephelium lappaceum* seed was carried out using various dosages (0.5–15 g/L) of *Nephelium lappaceum* on raw leachate, whereas for $SnCl_4$, dosages between 3.5 to 17.5 g/L were used. Then, the pre-determined dosages obtained were applied in varied pH landfill leachate from 2 to 12 to determine the best pH value for both *Nephelium lappaceum* seed and $SnCl_4$. Together with this experiment, a pH control experiment was also conducted by adjusting the pH of the sample without any coagulant. This control process was carried out to ensure that the removal of the pollutant was not made solely by acids or alkalis.

Once the optimum pH was determined, a new set of experiments were carried out to determine the optimum dosages of *Nephelium lappaceum* seed using different dosages varying from 0 (as a control) to 3 g/L. Meanwhile, different dosages of $SnCl_4$ varying from 0 (as control) to 17.5 g/L have been applied to the samples with the optimum pH value of $SnCl_4$.

A jar test experiment was conducted to evaluate the use of $SnCl_4$ as a coagulant in conjunction with *Nephelium lappaceum* seed as a flocculant. These were undertaken at different $SnCl_4$ and *Nephelium lappaceum* seed dosages. Different dosages of $SnCl_4$ were used, ranging from 5.60 g/L to 8.75 g/L. For *Nephelium lappaceum* seed, the dosage was from 0 g/L to 4 g/L. The same operating conditions for a jar test as before were applied.

3. Results and Discussion

3.1. Landfill Leachate Compositions

Table 1 shows the main characteristics of the raw leachate of APLS. The leachate has a temperature in the range of 26.48 °C to 32.07 °C with an average of 29.47 °C, which is within the permissible limit of below 40 °C as stated by the Malaysia Environmental Quality Act 1974. The leachate has a pH of 8.04 to 8.90, with an average of pH 8.59. This represents an old type of leachate. A young landfill (up to 5 years) has a pH from 3.7 to 6.5, whereas matured landfill has a pH higher than 7. This is because young landfill leachate contains carboxylic acid and bicarbonate ions that contribute to the low pH [26]. The concentration of the dissolved oxygen was between 2.14 and 4.38 mg/L with an average of 3.17 mg/L. The amount of DO in landfill leachate is usually low due to the various chemical pollutants as well as the high temperature that may restrain the concentration of oxygen in the landfill leachate [27]. COD value was between 2532 and 4215 mg/L with an average of 3440 mg/L. This far exceeded the standard discharge limit of 400 mg/L [22]. COD level in APLS landfill leachate was slightly lower than that of the previous study [28]. This is due to numerous factors that may affect the COD level of the landfill leachate, like; landfill age, the solid waste compositions, site characteristics, and climate conditions [29]. Besides that, APLS is a matured and stabilised landfill as the leachate contains a higher COD level, which is below 4000 mg/L [21]. BOD_5 of landfill leachate from APLS ranged from 160 to 333 mg/L with an average of 241 mg/L., also exceeding the standard discharge limit of 20 mg/L. It can be concluded that the APLS is a stabilised landfill site as the BOD_5 is less than 4000 mg/L [30]. The BOD_5/COD ratio in leachate was about 0.070 within the range of 0.057 to 0.080. In general, a mature and stabilised landfill has a lower fraction of BOD_5/COD, which is less than 0.1 due to the lower biodegradable fraction of organic pollutants in mature landfill leachate [21]. The low BOD_5/COD ratio indicates that it would be difficult to treat the leachate using a biological process. NH_3-N in APLS leachate was in the range of 1040 to 1357 mg/L and is slightly lower than the previous work [31]. The SS ranged from 350 to 604 mg/L and fell within the range of mature leachate. The leachate is negatively charged with a zeta potential reading of −21.5 mV; hence the majority of particles in leachate have the tendency to repel each other. Therefore, a coagulant with zeta potential higher than +21 mV is needed to neutralise the charges of particles in leachate and aggregate them into large flocs [12]. A recent study by Ramli et al. [32] showed similar results for the leachate characteristics. Figure 1 shows the particle size distribution of the leachate sample.

Table 1. Alor Pongsu Landfill Site (APLS) raw leachate characteristics.

Parameter	Min	Max	Average *	Permissible Limit
Temperature (°C)	26.48	32.07	29.47	40
pH	8.04	8.90	8.59	6.0–9.0
Total Dissolved Solids (g/L)	8.532	9.950	9.440	
Dissolved Oxygen (mg/L)	2.14	4.38	3.17	
BOD5 (mg/L)	160	333	241	20
COD (mg/L)	2533	4215	3440	400
Ratio BOD_5/COD	0.057	0.080	0.070	
Suspended Solids (mg/L)	350	604	469	50
Colour (PtCo)	13750	18733	16829	100
NH_3-N	1040	1357	1227	5
Zeta Potential (mV)	−22.4	−20.7	−21.5	

* Average value for six samples taken from January to April 2020.

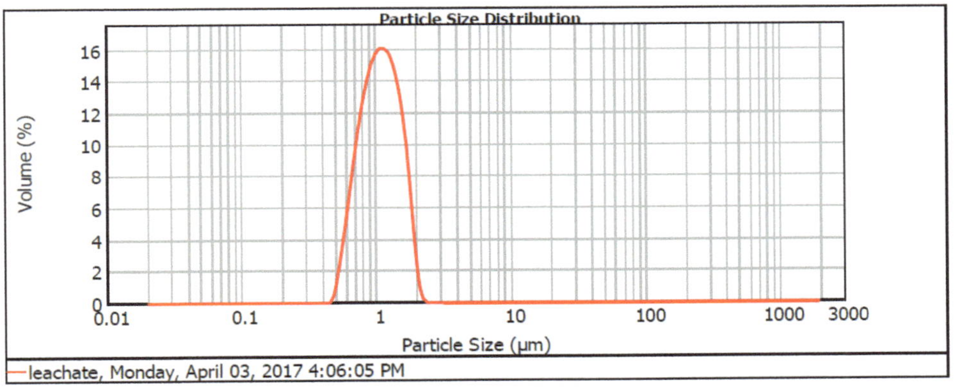

Figure 1. Particle size distribution for (a) landfill leachate.

3.2. Characteristic of *Nephelium lappaceum* Seeds

Nephelium lappaceum seed has an acidic condition (ranging between 3.56 and 5.36) with an average pH of 4.45 ± 0.71. Generally, *Nephelium lappaceum* is an acidic fruit. According to Arenas et al. [33], *Nephelium lappaceum* has a pH range from 4.6 to 5.5, of which the pH was similar to its seed. The Zeta potential of *Nephelium lappaceum* seed was recorded to be about 5.46 ± 0.615 mV at its natural pH. By knowing its zeta potential, the *Nephelium lappaceum* seed is a cationic coagulant which means that it has a positively charged particle surface. Figure 2 shows the particle size distribution of *Nephelium lappaceum*.

The molecular weight of *Nephelium lappaceum* seed was measured at 3850 kDa. It was slightly higher if compared to Longan seed, where its molecular weight was 3280 kDa [16]. The increase in molecular weight of the coagulant, according to Ebeling et al. [34], encourages more particles to adhere to each other by providing extended branches and channels, resulting in larger flocs in the coagulation–flocculation process. Nevertheless, different coagulants may have their own coagulation–flocculation behaviour despite their high molecular weight [6].

The functional group of *Nephelium lappaceum* seed was determined using an FTIR Spectrometer, and the results are given in Figure 3 and Table 2. It is shown that *Nephelium lappaceum* seed contains many functional groups such as alcohol (O-H), amide (N-H), and carboxylic groups (C=O). However, only certain functional groups play an important role in coagulation–flocculation processes. According to Zafar et al. [35], natural coagulant neutralises the particle charges and aggregates flocs through bridging mechanisms by biopolymers such as polysaccharides which can be found naturally in fruits and vegetables. Polysaccharides are one of the carbohydrate subgroups that contain carbon, hydrogen, and oxygen [36]. Furthermore, Aziz and Sobri [37] reported that carboxyl (C=O), hydroxyl (O-H), and amino (amine or amide) grouped together with polymeric chains have the capability to enhance flocculation through bridging. From the results, the content of (O-H), amide (N-H), and carboxylic groups (C=O) in *Nephelium lappaceum* seed are expected to serve as ion bridge or binding sites in coagulation–flocculation processes.

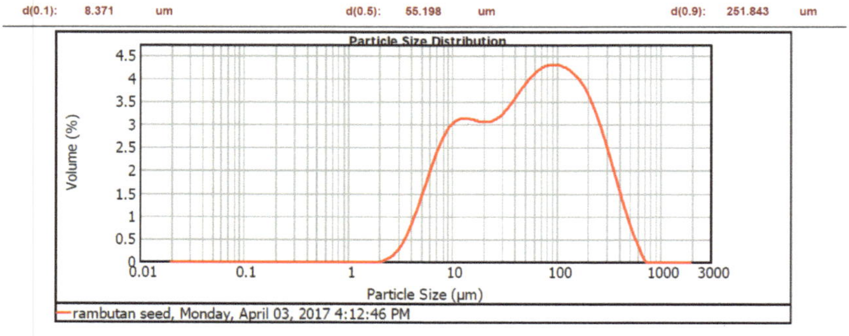

Figure 2. Particle size of *Nephelium lappaceum* seeds.

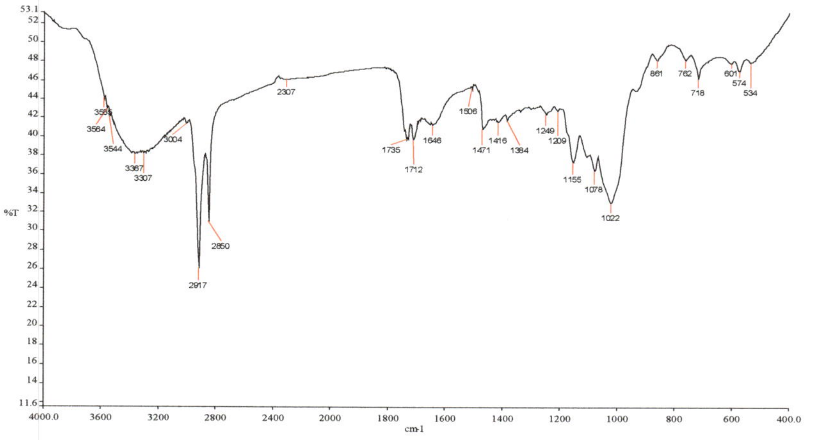

Figure 3. FTIR spectra of *Nephelium lappaceum* seed.

Table 2. Functional groups of *Nephelium lappaceum* seed.

Wavelength (cm^{-1})	Wavenumber Range	Intensity	Bond	Group Vibration	Functional Group
3585	3700–3584	Medium	O–H	Stretching	Alcohol
3564	3700–3500	Medium	N–H	Stretching	Amide
3544	3550–3200	Strong	O–H	Stretching	Alcohol
3367	3400–3300	Medium	N–H	Stretching	Aliphatic primary amine
3307	3300–2500	Strong	O–H	Stretching	Carboxylic acid

Table 2. *Cont.*

Wavelength (cm^{-1})	Wavenumber Range	Intensity	Bond	Group Vibration	Functional Group
3004	3000–2800	Strong	N–H	Stretching	Amine salt
	3100–3000	Medium	C–H	Stretching	Alkene
2917	3000–2800	Strong	N–H	Stretching	Amine salt
2850	3000–2840	Medium	C–H	Stretching	Alkane
2307	2280–2440	Medium	P–H	Stretching	Phosphine
1735	1750–1735	Strong	C=O	Stretching	δ–lactone
1712	1725–1705	Strong	C=O	Stretching	Aliphatic ketone
	1720–1706	Strong	C=O	Stretching	Carboxylic acid
	1650–1580	Medium	N–H	Bending	Amine
	1650–1566	Medium	C=C	Stretching	Cyclic alkene
1506	1550–1500	Strong	N–O	Stretching	Nitro compound
1471	1400–1480	Strong	C–O	Stretching	Methylene
1416	1440–1395	Medium	O–H	Bending	Carboxylic acid
	1420–1330	Medium	O–H	Bending	Alcohol
1384	1415–1380	Strong	S=O	Stretching	Sulfate
	1410–1380	Strong	S=O	Stretching	Sulfonyl chloride
	1400–1000	Strong	C–F	Stretching	Fluoro compound
1249	1342–1266	Strong	C–N	Stretching	Aromatic amine
	1250–1020	Medium	C–N	Stretching	Amine
1209	1275–1200	Strong	C–O	Stretching	Alkyl aryl ether
	1225–1200	Strong	C–O	Stretching	Vinyl ether
	1210–1163	Strong	C–O	Stretching	Ester
1155	1170–1155	Strong	S=O	Stretching	Sulphonamide
	1165–1150	Strong	S=O	Stretching	Sulfonic acid
	1160–1120	Strong	S=O	Stretching	Sulfone
	1205–1124	Strong	C–O	Stretching	Tertiary alcohol
1078	1085–1050	Strong	C–O	Stretching	Primary alcohol
1022	1050–1040	Strong	CO–O–CO	Stretching	Anhydride
861	880 ± 20	Strong	C–H	Bending	1,2,4—trisubstituted
	880 ± 20	Strong	C–H	Bending	1,3—disubstituted
762	850–550	Strong	C–Cl	Stretching	Halo compound
	780 ± 20	Strong	C–H	Bending	1,2,3—trisubstituted
	755 ± 20	Strong	C–H	Bending	1,2—disubstituted
	750 ± 20	Strong	C–H	Bending	Monosubstituted benzene derivative
718	730–665	Strong	C=C	Bending	Alkene
601	850–550	Strong	C–C	Stretching	Halo compound
	690–515	Strong	C–Br	Stretching	Halo compound
574	850–550	Strong	C–Cl	Stretching	Halo compound
534	850–550	Strong	C–Cl	Stretching	Halo compound
	600–500	Strong	C–I	Stretching	Halo compound

The surface morphology of *Nephelium lappaceum* seed is shown in Figure 4. According to FESEM micrographs, it can be seen that the *Nephelium lappaceum* seed surface has non-porous traits and mostly appears to be in globular shape. The surface morphology of *Nephelium lappaceum* seed appears similar to the Cassava peel, whereby it is smooth and globular in shape that covers the surface of Cassava peel [38]. It appears that most of the surfaces that contain starch are globular in shape due to the starch granules. Polysaccharides in the *Nephelium lappaceum* seed usually can be hydrolysed to hexose component, which is also known as starch. Moreover, *Nephelium lappaceum* seed has traces of alkaloids, sugar (1.25%), starch (25%), and ash (2%). Starch is considered as one of the biopolymers that are needed in coagulation–flocculation as it comprises a mixture of two polymers of anhydrous glucose units and amylopectin. These biopolymers can enhance the mechanism of adsorption, charge neutralisation, as well as perform inter-particle bridging in coagulation–flocculation.

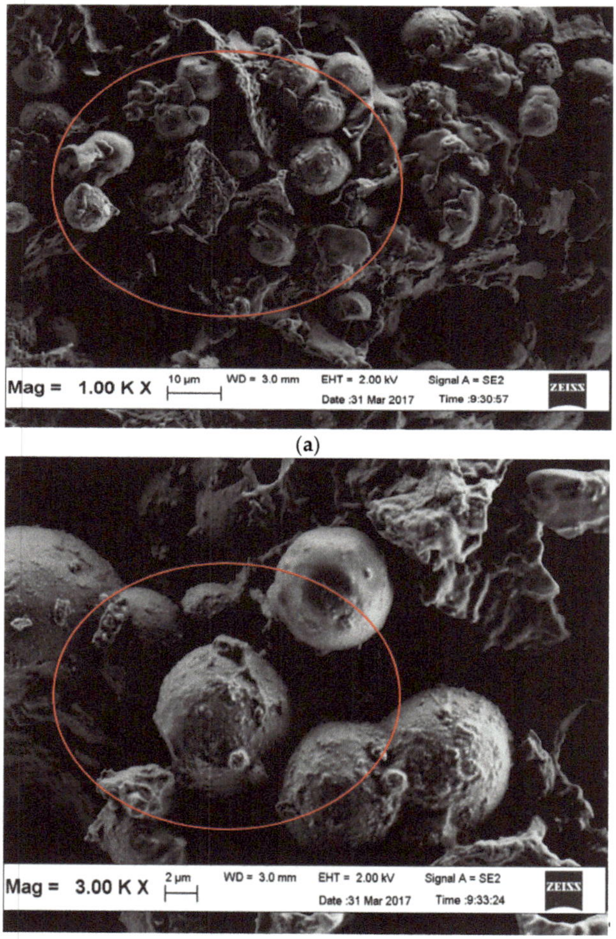

(b)

Figure 4. FESEM of *Nephelium lappaceum* seed. (**a**) 1000× magnification; (**b**) 3000× magnification.

3.3. Zeta Potential and Particle Size

According to Omar et al. [39], pH affects the zeta potential and aggregation of particles. For this, the zeta potential and particle size of leachate and *Nephelium lappaceum* seed solution were studied throughout a pH range of 2 to 12. Figure 5 shows the zeta potential of leachate, which was decreased from −9.53 mV to −29.5 mV as pH increased from pH 2 to 12, respectively. This trend indicates that landfill leachate was completely affected by pH. This observation is possibly due to the swift ionisation of both OH^- and H^+ on the surface of particle size when the leachate pH was adjusted using acid (HCl) and alkali (NaOH) [39]. At pH 8, the zeta potential was about −22.4 ± 1.95 mV, and this signified that the particle surface of leachate was naturally negatively charged. Thus, to neutralise the leachate surface charge, a positively charged coagulant with almost the same value as leachate particles was needed, and thus, the coagulation–flocculation process will then be highly effective [40,41].

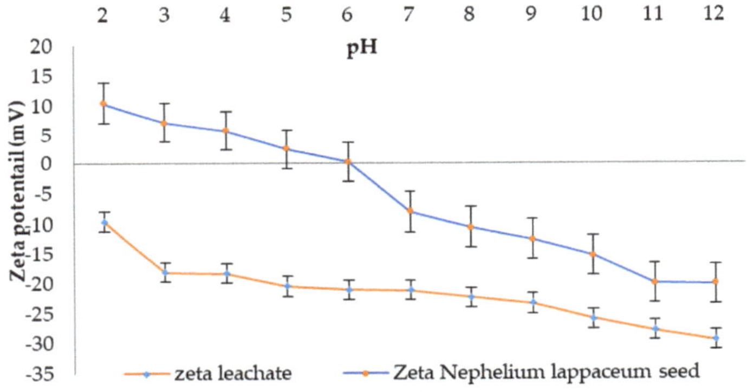

Figure 5. pH effect on Zeta potential of leachate.

Figure 6 shows the relationship between the particle size of leachate and pH. Similar to zeta potential, leachate particle size also dropped as the pH increased. However, when the landfill leachate was added with 0.5 M NaOH in order to increase the pH from pH 8 to pH 12, the particle size subsequently decreased from 228.1 to 207.1 d.nm. According to Cao, et al. [39], the reaction of negatively charged ion, OH^- from NaOH, has interfered with the destabilisation of particles in the leachate by repelling each other and thus, made it impossible to form larger flocs. The particle size of landfill leachate increased from 228.1 d.nm to 9520 d.nm when the pH of the leachate was altered to lower pHs, which varied from pH 8 to pH 2. This is most likely due to the fact that when the solution was acidic, the density of H^+ hydrolysates was extremely high, thus, dominate the charge neutralisation of the particles and forming a larger floc [39].

Kruszewski and Cyrankiewicz [42] stated that the aggregating agents (KCl or HCl) caused a substantial increase in particle size intensity as chloride was presented in these aggregating agents, which is capable of promoting the aggregation process and thus enhancing the flocculation process. Hence, when the pH of leachate increased, the particle size also increased. Characterisation of *Nephelium lappaceum* seed in terms of zeta potential and particle size as a function of pH were performed to evaluate the suitability of *Nephelium lappaceum* seed to be used as a natural coagulant. The association between pH and zeta potential, as well as particle size of *Nephelium lappaceum* seed, are depicted in Figures 1 and 2, respectively. It can be seen that the trend of zeta potential of *Nephelium lappaceum* seed was also decreased as pH increased, which was identical to the behaviour of the leachate zeta potential trend. Between pH 2 to pH 6, the zeta potential of *Nephelium lappaceum* seed was observed to be in the positively charged surface, which was between 10.47 to

0.20 mV, respectively, and it was near to the point of zero charges (PZC). With a further increase of pH, which was from pH 7 to pH 12, the zeta potential decreased drastically from −8.00 to −20.1 mV. This signifies that *Nephelium lappaceum* seed became negatively charged surface after pH 6. During the coagulation–flocculation process, neutralisation of negatively charged surface of landfill leachate required positively charged coagulant.

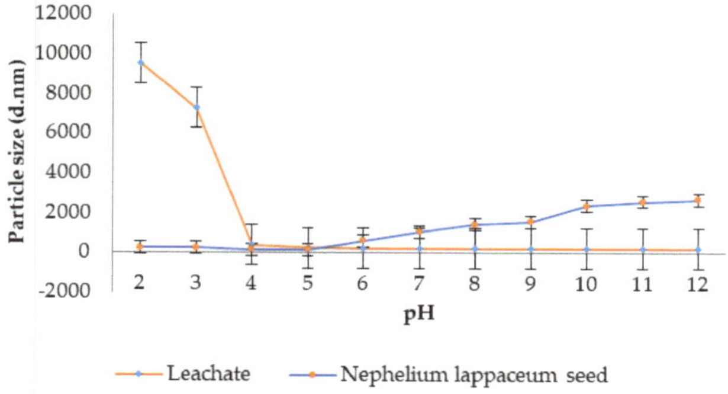

Figure 6. pH effect on particle size of leachate.

The trend of *Nephelium lappaceum* seed particle size with respect to pH was opposite to landfill leachate particles size. Overall, the particle size of Nephelim lappaceum seed increased as the pH increased. However, at pH 2 to 5, the *Nephelium lappaceum* seed particle size decreased from 260.0 to 162.7 d.nm. This might be due to the fact that some negatively charged surface in the solution attracted with the positively charged ions that were hydrolysed from HCl acid [42], which was added in the *Nephelium lappaceum* solution to decrease the pH and thus, this had promoted the flocculation of particles. Furthermore, when pH increased from its natural pH until pH 12, the particle size was observed to increase from 162.7 to 2690 d.nm. Supposedly, the particles would scatter when NaOH was added into the sample due to the negatively charged ions of OH-. However, in *Nephelium lappaceum* seed with a natural positively charged surface, it would promote aggregation as NaOH has strong OH-ions that would help to form the inter-bridge between particles [43]. Based on the results of zeta potential and particle size of *Nephelium lappaceum* seed solution, it can be concluded that it is wise to choose *Nephelium lappaceum* seed at pH less than 6 because when the pH is high (pH 7–12), the *Nephelium lappaceum* seed has negatively charged surface and large particle size. This will affect the coagulation–flocculation process as the negatively charged surface of *Nephelium lappaceum* seed will repel the negatively charged surface of landfill leachate.

3.4. Optimum Operating Conditions of Nephelium lappaceum Seed as a Sole Coagulant

3.4.1. Optimum pH

The effect of pH of *Nephelium lappaceum* seed as a natural coagulant in the coagulation–flocculation process was evaluated. A preliminary test was carried out first on raw leachate by using different dosages (0.5–15 g/L) of *Nephelium lappaceum* seed to find the pre-determined dosage used in determining the optimum pH of *Nephelium lappaceum* seed. Figure 7 demonstrates the effect of varying pH on the coagulation–flocculation process using a pre-determined dose of *Nephelium lappaceum* seed (1.5 g/L) and a pre-determined dosage of *Nephelium lappaceum* seed (1.5 g/L). At pH 2, the removal efficiency of SS, colour, and COD was 96.4%, 90.3%, and 56.4%, respectively. According to Bruice [43], acids are substances that can transfer a proton to a base and are known as proton donors. Therefore, at lower pH, a high concentration of H^+ due to the hydrolysation of HCl has neutralised negatively charged particles in the landfill leachate. The charge neutralisation and complex reaction in the process were dominated by hydrogen ions from HCl, and hence, the suspension, together with pollutants, began to flocculate [44]. Beyond pH 4, the removals of all parameters were fairly low (less than 30%). This could be explained by the fact that *Nephelium lappaceum* seed naturally has a milky colour. Beyond pH 3, *Nephelium lappaceum* seeds also lose the effectiveness to remove SS. This observation could be explained by the fact that the *Nephelium lappaceum* seed solution has a large particle size and higher SS in the solution, and thus, this has added the quantity of SS in leachate. In conclusion, the *Nephelium lappaceum* was ineffective when used as the sole coagulant. Nonetheless, pH 6 was observed to be better for all the parameters. In conclusion, the best and optimum pH value for *Nephelium lappaceum* seed to remove SS, colour, and COD was at pH 6 whereby it removed 19.2% COD and 21.8% colour. Table 3 shows a comparison of previous studies that used natural coagulants in landfill leachate treatment with the current studies.

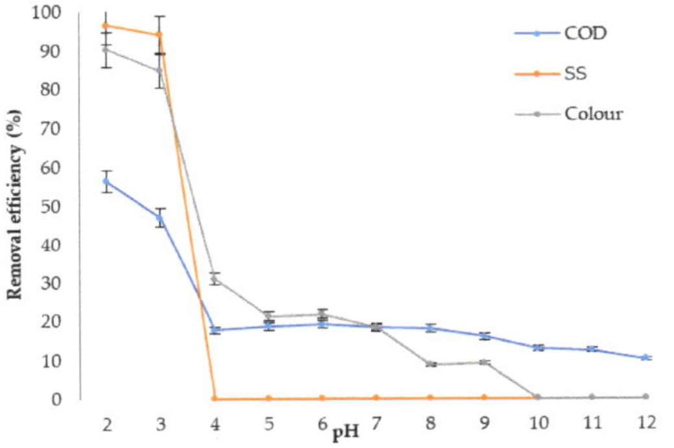

Figure 7. Effect of pH on pollutant removal using 1.5 g/L of *Nephelium lappaceum* seed dosage.

Table 3. Comparison of previous and current studies for best pH values of various natural coagulant.

Natural Coagulant	Optimum pH	Pollutant Removal Rate (%)			Reference
		SS	Colour	COD	
Commercial sago starch (CSS)	4	29.5	15.1	28.0	[37]
Chitosan	4	-	14.7	-	[32]
Durian seed starch	6	-	34.0	-	[45]
Tamarindus indica seed (TiS)	4	-	30.2	7.5	[46]
Longan seed	4	-	10.3	5.2	[19]
Nephelium lappaceum seed	**6**	**-**	**21.8**	**19.2**	**Current Study**

The bold is to show the results of the current study.

3.4.2. Optimum Dosage

After the optimum pH value was obtained, the optimum dosage of *Nephelium lappaceum* seed was determined. In this study, various dosages of *Nephelium lappaceum* seed within the range of 0 to 3 g/L were employed to attain the optimum dosage at optimum pH value, which was pH 6 that was determined earlier. Figure 8 depicts the removal effectiveness for all the parameters. The results demonstrated that COD was the best pollutant removed (35% reduction at 2 g/L) by *Nephelium lappaceum* seed compared to others. The overall performance of *Nephelium lappaceum* seed alone as a coagulant was considered to be fairly low for all the pollutants. In summary, COD and colour removal were at their maximum performance at dosage 2 g/L, but *Nephelium lappaceum* seed could not remove SS from landfill leachate at this dosage, although it could remove 5.6% of SS at dosage 1 g/L. Therefore, the best and optimum dosage of *Nephelium lappaceum* seed to remove the pollutants from landfill leachate was at dosage 2 g/L. Table 4 displays the comparison of *Nephelium lappaceum* seed with other natural coagulants from previous studies.

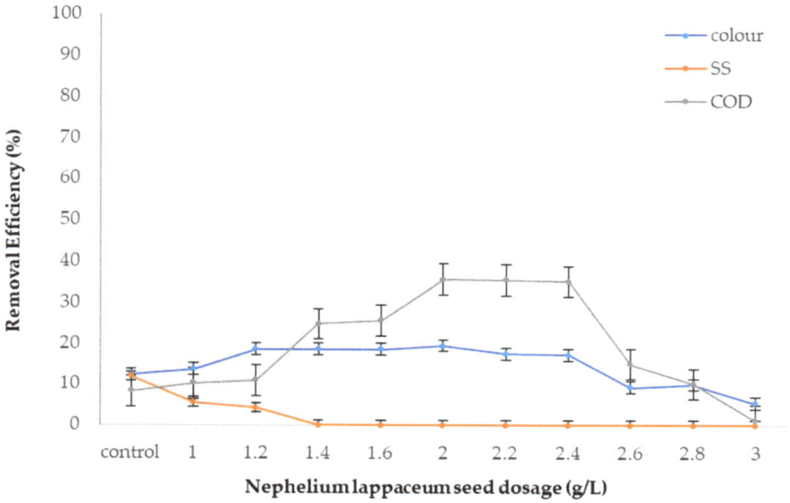

Figure 8. SS, colour, and COD removal efficiency with different dosages of *Nephelium lappaceum* seed at pH 6.

Table 4. Comparison of optimum dosage of *Nephelium lappaceum* with other natural coagulants from previous studies.

Natural Coagulant.	Optimum Dosage (g/L)	Pollutant Removal Rate (%)			Source
		SS	Colour	COD	
Commercial sago starch	6	29.5	15.1	28	[37]
Chitosan	0.06	-	14.7	-	[32]
Durian seed starch	4	-	34	-	[45]
Tamarindus indica seed	5.00	-	41.90	5.90	[46]
Longan seed	2.00	29.5	15.10	28.00	[19]
Nephelium lappaceum seed	**2.00**	**-**	**19.48**	**35.62**	**Current Study**

The bold is to show the results of the current study.

From the comparison above, it can be seen that *Nephelium lappaceum* seed has the same dosage as Longan seed in removing pollutants, i.e., SS, colour, and COD effectively with medium dosage, whereas other coagulants such as commercial sago starch, durian seed starch, and *Tamarindus indica* seed needs slightly higher dosages. Nevertheless, among all-natural coagulants, *Nephelium lappaceum* seed demonstrates the best COD removal in landfill leachate. This is probably due to its positive zeta potential, as well as the presence of biopolymer in the seed that may enhance the inter-particles bridging between the seed and landfill leachate that ensures excellent removal of COD excellently.

3.5. Optimum Operating Conditions of $SnCl_4$ a Sole Coagulant

3.5.1. Optimum pH

The effectiveness of $SnCl_4$ at different pH was investigated with the same procedures as before. Various dosages of $SnCl_4$ (3.5–17.5 g/L) were examined. The pre-determined dosage of $SnCl_4$, which was 10.5 g/L was employed at a broad range of landfill leachate pH from 2–12. Based on the results, the highest removal of SS, colour, and COD was noted at pH 7, whereby it removed 98.4% of SS, 96.8% of colour, and 8.0% of COD. According to Coffman [47], inorganic metal salts were normally found between pH 5 to 7. Yong and Aziz [46] found out that the optimum pH of PAC was pH 6, and Coffman [47] found that the optimum pH for titanium dioxide (TiO_2) was pH 5 for landfill leachate treatment.

3.5.2. Optimum Dosage

After obtaining the optimum pH value for $SnCl_4$, the dosage of $SnCl_4$ was then determined by using various dosages of $SnCl_4$ within the range of 0 to 17.50 g/L, and the results are presented in Figure 9. The removals for all the pollutants sharply increased from low dosages until they reached a steady state with 98%, 85% and 83% reductions of SS, colour and COD, respectively, at 10.5 g/L $SnCl_4$. Beyond this point, the removals dropped slightly. This is due to charge reversal, re-dispersal, and re-stabilisation of colloidal particles [43–45] and the overdosage phenomenon where the $SnCl_4$ would cause the particle to re-disperse.

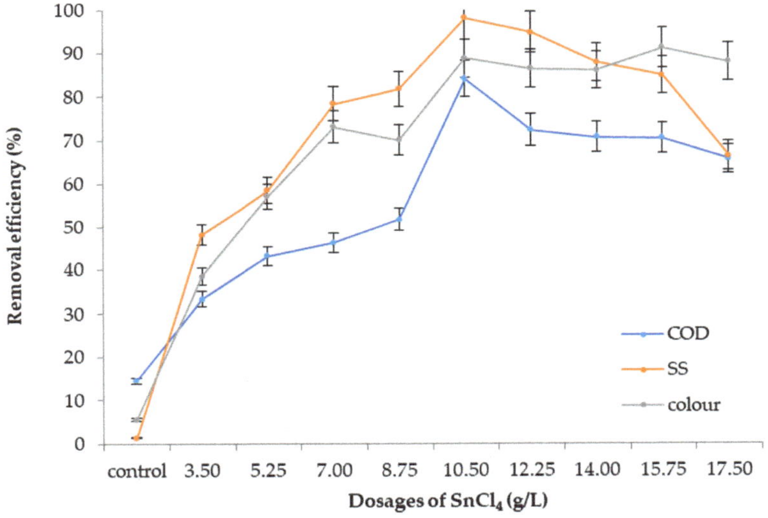

Figure 9. SS, colour, and COD removal efficiency with varied $SnCl_4$ dosages at optimum pH 7.

3.6. Performance of $SnCl_4$ as a Coagulant and Nephelium lappaceum Seeds as a Flocculant

$SnCl_4$ was used as a coagulant in this study, with the help of *Nephelium lappaceum* as a flocculant/coagulant aid. At the same time, the potential of *Nephelium lappaceum* seed assisting in reducing the coagulant dosages was also tested at a reducing rate from 8.75 g/L to 1.0 g/L of $SnCl_4$. The efficiency of $SnCl_4$ as a coagulant and *Nephelium lappaceum* seed as a flocculant was compared to the optimum and best dosage of $SnCl_4$ as a sole coagulant, which was 10.50 g/L.

Figure 10a illustrates the colour removal efficiency. At 8.40 g/L of $SnCl_4$ and 3 g/L of *Nephelium lappaceum* the removal for colour was about 89%. This is just slightly lower than 92% reduction when $SnCl_4$ was used alone at higher concentrations (10.5 g/L). Not much significant difference was noted for SS (Figure 10b), where nearly 90% reduction was observed compared with 96% when $SnCl_4$ was used alone. A slight reduction was due to the fact that *Nephelium lappaceum* seed particles are large in size [48,49], which renders them less effective in removing SS from landfill leachate [50]. Moreover, as discussed earlier, *Nephelium lappaceum* seed becomes negatively charged after pH 6. This might affect the particles charge neutralisation. Therefore, the removal of SS from landfill leachate mostly depended on $SnCl_4$. Further lowering of $SnCl_4$ dosages would only cause deterioration in treatment performance [51].

However, for the case of COD (Figure 10c), the performance of *Nephelium lappaceum* as coagulant aid was not so significant. The removal with and without *Nephelium lappaceum* was 76% and 89%, respectively. Generally, introducing *Nephelium lappaceum* seed as a flocculant with increasing dosage resulted in further reduction of COD removal efficiency [52]. This is because *Nephelium lappaceum* is an organic matter that contributes to the COD readings.

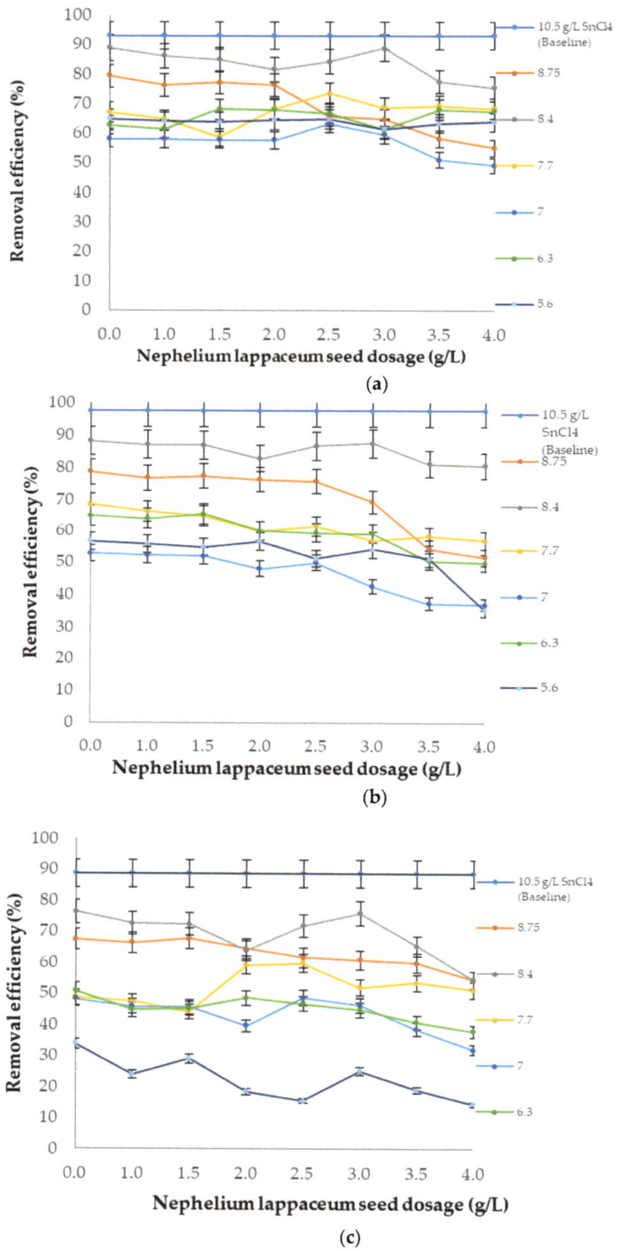

Figure 10. Removal efficiencies of pollutants at pH 7 with varied dosages of $SnCl_4$ as coagulant and *Nephelium lappaceum* seed as a flocculant. (**a**) Colour; (**b**) Suspended solids (SS); (**c**) COD.

The summary of the findings is plotted in Figure 11.

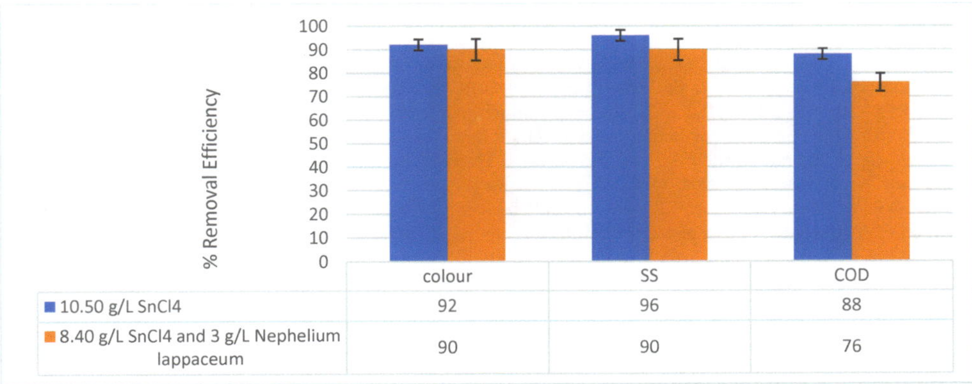

Figure 11. A summary of the removal efficiencies of pollutants conducted at pH 7 with varied dosages of $SnCl_4$ as coagulant and *Nephelium lappaceum* seed as a flocculant.

4. Conclusions

In this study, *Nephelium lappaceum* seeds were used as an alternative coagulant for leachate treatment and compared with *Nephelium lappaceum* seeds as a flocculant associated with Tin (IV) chloride ($SnCl_4$) as a main coagulant to remove suspended solids (SS), colour, and COD from landfill leachate. Results showed that *Nephelium lappaceum* was not effective when it was used as a sole coagulant. The best-operating conditions in terms of pH and dosage were found at pH 6 and 2 g/L, respectively. The removal of colour and COD were 19.5% and 35.7%, respectively. At pH 7, 10.5 g/L of $SnCl_4$ exhibited a good performance as a sole coagulant with 98%, 85%, and 63% reductions respectively for SS, colour and COD. At 8.40 g/L of $SnCl_4$ and 3 g/L of *Nephelium lappaceum* as flocculant, the removal for colour was about 89%, just slightly lower than 92% reduction when $SnCl_4$ was used alone. No significant difference was noted for SS where nearly 90% reduction was observed compared with 96% when $SnCl_4$ was used alone. However, *Nephelium lappaceum* as coagulant aid was not so significant for COD. With further investigations and optimisation work, *Nephelium lappaceum* could be a potential option to be an alternative to metal salt coagulant in treating landfill leachate.

Author Contributions: Conceptualization, methodology, supervision, validation, H.A.A.; data curation, investigation, formal analysis, N.S.R.; writing—original draft preparation, writing—review and editing, M.Y.D.A. All authors have read and agreed to the published version of the manuscript.

Funding: The authors would like to express the highest gratitude to Universiti Sains Malaysia (USM) for the provision of facilities required for this study. This work was also partly funded by the FRGS grant No. FRGS/1/2018/TK01/USM/01/1 and USM RUI grant No. 1001/PAWAM/8014081.

Institutional Review Board Statement: Not applicable.

Informed Consent Statement: Not applicable.

Data Availability Statement: The data presented in this study are available on request from the corresponding author.

Acknowledgments: Thank you is also extended to the Majlis Daerah Kerian (MDK), Perak, for their cooperation during the study and also to the laboratory technical staff who also contributed to the leachate sampling and laboratory works.

Conflicts of Interest: The authors declare no conflict of interest.

References

1. Aziz, H.A.; Alias, S.; Assari, F.; Adlan, M.N. The use of alum, ferric chloride and ferrous sulphate as coagulants in removing suspended solids, colour and COD from semi-aerobic landfill leachate at controlled pH. *Waste Manag. Res.* **2007**, *25*, 556–565. [CrossRef]
2. Yusoff, M.S.; Aziz, H.A.; Alazaiza, M.Y.; Rui, L.M. Potential use of oil palm trunk starch as coagulant and coagulant aid in semi-aerobic landfill leachate treatment. *Water Qual. Res. J.* **2019**, *54*, 203–219. [CrossRef]
3. Zainal, S.F.F.S.; Aziz, H.A.; Omar, F.M.; Alazaiza, M.Y. Influence of Jatropha curcas seeds as a natural flocculant on reducing Tin (IV) tetrachloride in the treatment of concentrated stabilised landfill leachate. *Chemosphere* **2021**, *285*, 131484. [CrossRef]
4. Aziz, H.A.; Yusoff, M.S.; Adlan, M.N.; Adnan, N.H.; Alias, S. Physico-chemical removal of iron from semi-aerobic landfill leachate by limestone filter. *Waste Manag.* **2004**, *24*, 353–358. [CrossRef]
5. Segundo, I.D.B.; Moreira, F.C.; Silva, T.F.; Webler, A.D.; Boaventura, R.A.; Vilar, V.J. Development of a treatment train for the remediation of a hazardous industrial waste landfill leachate: A big challenge. *Sci. Total Environ.* **2020**, *741*, 140165. [CrossRef]
6. Umar, M.; Aziz, H.A.; Yusoff, M.S. Trends in the use of Fenton, electro-Fenton and photo-Fenton for the treatment of landfill leachate. *Waste Manag.* **2010**, *30*, 2113–2121. [CrossRef]
7. Nai, C.; Tang, M.; Liu, Y.; Xu, Y.; Dong, L.; Liu, J.; Huang, Q. Potentially contamination and health risk to shallow groundwater caused by closed industrial solid waste landfills: Site reclamation evaluation strategies. *J. Clean. Prod.* **2021**, *286*, 125402. [CrossRef]
8. Zainal, S.F.F.S.; Abdul Aziz, H.; Mohd Omar, F.; Alazaiza, M.Y. Sludge performance in coagulation-flocculation treatment for suspended solids removal from landfill leachate using Tin (IV) chloride and Jatropha curcas. *Inter. J. Environ. Analyt. Chem.* **2021**, 1–15. [CrossRef]
9. Wei, Y.; Ye, Y.; Ji, M.; Peng, S.; Qin, F.; Guo, W.; Ngo, H.H. Microbial analysis for the ammonium removal from landfill leachate in an aerobic granular sludge sequencing batch reactor. *Bioresour. Technol.* **2021**, *324*, 124639. [CrossRef] [PubMed]
10. Aziz, H.A.; Daud, Z.; Adlan, M.N.; Hung, Y.-T. The use of polyaluminium chloride for removing colour, COD and ammonia from semi-aerobic leachate. *Inter. J. Environ. Eng.* **2009**, *1*, 20–35. [CrossRef]
11. Bashir, M.J.; Aziz, H.A.; Yusoff, M.S.; Huqe, A.; Mohajeri, S. Effects of ion exchange resins in different mobile ion forms on semi-aerobic landfill leachate treatment. *Water Sci. Technol.* **2010**, *61*, 641–649. [CrossRef]
12. Al-Hamadani, Y.A.; Yusoff, M.S.; Umar, M.; Bashir, M.J.; Adlan, M.N. Application of psyllium husk as coagulant and coagulant aid in semi-aerobic landfill leachate treatment. *J. Hazard. Mater.* **2011**, *190*, 582–587. [CrossRef] [PubMed]
13. Yusoff, M.S.; Adam, N.H.; Watalinggam, K.; Aziz, H.A.; Alazaiza, M.Y. Effectiveness of oil palm frond activated carbon for removing COD, color and Fe from landfill leachate. *J. Eng. Technol. Sci.* **2021**, *53*, 512–521. [CrossRef]
14. Bashir, M.J.; Aziz, H.A.; Yusoff, M.S. New sequential treatment for mature landfill leachate by cationic/anionic and anionic/cationic processes: Optimisation and comparative study. *J. Hazard. Mater.* **2011**, *186*, 92–102. [CrossRef] [PubMed]
15. Aziz, H.A.; AlGburi, H.R.; Alazaiza, M.Y.D.; Noor, A.F.M. Sequential treatment for stabilised landfill leachate by ozonation–adsorption and adsorption–ozonation methods. Inter. *J. Environ. Sci. Technol.* **2021**, *18*, 861–870.
16. Chiang, L.-C.; Chang, J.-E.; Chung, C.-T. Electrochemical oxidation combined with physical–chemical pre-treatment processes for the treatment of refractory landfill leachate. *Environ. Eng. Sci.* **2001**, *18*, 369–379. [CrossRef]
17. Duan, J.; Gregory, J. Coagulation by hydrolysing metal salts. *Adv. Coll. Interface Sci.* **2003**, *100*, 475–502. [CrossRef]
18. Kurniawan, T.A.; Lo, W.-H.; Chan, G.Y. Physico-chemical treatments for removal of recalcitrant contaminants from landfill leachate. *J. Hazard. Mater.* **2006**, *129*, 80–100. [CrossRef]
19. Aziz, H.A.; Rahim, N.A.; Ramli, S.F.; Alazaiza, M.Y.; Omar, F.M.; Hung, Y.T. Potential use of Dimocarpus longan seeds as a flocculant in landfill leachate treatment. *Water* **2018**, *10*, 1672. [CrossRef]
20. Zurina, A.Z.; Mohd Fadzli, M.; Ghani, A.; Abdullah, L. Preliminary study of rambutan (*Nephelium lappaceum*) seed as potential biocoagulant for turbidity removal. In *Advanced Materials Research*; Trans Tech Publications Ltd.: Bäch, Switzerland, 2014; Volume 12, pp. 96–105.
21. Aziz, H.A.; Ling, T.J.; Haque, A.A.M.; Umar, M.; Adlan, M.N. Leachate treatment by swim-bed bio fringe technology. *Desalination* **2011**, *276*, 278–286. [CrossRef]
22. APAH. *Standard Methods for the Examination of Water and Waste Water*, 21th ed.; American Public Health Association: Washington, DC, USA, 2005.
23. Thanki, A.; Padhiyar, H.; Kathiriya, T.; Shah, D.; Singh, N.K. Applications of Moringa Oleifera for wastewater treatment: Concepts and approaches. In *Green Innovation, Sustainable Development, and Circular Economy*; CRC Press: Boca Raton, FL, USA, 2020; Volume 11, pp. 129–140.
24. Chekli, L.; Phuntsho, S.; Kim, J.E.; Kim, J.; Choi, J.Y.; Choi, J.-S. A comprehensive review of hybrid forward osmosis systems: Performance, applications and future prospects. *J. Memb. Sci.* **2016**, *497*, 430–449. [CrossRef]
25. Awang, N.A.; Aziz, H.A. Hibiscus rosa-sinensis leaf extract as coagulant aid in leachate treatment. *Appl. Water Sci.* **2012**, *2*, 293–298. [CrossRef]
26. Sackey, L.N.A.; Meizah, K. Assesement of the quality of leachate at Sarbah landfill site at Weija in Accra. *J. Environ. Chem. Ecotox.* **2015**, *7*, 56–61.
27. Maqbool, F.; Bhatti, Z.; Malik, A.; Pervez, A.; Mahmood, Q. Effect of landfill leachate on the stream water quality. *Inter. J. Environ. Res.* **2011**, *5*, 491–500.

28. Aziz, H.; Yii, Y.; Syed Zainal, S.; Ramli, S.; Akinbile, C. Effects of using Tamarindus indica seeds as a natural coagulant aid in landfill leachate treatment. *Global NEST J.* **2018**, *20*, 373–380.
29. Aziz, S.Q.; Aziz, H.A.; Yusoff, M.S.; Bashir, M.J.; Umar, M. Leachate scharacterisation in semi-aerobic and anaerobic sanitary landfills: A comparative study. *J. Environ. Manag.* **2010**, *91*, 2608–2614. [CrossRef] [PubMed]
30. Guo, J.-S.; Abbas, A.A.; Chen, Y.-P.; Liu, Z.-P.; Fang, F.; Chen, P. Treatment of landfill leachate using a combined stripping, Fenton, SBR, and coagulation process. *J. Hazard. Mater.* **2010**, *178*, 699–705. [CrossRef] [PubMed]
31. Zakaria, S.N.F.; Abdul Aziz, H.; Amr, A.; Salem, S. Performance of ozone/$ZrCl_4$ oxidation in sstabilised landfill leachate treatment. In *Applied Mechanics and Materials*; Trans Tech Publications Ltd.: Bäch, Switzerland, 2015; pp. 501–506.
32. Ramli, S.F.; Aziz, H.A.; Omar, F.M.; Yusoff, M.S.; Halim, H.; Kamaruddin, M.A.; Ariffin, K.S.; Hung, Y.T. Reduction of COD and highly coloured mature landfill leachate by tin tetrachloride with rubber seed and polyacrylamide. *Water* **2021**, *13*, 3062. [CrossRef]
33. Arenas, M.G.H.; Angel, D.N.; Damian, M.T.M.; Ortiz, D.T.; Díaz, C.N.; Martinez, N.B. Characterization of rambutan (*Nephelium lappaceum*) fruits from outstanding mexican selections. *Rev. Bras. Frutic.* **2010**, *32*, 1098–1104. [CrossRef]
34. Ebeling, J.M.; Rishel, K.L.; Sibrell, P.L. Screening and evaluation of polymers as flocculation aids for the treatment of aquacultural effluents. *Aquacultur. Eng.* **2005**, *33*, 235–249. [CrossRef]
35. Zafar, M.S.; Tausif, M.; Mohsin, M.; Ahmad, S.W.; Zia-ul-Haq, M. Potato starch as a coagulant for dye removal from textile wastewater. *Water Air Soil Poll.* **2015**, *226*, 244. [CrossRef]
36. Solidum, J. Potential nutritional and medicinal sources from fruit peels in Manila, Philippines. *Inter. J. Biosci. Biochem. Bioinform.* **2012**, *2*, 270–274. [CrossRef]
37. Aziz, H.A.; Mohamad Sobri, N.I. Extraction and application of starch-based coagulants from sago trunk for semi-aerobic landfill leachate treatment. *J. Environ. Sci. Poll. Res.* **2015**, *22*, 16943–16950. [CrossRef]
38. Mohd-Asharuddin, S.; Othman, N.; Mohd Zin, N.; Tajarudin, H.A. A chemical and morphological study of cassava peel: A potential waste as coagulant aid. In *MATEC Web of Conferences*; EDP Sciences: Les Ulis, France, 2016; Volume 103.
39. Omar, F.M.; Aziz, H.A.; Stoll, S. Aggregation and disaggregation of ZnO nanoparticles: Influence of pH and adsorption of Suwannee River humic acid. *Sci. Total Environ.* **2014**, *468*, 195–201. [CrossRef]
40. Ab Ghani, Z.; Yusoff, M.S.; Zaman, N.Q.; Zamri, M.F.M.A.; Andas, J. Optimization of preparation conditions for activated carbon from banana pseudo-stem using response surface methodology on removal of color and COD from landfill leachate. *Waste Manag.* **2017**, *62*, 177–187. [CrossRef] [PubMed]
41. Cao, B.; Gao, B.; Xu, C.; Fu, Y.; Liu, X. Effects of pH on coagulation behavior and floc properties in Yellow River water treatment using ferric based coagulants. *Chin. Sci. Bulletin.* **2010**, *55*, 1382–1387. [CrossRef]
42. Kruszewski, S.; Cyrankiewicz, M. Aggregated silver sols as SERS substrates. *Acta Phys. Pol. Ser. A Gen. Phys.* **2012**, *121*, A68. [CrossRef]
43. Borja, K.; Mercado, J.; Combatt, E. Methods of mechanical dispersion for determining granulometric fractions in soils using four dispersant solutions. *Agron. Colomb.* **2015**, *33*, 253–260. [CrossRef]
44. Bruice, P.Y. *Organic Chemistry: Pearson New International Edition*; Pearson Education Limited: New York, NY, USA, 2014.
45. Mishra, A.; Bajpai, M. The flocculation performance of Tamarindus mucilage in relation to removal of vat and direct dyes. *Bioresour. Technol.* **2006**, *97*, 1055–1059. [CrossRef] [PubMed]
46. Muaz, M.; Yusof, M.S.; Aziz, H.A. The study of flocculant characteristics for landfill leachate treatment using starch based flocculant from Durio zibethinus seed. *Adv. Environ. Biol.* **2014**, *8*, 129–135.
47. Yong, C.; Aziz, H. Utilisation of Tamarindus Indica Seed as Natural Coagulant/Flocculant in Landfill Leachate Treatment. Master's Thesis, Universiti Sains Malaysia, Nibong Tebal, Malaysia, 2016. (unpublished).
48. Coffman, N. *Recovering Titanium Dioxide (TiO_2) after Its Useto Treat Leachate for Reuse on Future Leachate Flows*; Florida Atlantic University: Boca Raton, FL, USA, 2015.
49. Badrus, Z. Potential of natural flocculant in coagulation-flocculation wastewater treatment process. In *E3S Web of Conferences*; EDP Sciences: Les Ulis, France, 2018; p. 05006.
50. Baghvand, A.; Zand, A.D.; Mehrdadi, N.; Karbassi, A. Optimising coagulation process for low to high turbidity waters using aluminum and iron salts. *Am. J. Environ. Sci.* **2010**, *6*, 442–448. [CrossRef]
51. Osei, J.A. Utilization of Agricultural Food Waste Products for Bioethanol Generation, Kiambu County, Kenya. Ph.D. Thesis, Kenyatta University, Nairobi, Kenya, 2021.
52. Kalyanasundaram, K. Photochemical applications of solar energy: Photocatalysis and photodecomposition of. *Photochemistry* **2013**, *41*, 182.

International Journal of *Environmental Research and Public Health*

MDPI

Article

Seasonal and Technological Shifts of the WHO Priority Multi-Resistant Pathogens in Municipal Wastewater Treatment Plant and Its Receiving Surface Water: A Case Study

Łukasz Jałowiecki [1,*], Jakub Hubeny [2], Monika Harnisz [2] and Grażyna Płaza [1]

1 Environmental Microbiology Unit, Institute for Ecology of Industrial Areas, 40-844 Katowice, Poland; g.plaza@ietu.pl

2 Department of Engineering of Water Protection and Environmental Microbiology, Faculty of Geoengineering, University of Warmia and Mazury, 10-719 Olsztyn, Poland; jakub.hubeny@uwm.edu.pl (J.H.); monikah@uwm.edu.pl (M.H.)

* Correspondence: l.jalowiecki@ietu.pl

Abstract: The present study was focused on the identification of multi-resistant bacteria from the WHO priority pathogens list in the samples taken from different stages of the full-scale municipal wastewater treatment plant and receiving water. Additionally, the seasonal variations of the selected multi-resistant pathogens were analyzed in the samples. In order to the aim of the study, the metagenomic DNA from the collected samples was isolated and sequenced. The samples were collected in three campaigns (spring, summer, autumn). Metagenomic DNA was isolated by the commercial kits, according to the manufacturer's instruction. Illumina sequencing system was employed, and the R program was used to metagenomic analysis. It was found that the wastewater samples and receiving water contained the multi-resistant bacteria from the WHO priority pathogens list. The seasonal and technological variations affected the distribution of the pathogens in the wastewater. No effect of the effluent on the pathogens in the receiving water was observed. The results indicated that antibiotic-resistant "priority pathogens" from the WHO list are there in the waste- and receiving water. Technological process and seasons effected their distribution in the environment. Metagenomic analysis can be used as sufficient tool in microbiological and human health risk assessment.

Keywords: wastewater; multi-antibiotic resistant pathogens; WHO priority pathogens list; metagenome analysis

Citation: Jałowiecki, Ł.; Hubeny, J.; Harnisz, M.; Płaza, G. Seasonal and Technological Shifts of the WHO Priority Multi-Resistant Pathogens in Municipal Wastewater Treatment Plant and Its Receiving Surface Water: A Case Study. *Int. J. Environ. Res. Public Health* **2022**, *19*, 336. https://doi.org/10.3390/ijerph19010336

Academic Editors: Yung-Tse Hung, Hamidi Abdul Aziz and Issam A. Al-Khatib

Received: 22 November 2021
Accepted: 20 December 2021
Published: 29 December 2021

1. Introduction

Hundreds of antibiotics have been discovered or developed over 70 years, starting their huge application in medicine, veterinary, and agriculture [1,2]. Due to unmonitored and overuse of antibiotics, antimicrobial resistance (AMR) has been recognized by WHO as a major threat to global health [3,4]. The main reasons of AMR increase are the growth of various microbial infections, as well as the overuse and over-prescription of antimicrobials [5]. Human activities are mainly responsible for high levels and prevalence of antimicrobial resistance, which is now considered a modern phenomenon [6,7]. As calculated, around 10 million deaths are suspected by 2050 to occur annually due to AMR, and the cost of the situation is estimated ~US $100 trillion in total [8].

Among the anthropogenic sources, effluents from urban wastewater treatment plants (WWTPs) are suspected to be the main source of antibiotics resistance [9]. In recent years, wastewater plants were considered "hot spots" not only for antibiotic resistance but also for bacterial pathogens [10–14]. In addition, WWTPs are places where human activities and the environment are linked, and the horizontal transfer of resistance determinants among environmental microorganisms and clinically relevant pathogens is facilitated.

Many new techniques in treatment plants (biological, chemical, and physical) or their combinations have been now adopted to remove microbes from wastewater [15]. Despite the new approaches used, the pathogenic microbes exist in treated wastewater, and they are still considered a potential hazard to human health and the environment [16–18]. At present, biological indicators, such as the total coliforms, fecal coliforms, and *Escherichia coli*, are used to assess the quality of water and evaluate potential health risks, neglecting the problem of multi-antimicrobial resistance and multi-resistant pathogens [19].

WHO has developed a priority pathogens list (PPL) of antibiotic-resistant bacteria that pose the greatest threat to human health [20]. The PPL defines the priority of 12 pathogens, based on resistance to the most popular and widely applicable antibiotics for treating multi-drug resistant bacteria, such as carbapenems, third-generation cephalosporins, vancomycin, methicillin, penicillins, or fluoroquinolones. Twelve multi-resistant bacteria, posing the greatest threat to human health, are categorized to three priority tiers: critical, high, and medium, in terms of their resistance to the selected antimicrobials [21]. The pathogens are listed in Table 1 by group and antibiotic resistance characteristics.

Table 1. WHO priority pathogens list.

Categories	Bacteria	Antibiotic-Resistance
Priority 1: Critical	*Acinetobacter baumannii*	Carbapenem-resistant
	Pseudomonas aeruginosa	Carbapenem-resistant
	Enterobacteriaceae	Carbapenem-resistant, ESBL-producing
Priority 2: High	*Enterococcus faecium*	Vancomycin-resistant
	Staphylococcus aureus	Methicillin-resistant, vancomycin-resistant
	Helicobacter pylori	Clarithromycin-resistant
	Campylobacter spp.	Fluoroquinolone-resistant
	Salmonella spp.	Fluoroquinolone-resistant
	Neisseria gonorrhoeae	Cephalosporin-resistant, fluoroquinolone-resistant
Priority 3: Medium	*Streptococcus pneumoniae*	Penicillin-non-susceptible
	Haemophilus influenzae	Ampicillin-resistant
	Shigella spp.	Fluoroquinolone-resistant

In this context, the purpose of this study was to identify the multi-resistant bacteria from the WHO priority pathogens list in the samples taken from different wastewater treatment unit processes and receiver surface water. Additionally, the seasonal variations of the selected pathogens were analyzed in the samples. Our intention was to answer the question of whether, and how, the technological process and seasons affected the distribution of pathogens. This research may help us in understanding the dissemination of pathogenic bacteria, in terms of their significance as microbiological indicators in microbiological risk assessment (MRA). In the future, the results could be useful in developing appropriate treatment systems and their proper management.

2. Materials and Methods

2.1. Description of WWTP and Sample Collection

The samples were collected in three seasons: summer (June 2018), autumn (November 2018), and spring (March 2019) from the full-scale municipal wastewater treatment plant in the south part of Poland (50°5′35.881 N; 19°3′32.202 E). The detailed description of technological process of WWTP was presented by Rolbiecki et al. [22]. In Table 2, some technological parameters are presented.

Table 2. Some technological parameters of wastewater treatment plant and meteorological indicators.

Technological Parameters *	Unit	Wastewater	June 2018	Autumn 2018	March 2019
Flow	m^3/month		960,077	722,516	793,234
Temperature	°C	Influent	19.5	10.5	14.5
		Effluent	21	19	13.5
pH		Effluent	7.2	7.2	7.2
COD	mg/L	Influent	963	672	970
		Effluent	36.5	30.5	35.0
BOD5	mg/L	Influent	435	290	340
		Effluent	4.8	4.7	6.0
Suspension	mg/L	Influent	525	310	455
		Effluent	5.8	6.3	7.2
N_{TOT}	mg/L	Influent	106.1	78.1	84.2
		Effluent	10.4	7.9	6.2
N_{NH4+}	mg/L	Influent	32.25	48.90	56.90
		Effluent	0.36	0.22	0.31
P_{TOT}	mg/L	Influent	8.88	14.9	11.4
		Effluent	0.74	1.1	0.5
SRT	d		17	19	20
HRT	h		9	9	9
SS	kg/m^3		4.5	5.0	5.5
Meterological parameters *					
Temperature	°C		20.4	4.5	6.1
Rainfall	mm		71	14	59

* Monthly average; abbreviations: HRT—hydraulic retention time; SS—suspended solids; SRT—solid retention time.

During the sampling campaigns, 30 grab samples were collected in the sampling points presented in Figure 1. The sampling and transportation procedures used are described by Rolbiecki et al. [22] and Płaza et al. [23].

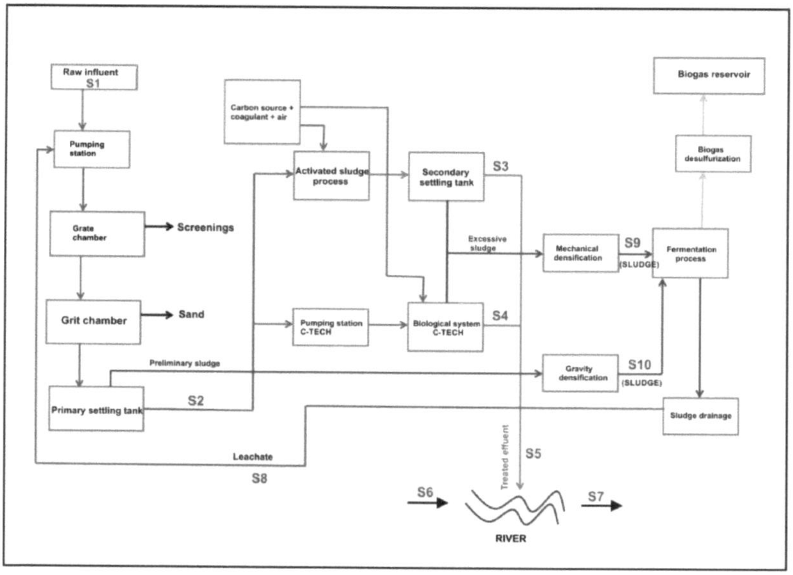

Figure 1. Scheme of wastewater treatment plant with sampling point.

2.2. DNA Extraction and Illumina Sequencing

The wastewater samples were filtered in triplicate through a 0.22 μm micropore membrane (Whatman, Merck, Germany) and kept at −80 °C until DNA extraction. The Power Water kit (MoBio Laboratories Inc., Carlsbad, CA, USA) was used for isolation of metagenomic DNA (met DNA) from wastewater and river waters. Metagenomic DNA (met DNA) from sewage sludge samples was isolated by the Power Soil kit (MoBio Laboratories Inc., Carlsbad, CA, USA). All isolations were performed according to the manufacturer's instruction. Quantity and quality of metDNA were determined by microspectrophotometry (BioSpectrometer, Eppendorf, Hamburg, Germany).

Extracted DNA samples were sent to Macrogen Inc. (Seoul, Korea) for library preparation and sequencing. An Illumina HiSeq sequencing system was used for sequencing.

2.3. Data Analysis

Sequencing results were uploaded to the MetaGenome Rapid Annotation Subsystems Technology (MG-RAST version 4.0.3) server as FASTQ files for analysis [24]. Details of the metagenomic analysis are presented in the paper by Płaza et al. [23]. The sequences were submitted to NCBI database, and they are under BioProject number ID: PRJNA666519.

Statistical analysis, in the form of PCA and Spearman rank correlation, were performed using Statistica v.13.3.

3. Results

In this paper, the technological and seasonal changes of antibiotic resistant priority pathogens from the WHO list were evaluated. Bacteria from the WHO priority pathogens list constituted from 2.15% of relative abundance of total identified bacteria in spring to 3.44% and 3.26% in autumn and summer, respectively. Twelve WHO priority pathogens were detected in all wastewater samples and the receiving surface water. The most abundant were pathogens belonging to priority 1, e.g., critical level in the waste- and receiving water. The dominant group of bacteria, from the WHO priority pathogens list, was *Enterobacteriaceae*. The average values of *Enterobacteriaceae* in all samples were 36.3%, 34.5%, and 23.7% in summer, autumn, and spring, respectively. *E. coli* and *Klebsiella pneumoniae* dominated in *Enterobacteriaceae*. The second most frequently isolated pathogen was *Acinetobacter baumannii*. The percentages of the bacteria were 18.6%, 17.5%, and 16.7% in summer, autumn, and spring, respectively. The third most frequently isolated pathogen was *Pseudomonas aeruginosa*. Among the rest of bacteria *Campylobacter* belonged to class of priority 2 (high), which was the most numerous.

The revealed information on the changes of the pathogens in the different technological steps and seasons is summarized in Figure 2. In Figure 3, changes in the *Enterobacteriaceae* family are presented. The distribution of all dominant species varied greatly in different seasons and technological steps. Treated wastewater did not have a significant influence on the structure and distribution of pathogens in the downstream surface water. In the Figure 4 the results from the principle component analysis are presented. There was a difference between distribution of WHO priority pathogens in raw sewage (influent) and treated wastewater. The weak correlations were detected between the samples. All the tested samples contained twelve bacteria from the WHO priority pathogens list. In Figure 5, the correlations between the seasons are presented. In all analyses, the values of Spearman correlation index between seasons were high. It is suggested that environmental parameters, such as temperature differences, effect pathogen distribution in various seasons.

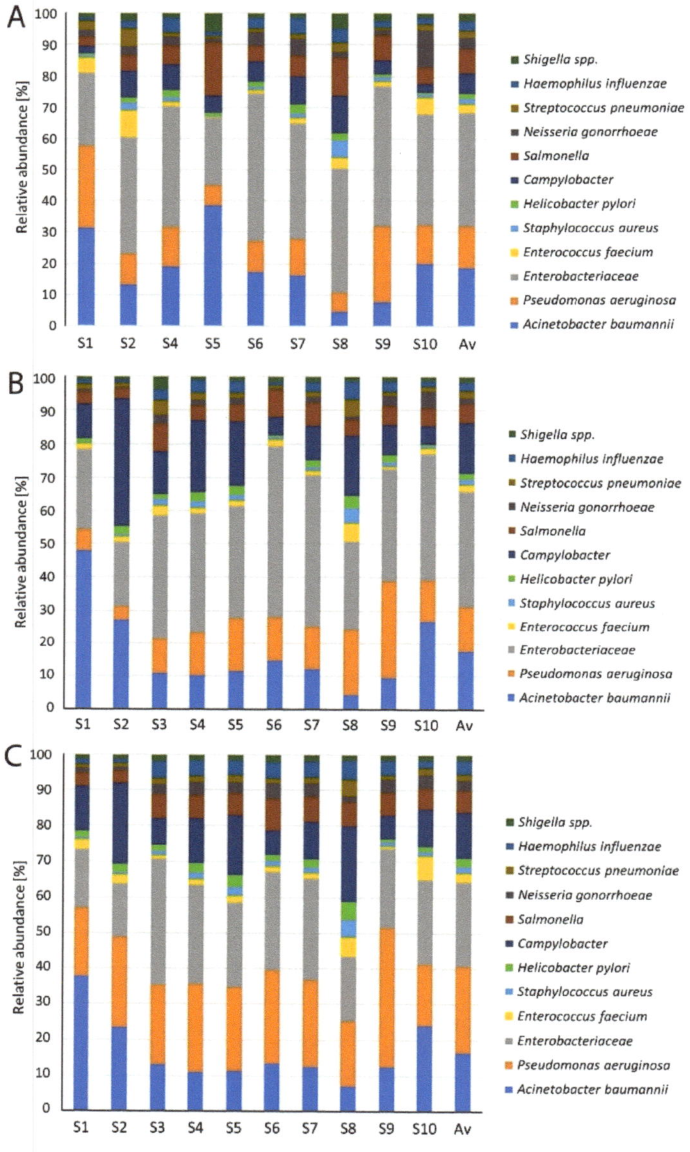

Figure 2. Technological and season changes of multi-antibiotic resistant pathogens from WHO list. (**A**) Summer (SUM); (**B**) autumn (AUT); (**C**) spring (SPR); Av—average.

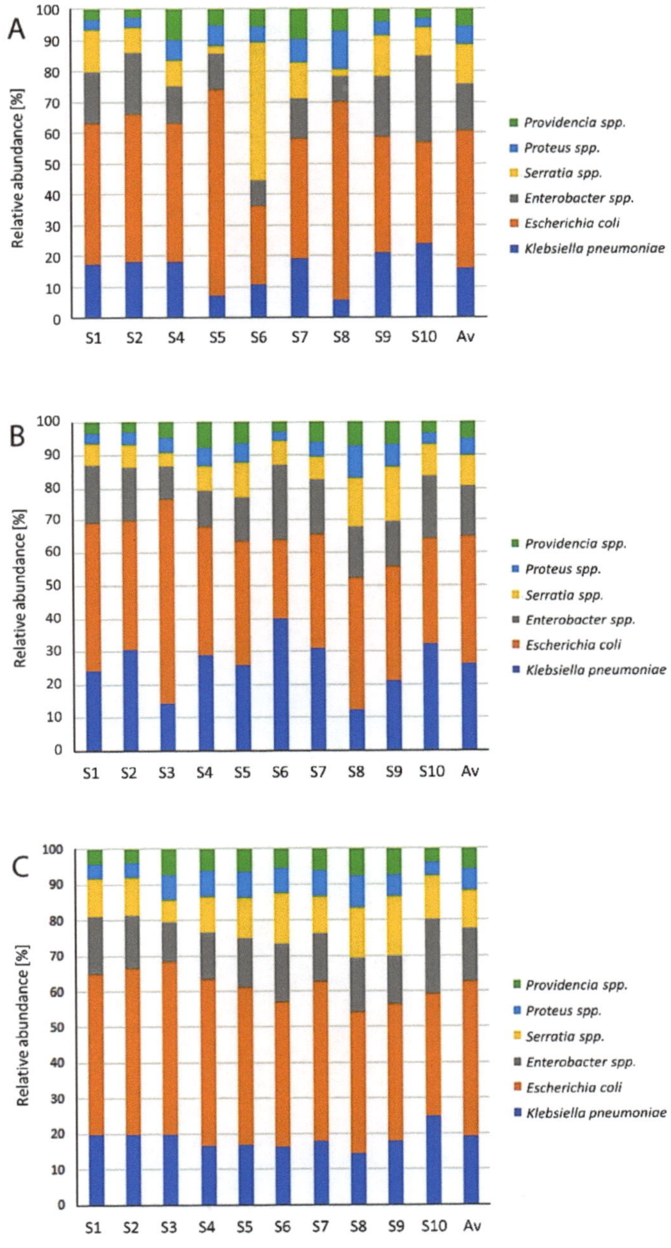

Figure 3. Technological and season changes of bacteria belonged to *Enterobacteriaceae*. (**A**) Summer (SUM); (**B**) autumn (AUT); (**C**) spring (SPR); Av—average.

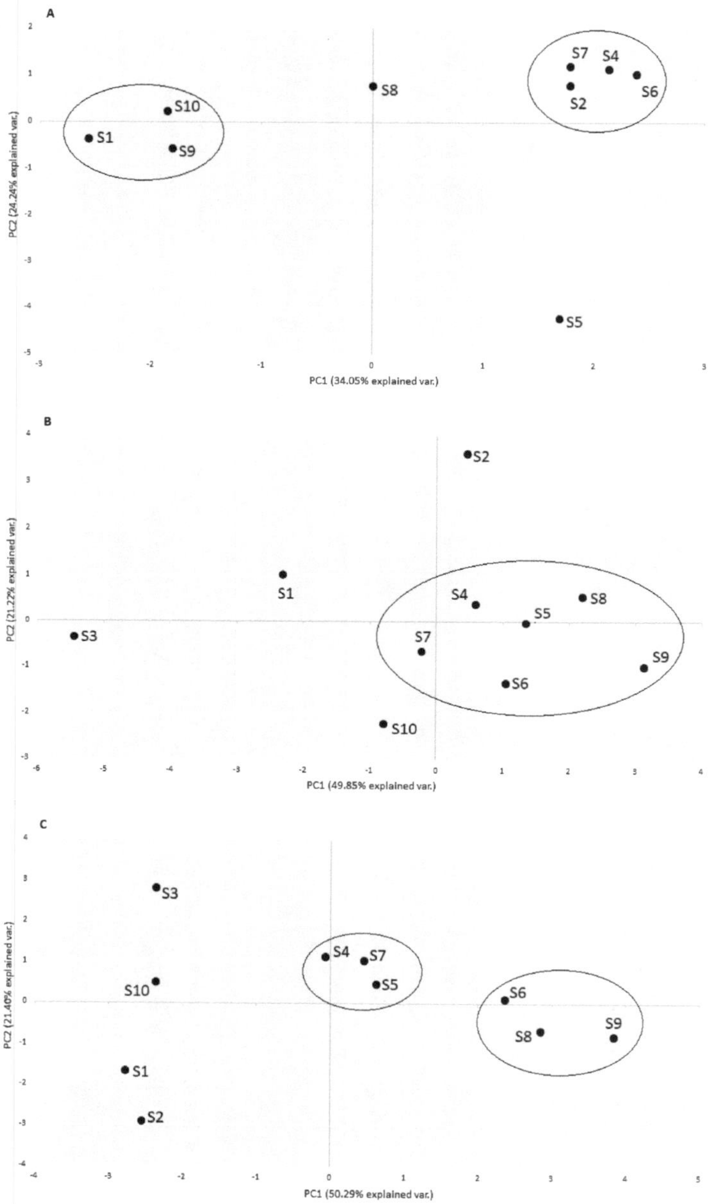

Figure 4. Principal component analysis (PCA) of the samples collected from different technological steps in various seasons on the variance-covariance matrix of the relative abundances of pathogens from the WHO multi-resistant pathogens list. The numbers in brackets describe the percentage of variance, explained by the first two components. (**A**) summer 2018; (**B**) autumn 2018; (**C**) spring 2019.

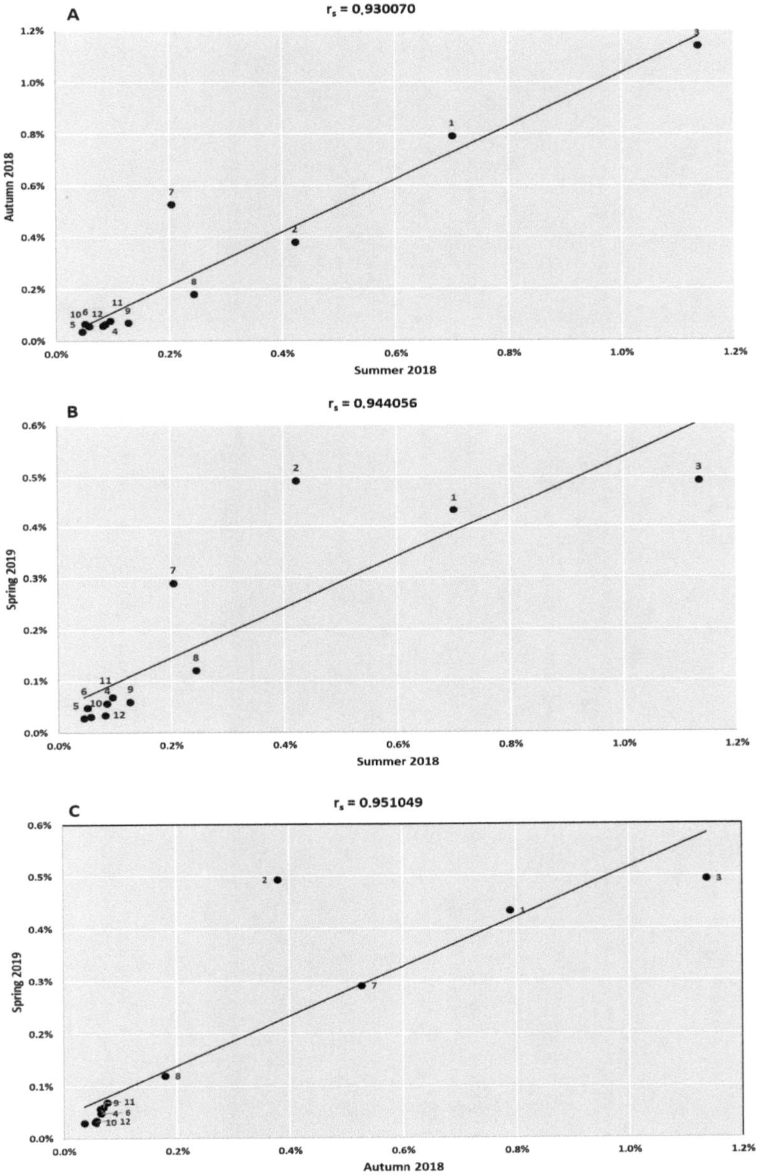

Figure 5. The Spearman's rank-order correlation coefficient between the bacterial pathogens from the WHO list, occurring in various seasons ($p < 0.01$). 1—*Acinetobacter baumannii*; 2—*Pseudomonas aeruginosa*; 3—*Enterobacteriaceae*; 4—*Enterococcus faecium*; 5—*Staphylococcus aureus*; 6—*Helicobacter pylori*; 7—*Campylobacter* spp.; 8—*Salmonella* spp.; 9—*Neisseria gonorrhoeae*; 10—*Streptococcus pneumoniae*; 11—*Haemophilus influenzae*; 12—*Shigella* spp. (**A**) summer 2018; (**B**) autumn 2018; (**C**) spring 2019.

In Figure 6A,B, the Venn diagrams for the selected technological steps and seasons are presented. *Venn diagram* worksheets present the relations between the abundance of pathogens in three seasons in the following technological steps. They graphically present how the distribution of pathogens in three seasons is different in the various technological steps.

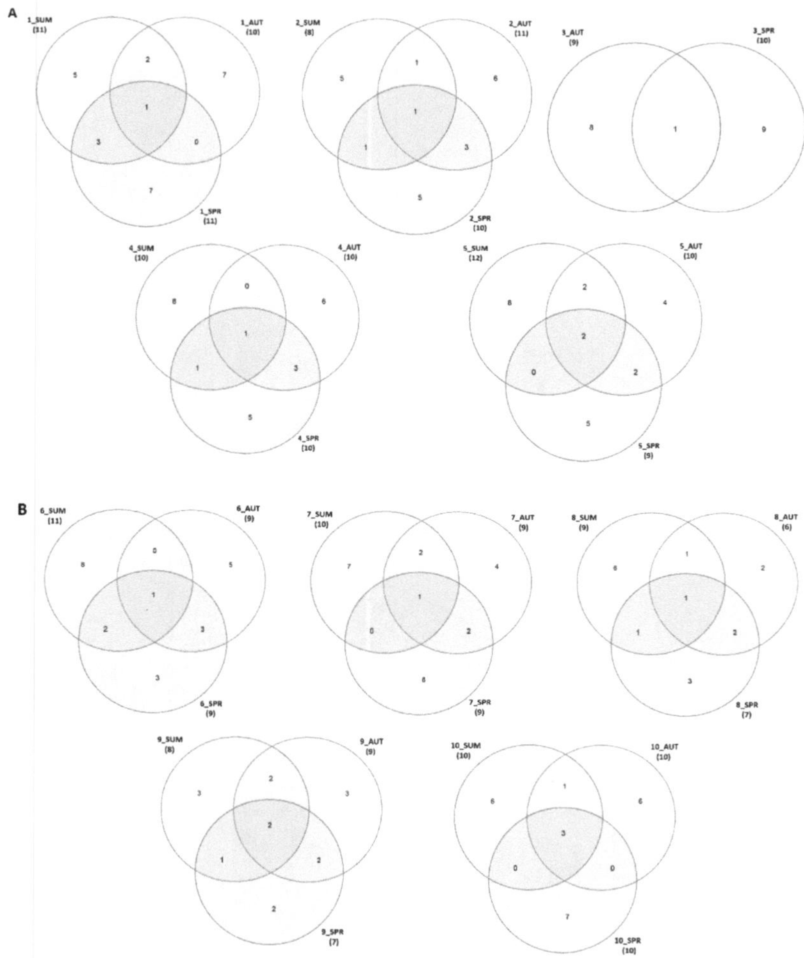

Figure 6. Venn diagrams illustrate the pathogens from WHO multi-resistant pathogens list in the samples collected from different technological steps. (**A**) Samples 1–5; (**B**) Samples 6–10; summer—SUM; autumn—AUT; spring—SPR.

4. Discussion

Based on the observed findings, the WHO priority multi-resistant pathogens were in wastewater and receiving surface waters. In the recent years, the number of pathogens with multi-drug resistance genes has significantly increased, and microbiological monitoring should be carried out in different environments, in order to prevent pollution and protect the environment and public health. Despite modern advances in wastewater technologies, treated effluents contain large amounts of various pollutants, including microbiological. The effluents are discharged into the environment, mainly into the receiving surface waters, consequently affecting public health directly or indirectly [25].

Generally, the microbial indicators are classified into three groups [26]: (i) general (process) microbial indicators, (ii) fecal indicators (such as *E. coli*), (iii) index organisms and model organisms. Presently, the following indicators of microbial contamination are used: total coliforms, enterococci, fecal streptococci, *Escherichia coli*, and *Clostridium perfringens*. Preferred indicators of fecal pollution are the enterococci [27]. These indicators are detected by the most of researches. Wen at al. [27] analyzed and compared water quality indicator systems in USA, Africa, and several countries in Asia and Europe. Currently, the bacterial indicators are the most popular in microbiological water quality monitoring, although several countries started to adopt the new methods for detecting other microbial pathogens, such as enteric viruses or protozoa [28].

Ajonina et al. [29] examined the microbiological quality of wastewater from the wastewater treatment plants in Hamburg City. As presented, the large amounts of coliform bacteria were found in treated wastewater and in River Elbe water. Marie and Lin [30] used the following bacterial indicators: *E. coli*, total coliforms, fecal coliforms, fecal streptococci, *Vibrio*, *Salmonella*, and *Shigella* for the evaluation of river water quality. Whereas Garrido-Perez et al. [31] evaluated the microbiological contamination of beach waters and sediments, using two indicators of fecal pollution: fecal coliforms and *Clostridium perfringens*. The occurrence of microbial indicators in waters is being reported by various researchers, but in most of the paper, there is no information about their antibiotic resistance. The characterization of the bacteria multi-antibiotic resistance has been neglected. Now, some epidemiological circumstances have changed this situation. Pérez-Rodríguez and Taban [32] reviewed the role of foods from animals (for example milk, meat, and poultry, etc.) as vehicles for multi-drug resistant pathogens and their role in the dissemination of antimicrobial resistances and novel characteristics, particularly multi-drug resistance.

However, most of the microbiological indicators are bacteria-specific, multi-antibiotic resistance; now, the multi-antibiotic resistance is a new characteristic of pathogens. Antimicrobial resistance (AMR) is huge public and health problem. As estimated, over 670,000 infections are caused annually by AMR pathogens in Europe. The cost of these infections is estimated to exceed 1 billion euros (ECDC, 2019). Most of the deaths are caused by pathogens, from which most of them are multi-drug resistant species, for example: *Escherichia coli*, *Acinetobacter* spp, *Pseudomonas aeruginosa*, *Klebsiella pneumoniae*, *Staphylococcus aureus*, and *Enterococcus* spp. Additionally, most of the infections are caused by these pathogens, and appropriate management, including antibiotic resistance of public health system, is essential.

In the review of Kakoullis et al. [33], the mechanisms of antibiotic resistance in pathogens, which are of great clinical important, were described. The author presented the resistance mechanisms of six pathogens, e.g., multi-drug resistant *Escherichia coli*, *Staphylococcus aureus*, *Pseudomonas aeruginosa*, *Enterococcus* spp., *Acinetobacter* spp., and *Klebsiella pneumoniae*. By the basic understanding the mechanisms of resistance, the clinicians can better comprehend and predict resistance patterns and subsequently select the most appropriate novel antimicrobial drugs for the pathogens or development of an effective vaccine [34,35].

In the study of Zaha et al. [36], the most commonly isolated pathogen was *Acinetobacter baumannii*, which were resistant to all the β–lactam antibiotics, including the carbapenems. Similar results were presented in the study of Handal et al. [37].

Fischbach and Walsh [38] have distinguished three classes of antibiotic-resistant pathogens, which are major threats to public health. First, methicillin-resistant *Staphylococcus aureus* (MRSA), which has high mortality rate. Pathogens from the second class, belonging to multidrug-resistant (MDR) and pan-drug-resistant (PDR) gram-negative bacteria. These strains of *Acinetobacter baumannii*, *Escherichia coli*, *Klebsiella pneumoniae*, and *Pseudomonas aeruginosa* are resistant to all antibiotic group: penicillins, cephalosporins, carbapenems, monobactams, quinolones, aminoglycosides, tetracyclines, and polymyxins. The third class belongs to strains of *Mycobacterium tuberculosis*.

Mhondoro et al. [39] noted that carbapenem-resistant *A. baumannii*, carbapenem-resistant *P. aeruginosa*, fluoroquinolone-resistant *Salmonella*, and ESBL producing *Enterobacteriaceae* were major health problems in Harare, Zimbabwe.

From the analysis conducted here, differential patterns in the distribution of WHO multi-resistant pathogens are clear, although the factors (technological steps and seasons) influencing the observed differences are yet to be fully explained. The effect of wastewater treatment processes on the fate of resistance is variable. As suggested by Karkman et al. [40], future research should examine whether WHO pathogens persist in downstream environments and urban water. Most of the WHO pathogens are important indicators for specific water purposes, such as agriculture and aquaculture wastewater and reuse.

Furthermore, the use of a metagenomic approach allowed us to identify pathogens in the collected samples. The sequence data can be also used to predict antibiotic resistance and virulence phenotypes. Detailed information on application of metagenomics approaches and other modern methods in pathogens detection is described in the literature [41–44]. So far, there are no reliable methods that could be used to detect pathogenic bacteria in the environmental samples, to the best of our knowledge. Despite the fact that numerous research has been done, new emerging pathogens or patterns and potential human risk from urban wastewater release are still uncertain. The lack of reporting standards of WHO pathogens also makes it is difficult to protect public health. Quantitative microbial risk assessment (QMRA) could be useful tool to evaluate the health risk assessment and describe scenarios for the spread of multi-resistant pathogens in the environment.

5. Conclusions

The study facilitated the evaluation of similarities and differences in composition of the pathogens from the WHO list, during the wastewater treatment process and seasons, and their distribution in effluent receiving water. Special attention was paid, in order to present the metagenomic analysis as a new tool in human health risk assessment. Twelve bacteria from the WHO priority pathogens list were detected in wastewater and receiving surface water. They disperse in the environment and disease transmission. WHO-listed bacteria pose a serious component of the threat of microbial contamination, in regard to public health and water quality, especially during water recycling and reuse processes. The process of the identification and detection of WHO-listed bacteria is an important step in the management of health and environmental risks associated with recycled water.

Author Contributions: Conceptualization, G.P. and M.H.; methodology and validation, G.P., M.H., Ł.J., and J.H.; formal analysis, Ł.J. and J.H.; writing—original draft preparation, G.P. and Ł.J.; writing—review and editing, Ł.J., J.H., M.H. and G.P.; supervision, G.P.; funding acquisition, M.H. and G.P. All authors have read and agreed to the published version of the manuscript.

Funding: This study was supported by National Science Centre, Poland (grant no. 2017/26/M/NZ9/00071).

Institutional Review Board Statement: Not applicable.

Informed Consent Statement: Not applicable.

Data Availability Statement: The metagenome data, used in submitted paper #738158, are deposited in the NCBI Sequence Read Archive (SRA), under accessions NCBI SRA BioProject: PRJNA666519—https://www.ncbi.nlm.nih.gov/bioproject/PRJNA6665196 (accessed on 20 February 2021).

Acknowledgments: The authors gratefully acknowledge the staff of WWTP for supporting the sampling campaigns and Dominka Głowacka (MSc) from Genomed S.A. for bioinformatic analysis.

Conflicts of Interest: The authors declare no conflict of interest.

References

1. Dias, D.A.; Urban, S.; Roessner, U. A historical overview of natural products in drug discovery. *Metabolites* **2012**, *2*, 303–336. [CrossRef] [PubMed]
2. Rex, J.H. ND4BB: Addressing the antimicrobial resistance crisis. *Nat. Rev. Microbiol.* **2014**, *12*, 231–242. [CrossRef]
3. WHO. Global Priority List of Antibiotic-Resistant Bacteria to Guide Research, Discovery, and Development of New Antibiotics. 2017. Available online: https://www.who.int/medicines/publications/global-priority-list-antibiotic-resistant-bacteria/en/ (accessed on 14 March 2021).

4. WHO. Antimicrobial Resistance. 2020. Available online: https://www.who.int/news-room/fact-sheets/detail/antimicrobial-resistance (accessed on 16 March 2021).
5. Michael, C.A.; Dominey-Howes, D.; Labbate, M. The antimicrobial resistance crisis: Causes, consequences, and management. *Front. Public Health* **2014**, *2*, 145. [CrossRef]
6. Davies, J.; Davies, D. Origins and evolution of antibiotic resistance. *Microbiol. Mol. Biol. Rev.* **2010**, *74*, 417–433. [CrossRef] [PubMed]
7. D'Costa, V.M.; King, C.E.; Kalan, L.; Morar, M.; Sung, W.W.L.; Schwarz, C.; Froese, D.; Zazula, G.; Calmels, F.; Debruyne, R.; et al. Antibiotic resistance is ancient. *Nature* **2011**, *477*, 457–461. [CrossRef] [PubMed]
8. Trotter, A.J.; Aydin, A.; Strinden, M.J.; O'Grady, J. Recent and emerging technologies for the rapid diagnosis of infection and antimicrobial resistance. *Curr. Opin. Microbiol.* **2019**, *51*, 39–45. [CrossRef] [PubMed]
9. Rizzo, L.; Manaia, C.; Merlin, C.; Schwartz, T.; Dagot, C.; Ploy, M.C.; Michael, I.; Fatta-Kassinos, D. Urban wastewater treatment plants as hotspots for antibiotic resistant bacteria and genes spread into the environment: A review. *Sci. Total Environ.* **2013**, *447*, 345–360. [CrossRef] [PubMed]
10. Cai, L.; Zhang, T. Detecting human bacterial pathogens in wastewater treatment plants by a high-throughput shotgun sequencing technique. *Environ. Sci. Technol.* **2013**, *47*, 5433–5441. [CrossRef]
11. Kumaraswamy, R.; Amha, Y.M.; Anwar, M.Z.; Henschel, A.; Rodriguez, J.; Ahmad, F. Molecular analysis for screening human bacterial pathogens in municipal wastewater treatment and reuse. *Environ. Sci. Technol.* **2014**, *48*, 11610–11619. [CrossRef] [PubMed]
12. Chu, B.T.T.; Petrovich, M.L.; Chaudhary, A.; Wright, D.; Murphy, B.; Wells, G.; Poretsky, R. Metagenomics reveals the impact of wastewater treatment plants on the dispersal of microorganisms and genes in aquatic sediments. *Appl. Environ. Microbiol.* **2018**, *84*, e02168-17. [CrossRef]
13. Huang, K.; Zhao, F.; Zhang, X.-X.; Ye, L.; Ren, H.; Zhang, T.; Mao, Y.; Ju, F.; Wang, Y.; Li, B. Free-living bacteria and potential bacterial pathogens in sewage treatment plants. *Appl. Microbiol. Biotechnol.* **2018**, *102*, 2455–2464. [CrossRef]
14. Osunmakinde, C.O.; Selvarajan, R.; Mamba, B.B.; Msagati, T.A.M. Profiling bacterial diversity and potential pathogens in wastewater treatment plants using high-throughput sequencing analysis. *Microorganisms* **2019**, *7*, 506. [CrossRef]
15. Rajasulochana, P. Comparison on efficiency of various techniques in treatment of waste and sewage water—A comprehensive review. *Resour. Technol.* **2016**, *2*, 175–184. [CrossRef]
16. Okoh, A.I.; Odjadjare, E.E.; Igbinosa, E.O.; Osode, A.N. Wastewater treatment plants as a source of microbial pathogens in receiving watersheds. *Afr. J. Biotechnol.* **2007**, *6*, 2932–2944. [CrossRef]
17. Naidoo, S.; Olaniran, A.O. Treated wastewater effluent as a source of microbial pollution of surface water resources. *Int. J. Environ. Res. Public Heal.* **2013**, *11*, 249–270. [CrossRef]
18. Lu, X.; Zhang, X.; Wang, Z.; Huang, K.; Wang, Y.; Liang, W. Bacterial pathogens and community composition in advanced sewage treatment systems revealed by metagenomics analysis based on high-throughput sequencing. *PLoS ONE* **2015**, *5*, e0125549. [CrossRef] [PubMed]
19. Lucena, F.; Duran, A.E.; Morón, A.; Calderón, E.; Campos, C.; Gantzer, C.; Skraber, S.; Jofre, J. Reduction of bacterial indicators and bacteriophages infecting fecal bacteria in primary and secondary wastewater treatments. *J. Appl. Microbiol.* **2004**, *97*, 1069–1076. [CrossRef] [PubMed]
20. WHO. Global Action Plan on Antimicrobial Resistance. 2015. Available online: http://www.who.int/antimicrobial-resistance/global-action-plan/en/ (accessed on 20 February 2021).
21. WHO. *Prioritization of Pathogens to Guide Discovery, Research and Development of New Antibiotics for Drug Resistant Bacterial Infections, Including Tuberculosis*; WHO: Geneva, Switzerland, 2017.
22. Rolbiecki, D.; Harnisz, M.; Korzeniewska, E.; Jałowiecki, Ł.; Płaza, G. Occurrence of fluoroquinolones and sulfonamides resistance genes in wastewater and sludge at different stages of wastewater treatment: A preliminary case study. *Appl. Sci.* **2020**, *10*, 5816. [CrossRef]
23. Płaza, G.; Jałowiecki, Ł.; Głowacka, D.; Hubeny, J.; Harnisz, M.; Korzeniewska, E. Insights into the microbial diversity and structure in a full-scale municipal wastewater treatment plant with particular regard to Archaea. *PLoS ONE* **2021**, *16*, e0250514. [CrossRef]
24. Meyer, F.; Paarmann, D.; D'Souza, M.; Olson, R.; Glass, E.M.; Kubal, M.; Paczian, T.; Rodriguez, A.; Stevens, R.; Wilke, A.; et al. The metagenomics RAST server—A public resource for the automatic phylogenetic and functional analysis of metagenomes. *BMC Bioinform.* **2008**, *9*, 386. [CrossRef]
25. De Lima, I.R.; Dos Santos, L.U.; Tosetto, M.S.; Franco, R.M.; Guimarães, J.R. Urban water reuse: Microbial pathogens control by direct filtration and ultraviolet disinfection. *J. Water Health* **2014**, *12*, 465–473. [CrossRef]
26. Ashbolt, N.J.; Grabow, W.O.K.; Snozzi, M. Indicators of microbial water quality. In *Water Quality: Guidelines, Standards and Health*; Fewtrell, L., Bartram, J., Eds.; IWA Publishing: London, UK, 2001; pp. 127–145.
27. Wen, X.; Chen, F.; Lin, Y.; Zhu, H.; Yuan, F.; Kuang, D.; Jia, Z.; Yuan, Z. Microbial indicators and their use for monitoring drinking water quality—A review. *Sustainability* **2020**, *12*, 2249. [CrossRef]
28. Jurzik, L.; Hamza, I.A.; Puchert, W.; Uberla, K.; Wilhelm, M. Chemical and microbiological parameters as possible indicators for human enteric viruses in surface water. *Int. J. Hyg. Environ. Health* **2010**, *213*, 210–216. [CrossRef] [PubMed]

29. Ajonina, C.; Buzie, C.; Rubiandini, R.H.; Otterpohl, R. Microbial pathogens in wastewater treatment plants (WWTP) in Hamburg. *J. Toxicol. Environ. Health Part A* **2015**, *78*, 381–387. [CrossRef]
30. Marie, V.; Lin, J. Microbial indicators and environmental relationships in the Umhlangane River, Durban, South Africa. *Open Life Sci.* **2018**, *13*, 385–395. [CrossRef]
31. Garrido-Perez, M.C.; Anfuso, E.; Acevedo, A.; Perales-Vargas-Machuca, J.A. Microbial indicators of faecal contamination in waters and sediments of beach bathing zones. *Int. J. Hyg. Environ. Health* **2008**, *211*, 510–517. [CrossRef]
32. Pérez-Rodríguez, F.; Taban, B.M. A State-of-Art Review on Multi-Drug Resistant Pathogens in Foods of Animal Origin: Risk Factors and Mitigation Strategies. *Front. Microbiol.* **2019**, *10*, 2091. [CrossRef]
33. Kakoullis, L.; Papachristodoulou, E.; Chra, P.; Panos, G. Mechanisms of antibiotic resistance in important gram-positive and gram-negative pathogens and novel antibiotic solutions. *Antibiotics* **2021**, *10*, 415. [CrossRef] [PubMed]
34. ECDC. *Antimicrobial Resistance in the EU/EEA—AER for 2019*; ECDC: Stockholm, Sweden, 2020.
35. Banin, E.; Hughes, D.; Kuipers, O.P. Bacterial pathogens, antibiotics and antibiotic resistance. *FEMS Microbiol. Rev.* **2017**, *41*, 450–452. [CrossRef] [PubMed]
36. Zaha, D.C.; Bungau, S.; Aleya, S.; Tit, D.M.; Vesa, C.M.; Popa, A.R.; Pantis, C.; Maghiar, O.A.; Bratu, O.G.; Furau, C.; et al. What antibiotics for what pathogens? The sensitivity spectrum of isolated strains in an intensive care unit. *Sci. Total Environ.* **2019**, *687*, 118–127. [CrossRef] [PubMed]
37. Handal, R.; Qunibi, L.; Sahouri, I.; Juhari, M.; Dawodi, R.; Marzouqa, H.; Hindiyeh, M. Characterization of carbapenem-resistant *Acinetobacter baumannii* strains isolated from hospitalized patients in Palestine. *Int. J. Microbiol.* **2017**, *2017*, 8012104. [CrossRef]
38. Fischbach, M.A.; Walsh, C.T. Antibiotics for Emerging Pathogens. *Science* **2009**, *325*, 1089–1093. [CrossRef] [PubMed]
39. Mhondoro, M.; Ndlovu, N.; Donewell, B.; Juru, T.; Tafara, G.N.; Gerald, S.; Peter, N.; Mufuta, T. Trends in antimicrobial resistance of bacterial pathogens in Harare, Zimbabwe, 2012–2017: A secondary dataset analysis. *BMC Infect. Dis.* **2019**, *19*, 746. [CrossRef]
40. Karkman, A.; Pärnänen, K.; Larsson, D.G.J. Fecal pollution can explain antibiotic resistance gene abundances in anthropogenically impacted environments. *Nat. Commun.* **2019**, *10*, 80. [CrossRef] [PubMed]
41. Miller, R.R.; Montoya, V.; Gardy, J.L.; Patrick, D.M.; Tang, P. Metagenomics for pathogen detection in public health. *Genome Med.* **2013**, *5*, 81. [CrossRef] [PubMed]
42. Ibekwe, A.M.; Leddy, M.; Murinda, S.E. Potential human pathogenic bacteria in a mixed urban watershed as revealed by pyrosequencing. *PLoS ONE* **2013**, *8*, e79490. [CrossRef]
43. Ramírez-Castillo, F.; Loera-Muro, A.; Jacques, M.; Garneau, P.; Avelar-González, F.; Harel, J.; Guerrero-Barrera, A. Waterborne pathogens: Detection methods and challenges. *Pathogens* **2015**, *4*, 307–334. [CrossRef] [PubMed]
44. Ricchi, M.; Bertasio, C.; Boniotti, M.B.; Vicari, N.; Russo, S.; Tilola, M.; Bellotti, M.A.; Bertasi, B. Comparison among the quantification of bacterial pathogens by qPCR, dPCR, and cultural methods. *Front. Microbiol.* **2017**, *8*, 1174. [CrossRef]

MDPI AG
Grosspeteranlage 5
4052 Basel
Switzerland
Tel.: +41 61 683 77 34

International Journal of Environmental Research and Public Health Editorial Office
E-mail: ijerph@mdpi.com
www.mdpi.com/journal/ijerph

www.ingramcontent.com/pod-product-compliance
Lightning Source LLC
Chambersburg PA
CBHW041415210926
51850CB00013B/3281

* 9 7 8 3 7 2 5 8 3 5 0 6 5 *